**Tutorials, Schools, and Workshops
in the Mathematical Sciences**

This series will serve as a resource for the publication of results and developments presented at summer or winter schools, workshops, tutorials, and seminars. Written in an informal and accessible style, they present important and emerging topics in scientific research for PhD students and researchers. Filling a gap between traditional lecture notes, proceedings, and standard textbooks, the titles included in TSWMS present material from the forefront of research.

More information about this series at http://www.springer.com/series/15641

Sergio Cacciatori • Batu Güneysu • Stefano Pigola
Editors

Einstein Equations: Physical and Mathematical Aspects of General Relativity

Domoschool 2018

 Birkhäuser

Editors

Sergio Cacciatori
Department of Science and High
Technology
University Insubrien
Como, Italy

Batu Güneysu
Mathematisches Institut
Humboldt-Universität zu Berlin
Berlin, Germany

Stefano Pigola
Department of Science and High
Technology
University Insubrien
Como, Italy

ISSN 2522-0969   ISSN 2522-0977   (electronic)
Tutorials, Schools, and Workshops in the Mathematical Sciences
ISBN 978-3-030-18063-8   ISBN 978-3-030-18061-4   (eBook)
https://doi.org/10.1007/978-3-030-18061-4

Mathematics Subject Classification (2010): 83C05, 83C15, 83C20, 83C35, 83C40, 83C47, 83C57

This book is published under the imprint Birkhäuser, www.birkhauser-science.com, by the registered company Springer Nature Switzerland AG.
The registered company address is: Gewerbestrasse 11, 6330 Cham, Switzerland

Preface

This book presents four lecture courses and, in addition, nine talks given at the first edition of the *Domoschool*, the *International Alpine School in Mathematics and Physics*, held in Domodossola in July 2018 with the title "Einstein Equations: Physical and Mathematical aspects of General Relativity."

Domoschool is a 5-day summer school aiming to bring together young mathematicians and physicists working in the same field of physics. Four days are devoted to four lecture courses (each consisting of four lectures) given by prominent experts. Despite its theoretical nature, the school also includes a short course on experimental aspects of the chosen topic, aiming to motivate a further theoretical research. The additional day of Domoschool is devoted to social activities (such as a hiking tour and a dinner), aiming to show the participants the town of Domodossola and its surroundings. The school also includes a public meeting with the local community in order to explain to the citizens of Domodossola the contents of the summer school and the scientific relevance. In 2018, the topic was chosen to be General Relativity. Public lectures were given by *Daniela Bortoletto, Luigi Bignami*, and *Giancarlo Cella*. Alessandro Carlotto gave an additional, spontaneous public lecture, explaining the contents of this summer school to the broader audience.

General Relativity is perhaps the most spectacular creation of Albert Einstein. He started elaborating a new theory of gravitation immediately after his annus mirabilis (1905). However, creating a relativistic theory of gravity took 10 long years of hard and almost unsuccessful work, and the correct formulation suddenly appeared in the last few months of his works, written between the end of 1915 and the beginning of 1916. General Relativity immediately had many groundbreaking consequences, even though the theory was not initially accepted by the scientific community. For Einstein himself, the correct prediction of the secular precession of Mercury's perihelion was the smoking gun, but what brought the theory to the forefront of the world was the verification, in 1919, of the bending of light by the sun, predicted by this new theory. Today, we know hundreds of tests agreeing with the predictions of General Relativity while ruling out several alternative theories of gravity and adding very strong constraints.

However, the history of General Relativity is even more interesting. In 1917, in a famous paper where he introduced the cosmological constant, Einstein gave origin to modern cosmology, and in 1918, he predicted the existence of gravitational waves but considered it as just a theoretical speculation since the weakness of the phenomenon appeared to make it undetectable. However, despite its experimental verification before 1920, it is worth mentioning that General Relativity did not receive much attention in the physics community for quite some time. A new golden era started in the 1960s when black hole physics attracted the attention of mathematicians and physicists, like Roger Penrose, Stephen Hawking, Subrahmanyan Chandrasekhar, etc., even though the first black hole solution was already known in 1917, thanks to the works of Karl Schwarzschild. In the same years, the first observations of neutron stars, pulsars, and very compact objects, whose theoretical description required the use of a relativistic theory of gravity, contributed heavily to the acceptance of General Relativity as the true theory of gravity. In particular, in 1974, Russell Alan Hulse and Joseph Hooton Taylor Jr. observed the orbit of a pulsar around a neutron star measuring a decreasing of the orbiting period in perfect agreement with the prediction of the Einstein equations about the emission of gravitational waves. The direct detection of gravitational waves, however, was not before the twenty-first century. Starting from September 2015 up to August 2017, six events have been detected by LIGO and Virgo labs: GW150914, GW151226, GW170104, GW170608, GW170814, and GW170817. The first five events corresponded to the catastrophic merging of black hole pairs after tremendously violent dances. All of them agreed with the predictions of the Einstein equations as computed by the new methods introduced by Thibault Damour and collaborators. The sixth event was even more formidable: it was generated by the merging of two neutron stars, giving rise not only to gravitational waves but also to electromagnetic waves, allowing for comparisons and several new tests of the theories. This represented the beginning of a new era! Merging of neutron stars, called *multimessengers*, is a new instrument for looking at our universe as we have never seen it before.

But the surprises arising from the Einstein equations are not only at the experimental level. The Cauchy problem for the Einstein equations can be formulated by specifying, on a suitable spacelike surface, initial data that have to satisfy certain constraint equations. In their study of precisely these constraint equations, Alessandro Carlotto and Richard Schoen discovered certain localization phenomena having many consequences. Among them, as Piotr Chrusciel pointed out, there is a gravitation screening phenomenon, very different from the one that can happen in Newtonian gravity. After a century, Einstein's equations continue to provide new surprises! These are the reasons underlying our choice to organize a school on both theoretical and experimental recent achievements in General Relativity.

The book is divided into two parts. The first part is devoted to the courses given at the school by leading international experts.

The first course, given by *Alessandro Carlotto*, had the primary scope of introducing graduate students to the mathematics of gravitational isolated systems. It was divided into four lectures. The first lecture provided a nice introduction to Einstein equations expressed in terms of constraint equations for the Cauchy problem, starting from Newtonian gravity and passing through Lorentzian geometry. The second lecture was devoted to the positive Riemannian mass theorem, by Richard Schoen and Shing-Tung Yau, which states the positivity of the Arnowitt-Deser-Misner mass for asymptotically flat manifolds with Euclidean signature. In particular, he presented the very recent proof valid for dimensions $D \geq 8$. In lecture three, it was shown that a massless asymptotically flat Riemannian manifold with nonnegative scalar curvature is necessarily a flat Euclidean space. It also included an interesting discussion of similar rigidity phenomena. Finally, the fourth lecture course was devoted to the gluing methods used by Alessandro Carlotto and Richard Schoen in order to construct new localized solutions, which are asymptotically flat initial data sets, with positive ADM mass that are precisely trivial outside a cone of a given angle.

The second course was given by *Felix Finster* and concerns the linear stability for non-extremal Kerr black holes under perturbations. The course introduced to an advanced topic in a very didactic way. The author started with the description of the Kerr solution, examining its symmetries and the associated conserved quantities. He then discussed the main instability processes, like superradiance and the Penrose process for extracting energy, and the long-time behavior of the Dirac field. It was then shown that the Teukolsky equation, arising in the decoupling of waves of different spin, can always be fully separated. This allowed him to analyze the stability of the non-extremal Kerr spacetime after expressing the problem in terms of the decay of solutions of the Cauchy problem for the Teukolsky equation. An outline of the proof of the stability was finally given.

The third course was given by *Alexander Kamenshchik* and was devoted to cosmology. This course was rather atypical for cosmology in the sense that when solely looking at some symmetries of the Einstein equation completely, new phenomena can arise. The course started with a discussion of the symmetries of the Bianchi classification of three-dimensional Lie algebras and, in particular, of the homogeneous cosmologies. The author then took a closer look at the solutions of the Einstein equations for Bianchi-I and Bianchi-II universes and the geometry of Bianchi-IX universes. He also explained the oscillatory method introduced by V.A. Belinski, I.M. Khalatnikov, and E.M. Lifshitz for approaching the cosmological singularity and its Hamiltonian description, the Mixmaster universe, introduced by C. Misner. Finally, it was shown how chaos enters in cosmology and how infinite dimensional Lie algebras might play a crucial role in this context.

The fourth course was given by *Giancarlo Cella*. This course was very different from the other ones in the sense that it was devoted to experimental physics: the author explained the experiments of the LIGO-Virgo labs, which detected the gravitational wave events mentioned above. The lectures gave quite a complete panorama on the phenomenological and experimental aspects and showed the magnificent predictive power of General Relativity. After a historical introduction, the basic principles of the gravitational detectors were discussed. Then, the fundamental sources of noise were analyzed: seismic, thermal, and quantum noise, parametric instabilities, and gravity fluctuations. Next, the detections were described, both for the black holes coalescences and for the multimessenger event GW170817. The potential of multimessenger detection was also discussed. Finally, the author presented proposals for other possible sources beyond coalescences, like continuous sources, burst signals, and stochastic background detections.

The school also included two other courses not transposed in the present book. A course given by *Piotr Chrusciel*, "Introduction to the General Relativistic Constraint Equations," was not included here because the course aimed to introduce the physics students who attended Domoschool to the mathematical concepts of General Relativity. The presented material is available in the literature and can be found, for example, on his personal homepage: *homepage.univie.ac.at/piotr.chrusciel*. The other course not included here was a course given by *Eric Poisson*, "Advanced Newtonian Gravity," whose lectures have been based on his book [E. Poisson and Clifford M. Will, *Gravity: Newtonian, Post-Newtonian, Relativistic*, Cambridge University Press (2014) Cambridge, UK].

Although Domoschool is a school aiming to offer PhD students and young researchers high-level courses given by experts, it also provides space for young researchers to present their research results in short talks. Reflecting these talks, the second part of the book presents these talks in a conference proceeding-type manner. In Domoschool 2018, nine of the participants presented a short talk:

Gabriele Bozzola and Vasileios Paschalidis (University of Arizona, Tucson, USA): *Generation of Initial Data for General-Relativistic Simulations of Charged Black Holes*

Federico Capone (University of Southampton, UK): *BMS Symmetries and Holography: An Introductory Overview*

Vittorio De Falco (Silesian University in Opava, Czech Republic): *Relativity of Observer Splitting Formalism and Some Astrophysical Applications*

Davide Fermi (Università degli Studi di Milano): *Some Remarks on a New Exotic Spacetime for Time Travel by Free Fall*

Colin MacLaurin (University of Queensland, Brisbane, Australia): *Generalised Gullstrand-Painleve Slicings of Schwarzschild*

Jiří Ryzner and Martin Žofka (Charles University in Prague): *Crystal Spacetimes with Discrete Translational Symmetry*

Jiří Veselý and Martin Žofka (Charles University in Prague): *Electrogeodesics and Extremal Horizons in Kerr–Newman–(anti-)de Sitter*

Sebastian H. Völkel and Kostas D. Kokkotas (University of Tübingen, Germany): *Hearing the Nature of Compact Objects*

Adamantia Zampeli (Charles University in Prague): *Minisuperspace Quantisation via Conditional Symmetries*

Domoschool was a project initiated by Sergio Cacciatori, but its realization was made possible, thanks to the open-minded people and the enthusiastic support of the *Municipality of Domodossola*, which supported the school both financially and concretely in its organization. For this, we are indebted to the Mayor *Lucio Pizzi*; Councilor for Culture *Daniele Folino*, who was the first one welcoming our proposal and connecting us to the right people; and, especially, Deputy Mayor *Angelo Tandurella*, who was fully available to us, assisting step-by-step through the evolution, the logistics, the public events, and so on.

Additionally, we want to acknowledge the noprofit association *Ars.Uni.VCO* (Associazione per lo Sviluppo della Cultura di Studi Universitari e della Ricerca nel Verbano Cusio Ossola); its outgoing president and current vice president, *Giulio Gasparini*, who was present during the foundation of the school; and the incoming president, *Stefania Cerutti*, who inaugurated the school, together with *Angelo Tandurella*. We are particularly indebted to the secretary, *Andrea Cottini*, and the communication manager, *Federica Fili*, for their continuous support, organizing the school, assisting all participants, and, in particular, together with the Rosminian Fathers (especially *Padre Fausto*), making it possible to gain Collegio Rosmini as the location of the school.

Furthermore, we are particularly grateful to the following speakers of Domoschool who kindly accepted our invitations and whose participation brought high prestige to the school:

Alessandro Carlotto from ETH, Zurich, and IAS, Princeton

Felix Finster from Universität Regensburg, Germany

Alexander Kamenshchik from Università degli Studi di Bologna, Italy, and Landau Institute, Moscow

Giancarlo Cella, INFN, Pisa, Italy

Piotr Chrusciel, from the University of Vienna, Vienna, Austria

Eric Poisson, from the University of Guelph, Guelph, Ontario, Canada

We acknowledge the speakers who accepted our invitation for the public lectures given at Rovereto Square just outside the city hall of Domodossola. Thanks to them, the public event was a complete success. They are as follows:

Daniela Bortoletto, from the University of Oxford, Oxford, UK *Luigi Bignami*, Italian scientific journalist and popularizer

Giancarlo Cella, researcher at INFN

Alessandro Carlotto, who kindly spontaneously offered his contribution to the public lectures

Moreover, we extend our acknowledgments to the participants of the school, including two high school teachers, who, with their active participation, contributed to making the atmosphere of the school particularly pleasant. And, to other members of the scientific board, *Francesco Belgiorno*, *Simone Noja*, *Riccardo Re*, *Mauro Giudici*, and *Pietro Antonio Grassi*, as well as the members of the organizing committee, *Andrea Cottini* and *Giorgio Mantica*, we thank you for your support.

Last, but not least, we are grateful to the sponsors of Domoschool: *Città di Domodossola*, *GNFM-INDAM*, *Fondazione CRT*, and *Consiglio Regionale del Piemonte*.

Como, Italy   Sergio Cacciatori
Berlin, Germany   Batu Güneysu
Como, Italy   Stefano Pigola
February 28, 2019

Saluto del Sindaco di Domodossola Lucio Pizzi

Raccolgo molto volentieri l'invito a rivolgere un saluto da queste pagine.

La scorsa estate il prof. Sergio Cacciatori, unitamente ai Suoi valenti collaboratori, ha dato vita alla prima edizione di "Domoschool" e ha fondato nella nostra città la prima "Alpine School in Mathematics and Physics", una scuola internazionale di matematica e fisica teorica ai più alti livelli.

Dalla prima edizione di "Domoschool" nasce questo libro che raccoglie le lezioni svolte, nel luglio 2018, da prestigiosi docenti di fama internazionale convenuti a Domodossola. Uno di questi ha definito Domodossola "...a hidden gem not far from Milan, Zurich and Geneva...".

...Una gemma nascosta non lontana da Milano, Zurigo e Ginevra...

E' questa l'immagine da lanciare per la nostra Domodossola, una gemma che vorremmo sempre più centrale rispetto agli itinerari di viaggio, baricentrica rispetto alle grandi città della Svizzera e del Nord Italia.

Facile da raggiungere e bella da visitare, accogliente come l'intero territorio circostante.

Un'immagine che si costruisce attraverso la promozione e attraverso la creazione di eventi di richiamo, primi tra tutti quelli culturali e scientifici.

In quest'ottica stiamo andando ad esempio con VisitOssola, moderno sito web, con Domosofia, Festival delle idee e dei saperi, con le grandi mostre di Palazzo San Francesco e anche con "Domoschool", un'iniziativa che contribuisce ad impreziosire la nostra offerta e a nobilitarla.

Dopo l'evento questo libro che diventa memoria volta a diffondersi in una lingua che oltretutto ci invita ad uscire ulteriormente dai nostri confini e ad ampliare i nostri orizzonti, guardando dal nostro splendido Borgo della Cultura verso il mondo.

Buona lettura.

Lucio Pizzi
Sindaco di Domodossola

Contents

Part I
Main Lectures

Four Lectures on Asymptotically Flat Riemannian Manifolds

Alessandro Carlotto

Abstract After an introduction to the mathematical description of isolated gravitating systems, we present a proof of the positive mass theorem relying on the compactification trick that lies behind the argument proposed by Schoen and Yau in 2017 to handle the possible occurrence of high-dimensional singularities of minimizing cycles. From there, we describe some rigidity phenomena involving Riemannian manifolds of non-negative scalar curvature, and survey various recent results concerning the large-scale geometric structure of asymptotically flat spaces. In the last section, we outline the construction, due to Schoen and the author, of localized solutions of the Einstein constraints and discuss its implications both on the physical and the geometric side.

Keywords Asymptotically flat manifold · ADM mass · Positive mass theorem · Isoperimetric profile · Einstein constraint equations

Mathematics Subject Classification (2000) Primary 83C05; Secondary 53A10

1 Introduction

These lecture notes originated from a course taught by the author in the context of the summer school "*Einstein equations: physical and mathematical aspects of general relativity*" held in Domodossola in July 2018.

The primary scope of the lectures was to introduce graduate or advanced undergraduate students to the mathematical description of *isolated gravitational systems* as encoded, in the context of general relativity, by the mathematical

A. Carlotto (✉)
IAS, School of Mathematics, Princeton, NJ, USA

ETH, Department of Mathematics, Zürich, Switzerland
e-mail: alessandro.carlotto@math.ias.edu; alessandro.carlotto@math.ethz.ch

© Springer Nature Switzerland AG 2019   3
S. Cacciatori et al. (eds.), *Einstein Equations: Physical and Mathematical Aspects of General Relativity*, Tutorials, Schools, and Workshops in the Mathematical Sciences, https://doi.org/10.1007/978-3-030-18061-4_1

notion of *asymptotically flat manifold*. In particular, after recollecting some basic facts in Newtonian gravity, we presented the definition of ADM mass and discussed its positivity relying on the compactification lemma that is ancillary to the elegant proof presented by Schoen and Yau in 2017. While the goal of that approach was to effectively handle, via a recursive argument based on the notion of minimal k-slicing, the possible occurrence of high-dimensional singularities of area-minimizing cycles (an issue which we do not address here), we thought it would be useful for the audience to see this proof rather than the original one (dating back to 1979), to appreciate its simplicity and effectiveness. Motivated by the wealth of ideas related to this landmark theorem, we went on to discuss some aspects of the impressive body of research that it fostered over the last two decades, and remarkably so in the last few years. More specifically, we presented a few rigidity results concerning 3-manifolds of non-negative scalar curvature and described (without any technical details) the significant progress in our understanding of the isoperimetric structure of asymptotically flat Riemannian manifolds. In the fourth and last lecture, we then outlined the construction, due to Schoen and the author, of localized solutions of the Einstein constraint equations, whose geometry is in striking contrast with that exhibited by asymptotically Schwarzschildean manifolds.

The audience was, for this course, quite broad and inhomogeneous: for this reason the lectures, while advanced in many respects, only assume some basic knowledge in Riemannian geometry and some prior exposure to general relativity.

In preparing these notes, we have refrained from the strong temptation of expanding the content of the lectures much beyond what could actually be covered in class: as a result, a lot of significant and related themes, like, for instance, Penrose-type inequalities or notions of quasi-local mass have completely been omitted. Needless to say, the reader is encouraged, starting from these notes, to expand the narrow horizon we embraced here and study the original sources that contributed to the development of this exciting area of research at the borderline between geometric analysis and mathematical general relativity.

I wish to express my gratitude to Sergio Cacciatori and Stefano Pigola for inviting me to lecture in such a unique context, and to Piotr Chruściel for valuable feedback on various preliminary versions: these notes are much better as a result of his input.

2 First Lecture: From Newtonian to Einsteinian Gravity

The scope of these lectures is to provide a brief introduction to the mathematical description of isolated gravitating systems, in the context of general relativity. As a preliminary step in that direction, we recall some basic facts in Newtonian gravity and discuss how the notion of ADM mass can be related to its classical counterpart.

2.1 Newtonian Gravity and Isolated Systems

We can summarize the Newtonian description of gravitational forces as follows:

Step 1: The background for physical phenomena is provided by a spacetime that factors as a trivial bundle, more precisely as a Riemannian product of the form $\mathbb{R}_t \times \mathbb{R}_x^3$, where both factors are endowed with the flat metric; furthermore the motion of bodies is described by means of Galileian kinematics and is ruled by the Newtonian laws of dynamics.

Step 2: There is a well-defined notion of gravitational mass, which can (initially) be regarded as a positive-definite function

$$m : \{\text{pointwise bodies}\} \to \mathbb{R}$$

such that for any two couple of pointwise bodies one has a purely attractive force

$$F = -\frac{km_1 m_2}{|x|^3} x$$

where x is a vector from the first to the second body, and k is a physical constant, which we shall assume from now onwards to be unitary (a conversion table from geometrized to nongeometrized units is provided, for instance, in appendix F of [65]). Gravitational forces propagate at infinite speed and therefore give rise to global interactions. From this basic postulate, we can derive important consequences:

- *if* it is assumed that the inertial and gravitational mass, while conceptually distinct, are proportional (hence equal modulo renormalization), then one can describe this force in terms of a gravitational field (generated by a pointwise mass $m > 0$ placed at the origin), defined to be the vector field

$$G = -\frac{m}{|x|^3} x$$

  which is nothing but the acceleration measured for a unit mass playing the role of test mass and instantaneously located at $x \in \mathbb{R}^3$;
- the field above is conservative and thus (since the domain $\mathbb{R}^3 \setminus \{0\}$ is simply-connected) there exists a potential $\varphi : \Omega \to \mathbb{R}$ such that $G = -\nabla\varphi$. Such a potential is unique modulo additive constants, so it is fixed once we assign the value at a point or at infinity;
- the Gauss law holds, namely for $\Omega \subset \mathbb{R}^3$ an open, bounded domain with smooth boundary one has

$$\int_{\partial\Omega} (G \cdot v)\, d\mathscr{H}^2 = -4\pi m$$

in case Ω contains the origin, or else with zero right-hand side if that is not the case. Here ν denotes the outward-pointing, unit normal vector field to $\partial\Omega$. One can rewrite this equation as $\mathrm{div}(G) = -4\pi\mu$ where we have set $\mu = m\delta_0$ and the equation is understood in the standard weak sense of distributions.

Furthermore, we can rewrite the equation above in terms of φ as the Poisson equation

$$\Delta\varphi = 4\pi\mu.$$

Notice that given this sole equation one can in fact reconstruct the whole theory. Indeed, by considering the above equation for $\mu = \delta_x$ (a Dirac mass at $x \in \mathbb{R}^3$) we recover the fundamental solution of the Laplace operator

$$E(y) = -\frac{1}{|x-y|}$$

which in turn gives rise to the force law described above. This completes the description for pointwise masses.

Step 3: We can now consider extended bodies described by, say, a locally finite measure μ and describe their gravitational effects by means of a suitably normalized potential φ defined by solving the Poisson equation above, which we shall then regard as the field equation in the context of Newtonian gravity. If we restrict our attention to the case when the sources are described by a Borel measure that is finite and compactly supported, then it is standard (cf. [44]) to prove that such an equation has indeed a distributional solution, which coincides with a smooth function outside a compact set and is unique provided we require the potential to decay to zero at infinity. Furthermore, such a unique potential can be expanded at infinity as

$$\varphi(x) = -\frac{1}{|x|}\int_{\mathbb{R}^3} d\mu(y) - \frac{x}{|x|^3}\cdot\int_{\mathbb{R}^3} y\,d\mu(y) + O(|x|^{-3}).$$

Indeed, let $d > 0$ be such that $\mathrm{spt}(\mu) \subset B_d(0)$ so for $|x| > 2d$ we can certainly write

$$|x-y| = |x|\left(1 - 2\frac{x\cdot y}{|x|^2} + \frac{|y|^2}{|x|^2}\right)^{1/2} = |x|\left(1 - \frac{x\cdot y}{|x|^2} + O(|x|^{-2})\right) \quad \text{for } |x| \to +\infty$$

which holds uniformly for $|y| < d$. Thus, plugging this expansion in the formula

$$\varphi(x) = -\int_{\mathbb{R}^3}\frac{1}{|x-y|}\,d\mu(y)$$

gives the conclusion. This simple result justifies the following definitions:

$$m := \int_{\mathbb{R}^3} d\mu(y), \quad \text{mass of the system;}$$

$$c := \frac{1}{m} \int_{\mathbb{R}^3} y \, d\mu(y), \quad \text{center of mass of the system.}$$

Later and in the next lectures we will discuss the relativistic counterparts of these notions.

2.2 Lorentzian Geometry

The general theory of relativity reduces the study of gravitational forces to the analysis of certain classes of mathematical (in fact: geometric) objects called Lorentzian manifolds. Let us now introduce such objects.

Given a vector space V over the field $\mathbb{R}$ we know (by Sylvester's Theorem, see e.g. chapter XV of [49]) that any symmetric bilinear form $q : V \times V \to \mathbb{R}$ has a well-defined invariant called signature: there exists an orthogonal basis $\{v_0, \ldots, v_n\}$ such that $q(v_i, v_i) \in \{-1, 0, 1\}$ and a well-defined triple (s_-, s_0, s_+) with

$$s_\pm = card\,\{0 \le i \le n \, : \, q(v_i, v_i) = \pm 1\}, \ s_0 = card\,\{0 \le i \le n \, : \, q(v_i, v_i) = 0\}$$

which is in fact independent of the choice of the basis in question. Those forms satisfying $s_0 = 0$ shall be called nondegenerate.

Definition 2.1 Given a real vector space V of dimension $n+1 \ge 2$ we shall say that a symmetric bilinear form $q : V \times V \to \mathbb{R}$ is a Lorentzian form if it has signature $(1, 0, n)$. The couple (V, q) will be called a Lorentzian vector space.

It follows from the previous remark that any Lorentzian form q can always be written in canonical form, namely

$$q = -v_0^* \otimes v_0^* + \sum_{i=1}^{n} v_i^* \otimes v_i^*$$

for a suitable orthogonal basis $\{v_0, \ldots, v_n\}$ and where we have set, for any two vectors $v, w \in \{v_0, \ldots, v_n\}$

$$v^*(w) = \begin{cases} 1 & \text{if } v = w \\ 0 & \text{otherwise.} \end{cases}$$

Definition 2.2 Given a Lorentzian vector space (V, q) we shall say that $v \in V$ is

timelike if $q(v, v) < 0$
null   if $q(v, v) = 0$ and $v \neq 0$
spacelike if $q(v, v) > 0$ or $v = 0$.

Furthermore, v will be causal if it is either timelike or null. In this setting, we will call light cone the set

$$CV := \{v \in V \ : \ q(v, v) = 0\}.$$

Remark 2.3 In the setting above, the set of timelike vectors in V consist of two (open) connected components, in fact two open cones in V called timecones.

Let now M be a smooth[1] manifold and let SM be the total space of the vector bundle $\pi : SM \to M$ consisting of symmetric bilinear forms defined on the tangent space to M so that for $x \in M$ the fiber SM_x consists of $\mathrm{Sym}_2(T_x M; \mathbb{R})$.

Definition 2.4 Given a smooth manifold M (and defined SM as above) a Lorentzian metric γ is a section of the bundle SM having signature $(1, 0, n)$ at each point of M. The couple (M, γ) will be called Lorentzian manifold.

Remark 2.5 If (M, γ) is a Lorentzian manifold, and $x \in M$ we can always find a neighborhood U of x and a smooth frame $\{V_0, V_1, \ldots, V_n\}$ such that $\gamma(V_i, V_j) = 0$ for any $i \neq j$ and $\gamma(V_0, V_0) = -1$, $\gamma(V_i, V_i) = 1$ for any $1 \leq i \leq n$. We will call any such local frame a Lorentzian frame.

Given a Lorentzian manifold, we can distinguish three types of vectors at each tangent space $T_x M$, as encoded in the terminology that was presented above.

Definition 2.6 Given a Lorentzian manifold (M, γ) we shall say that a tangent vector $v \in T_x M$ is

timelike if $\gamma(v, v) < 0$
null   if $\gamma(v, v) = 0$ and $v \neq 0$
spacelike if $\gamma(v, v) > 0$ or $v = 0$.

Furthermore, v will be causal if it is either timelike or null. In this setting, we will call light cone of M at x the subspace $C_x M \subset T_x M$ given by

$$C_x M := \{v \in T_x M \ : \ \gamma(v, v) = 0\}.$$

---

[1]For the sake of clarity, we shall always tacitly assume the word *smooth* to mean C^∞ and leave the straightforward modifications needed to handle the case of finite regularity, as encoded by functional spaces like $C^{k,\alpha}$ or $W^{k,p}$, to the reader. Such spaces will be relevant at a later stage and we will recall their definitions at due course.

Remark 2.7 In the setting above, at each point $x \in M$ the set of timelike vectors in $T_x M$ consist of two (open) connected components, in fact two open cones in $T_x M$ called timecones.

Definition 2.8 We shall say that a Lorentzian manifold (M, γ) is time-orientable if there exists a globally defined, smooth vector field $V : M \to TM$ such that $V(x)$ is timelike at each $x \in M$.

One has the following simple fact:

Proposition 2.9 *A smooth manifold M can be endowed a time-orientable Lorentzian metric if and only if there exists a nowhere vanishing vector field $V : M \to TM$.*

As an application, we see at once that (by virtue of the Poincaré-Hopf theorem) manifolds like S^{2n}, $n \geq 1$ do not admit a time-orientable Lorentz structure. In fact, it is well-known (see e.g. chapter 3 of [43]) that a compact, connected manifold without boundary admits a nowhere vanishing vector field if and only if its Euler characteristic is zero, so that such a condition is necessary and sufficient for the manifold in question to be endowed with a time-orientable Lorentzian metric. On the contrary, the condition required by Proposition 2.9 (namely: the existence of a nowhere vanishing vector field) is satisfied by any non-compact manifold.

Keeping these facts in mind, we shall then introduce a fundamental concept:

Definition 2.10 A spacetime is a connected, time-orientable Lorentzian manifold (M, γ).

Remark 2.11 In a spacetime (M, γ), one can canonically (and globally) distinguish the two timecones of the tangent space at each given point. Indeed, given a time-orienting vector field V and set $v = V(p)$ for some $p \in M$ we shall say that a timelike vector $w \in T_p M$ is

$$\text{future-pointing if } \gamma(v, w) < 0$$
$$\text{past-pointing} \quad \text{if } \gamma(v, w) > 0.$$

Now, let $\sigma : I \to M$ be a smooth curve. In general, the type of its tangent vector σ' will depend on the point and will change from point to point. In a number of situations, one is mostly interested to those curves for which such a type is the same at all points.

Definition 2.12 Given a Lorentzian manifold (M, γ) we shall say that $\sigma : I \to M$ is a timelike curve (resp. a null curve or spacelike curve) if $\sigma'(s)$ is timelike (resp. null or spacelike) for all $s \in I$. Furthermore, we shall say that $\sigma : I \to M$ is a causal curve if $\sigma'(s)$ is causal for all $s \in I$. Lastly, we shall say that $\sigma : I \to M$ is future-pointing (resp. past-pointing) if $\sigma'(s)$ is future-pointing (resp. past-pointing) for all $s \in I$.

There are important geometric quantities, independent of the chosen parametrization, associated with such specific classes of curves.

Definition 2.13 Given a Lorentzian manifold (M, γ) and an interval $I \subset \mathbb{R}$ containing the origin, we shall define

1. for a spacelike curve $\sigma : I \to M$ the arclength parameter to be the function

$$l(\bar{s}) = \int_0^{\bar{s}} \sqrt{\gamma(\sigma'(s), \sigma'(s))} \, ds;$$

2. for a timelike curve $\sigma : I \to M$ the proper time to be the function

$$\tau(\bar{s}) = \int_0^{\bar{s}} \sqrt{-\gamma(\sigma'(s), \sigma'(s))} \, ds.$$

Timelike curves represent the spacetime positions of physical observers and the proper time of such a curve is nothing but a natural notion of time for the corresponding observer (naturally meaning that the notion in question is independent of the specific parametrization of the curve).

2.3 The Einstein Equation as an Initial Value Problem

Given a Lorentzian manifold (M, γ) we let ∇ stand for its Levi-Civita connection. Denoted the second covariant derivative by ∇^2, so that by definition

$$\nabla^2_{X,Y} = \nabla_X \nabla_Y - \nabla_{\nabla_X Y},$$

we let

$$
\begin{aligned}
Riem_\gamma(X, Y, Z, W) &= -\gamma(\nabla^2_{X,Y} Z - \nabla^2_{Y,X} Z, W) \\
&= -\gamma(\nabla_X \nabla_Y Z - \nabla_Y \nabla_X Z - \nabla_{[X,Y]} Z, W)
\end{aligned}
$$

denote the $(0, 4)$ Riemann curvature tensor. Correspondingly, the Ricci and scalar curvatures are defined by taking traces over local Lorentzian bases: given a local Lorentzian frame $\{V_0, V_1, \ldots, V_n\}$ (according to the definition given in the previous section) one sets

$$Ric_\gamma(X, Y) = -Riem_\gamma(X, V_0, Y, V_0) + \sum_{i=1}^n Riem_\gamma(X, V_i, Y, V_i)$$

and

$$R_\gamma = -Ric_\gamma(V_0, V_0) + \sum_{i=1}^{n} Ric_\gamma(V_i, V_i).$$

We shall adopt the very same notation for Riemannian metrics as well.

The Newtonian description of gravity is in patent contrast with the foundational principles of special relativity, and in fact poses experimental and theoretical issues. In order to overcome those issues and moving in the direction of a conceptually consistent description of physical phenomena, Einstein proposed in 1915 (see [34, 35]) a completely new framework for gravitational forces, which we will now start to describe. General relativity is built starting from two basic, dual postulates:

Postulate 1 The action of gravitational forces is encoded in the geometric properties of connected, time-oriented Lorentzian manifolds $(\mathbb{L}, \gamma)$, called spacetimes, satisfying the field equations

$$Ric_\gamma - \frac{1}{2}R_\gamma\gamma + \Lambda\gamma = 8\pi T$$

where $\Lambda \in \mathbb{R}$ is a physically determined number called cosmological constant and T is the stress–energy tensor describing matter (a suitable generalization of the stress tensor in classical mechanics);

Postulate 2 Free fall (namely: motion under the sole influence of gravity) is geodesic.

In the setting above, we call Einstein tensor the term

$$G_\gamma = Ric_\gamma - \frac{1}{2}R_\gamma\gamma.$$

and recall that it is a divergence-free tensor whose (Lorentzian) trace is equal to $(1 - n)/2$, so that in the physical case of $1 + 3$ spacetimes we get $G_\gamma = -Ric_\gamma$.

For various classes of relevant physical systems, and specifically as far as one is concerned with isolated gravitational systems (as we shall be here) the effect of the term involving Λ is essentially negligible, so that it is customary to study the above equations under the assumption that $\Lambda = 0$. This is the convention we will tacitly make for what concerns the rest of these lectures, unless otherwise explicitly stated. That being said, observe that when $n > 1$ (namely with at least two spatial dimensions) one has

$$G_\gamma = 0 \quad \Leftrightarrow \quad Ric_\gamma = 0$$

namely the vacuum Einstein equations, corresponding to $T = 0$, are satisfied by all and only Ricci-flat (connected, time-orientable) Lorentzian manifolds. This suggests that (differently from Newtonian gravity) non-trivial vacuum solutions should in fact exist in abundance.

Concerning the physical term T, we introduce some terminology that is motivated by the study of a number of specific, concrete examples.

Definition 2.14 Let $(\mathbb{L}, \gamma)$ be a Lorentzian manifold solving the Einstein field equations and let $\{E_0, E_1, \ldots E_n\}$ be a local Lorentzian frame. Then, considered the components $T_{ij} := T(E_i, E_j)$ we shall employ the following terminology:

T_{00}	is the energy density
$T_{i0}\ i \neq 0$	is the momentum density
$T_{ii}\ i \neq 0$	is the normal stress
$T_{ij}\ i \neq j, i \neq 0, j \neq 0$	is the shear stress

In practice, rather than restricting to a specific matter model, it is often useful to just introduce suitable energy conditions that place some general constraints on the structure of the stress–energy tensor. These come in the form of functional inequalities that are required to be satisfied pointwise by the classes of data one considers. We shall focus here on the one that is most commonly adopted.

Definition 2.15 We shall say that the stress–energy tensor T satisfies the dominant energy condition (often abbreviated DEC in the sequel) if for any future-pointing causal vector field U one has that the vector field $-T(U, \cdot)^\sharp$ is itself future-pointing causal.

The DEC can be physically interpreted as the requirement that the flow of mass-energy can never be measured (by any physical observer, see our comments after Definition 2.13) to happen at a speed larger than the speed of light.

We shall now set up the Einstein field equations as an hyperbolic initial value problem. In particular, we will present and discuss the constraint equations involving the first and second fundamental form for a spacelike hypersurface inside a spacetime solving the Einstein field equations.

Let us then assume that $(\mathbb{L}, \gamma)$ contains a spacelike hypersurface M and let g, h be respectively the first and second fundamental form of M in $\mathbb{L}$. The conventions adopted here are the following ones:

$$h(X, Y) = \gamma(\nabla_X V, Y) = -\gamma(\nabla_X Y, V)$$

for every couple of sections of TM and where V is a (locally defined) timelike, future-pointing, unit normal vector field to M. Before proceeding, let us recall two basic facts about the extrinsic geometry of submanifolds: in the setting above, if X, Y, Z, W denote smooth sections of the tangent vector bundle of TM then

$$Riem_g(X, Y, Z, W) = Riem_\gamma(X, Y, Z, W) - h(X, Z)h(Y, W) + h(X, W)h(Y, Z)$$

which is the Gauss equation for spacelike hypersurfaces in Lorentzian manifolds; furthermore

$$Riem_\gamma(X, Y, Z, V) = \nabla_X h(Y, Z) - \nabla_Y h(X, Z)$$

which is a special case of the general Codazzi–Mainardi equations, the general formula concerning the shape operator of submanifolds of arbitrary codimension.

Now, let us further recall that in the Euclidean space the two theorems above give the geometric constraints on any smooth immersion $F : U \subset \mathbb{R}^2 \to \mathbb{R}^3$, which are in fact necessary and sufficient conditions for its local realizability. This assertion is clarified by means of the following statement:

Theorem 2.16 (Bonnet) *The Gauss and Codazzi equations always locally determine one and only one Euclidean surface up to rigid motions: given g, h defined on an open set $\Omega \subset \mathbb{R}^2$ and satisfying the Gauss and Codazzi equations, and $q \in \Omega$ one can always find an open connected set $\Omega_0 \subset \Omega$ containing q and a smooth immersion $\varphi : \Omega_0 \to \mathbb{R}^3$ with first fundamental form g and second fundamental form h; furthermore if $\varphi_1, \varphi_2 : \Omega_0 \to \mathbb{R}^3$ are two such immersions, then there exist $\varrho \in SO(3)$ and $b \in \mathbb{R}^3$ such that $\varphi_2 = \varphi_1 \cdot \varrho + b$.*

This is sometimes referred to as *fundamental theorem of the local theory of surfaces*, see e.g. Section 4.9 in [1] for a complete proof. That important point being recalled, let us go back to our setting:

Theorem 2.17 *Let $(\mathbb{L}, \gamma)$ be a spacetime solving the Einstein equations. Then any spacelike hypersurface (M, g, h) satisfies the system*

$$\begin{cases} R_g - \|h\|_g^2 + (tr_g h)^2 = 16\pi\mu \\ div_g(h - (tr_g h)g) = 8\pi J \end{cases} \tag{2.1}$$

where we have set $\mu = T(V, V)$ and $J = T(V, \cdot)$ for V a future-pointing timelike unit normal vector field to M. These are known as Einstein constraint equations.

Definition 2.18 Let (M, g, h) be a triple where:

(M, g) is a smooth, complete Riemannian manifold

h is a smooth $(0, 2)$ tensor field.

We call (M, g, h) an initial data set if it satisfies the Einstein constraint equations. Furthermore, we shall say that (M, g, h) satisfies the dominant energy condition if $\mu \geq |J|_g$ at every point of M, where μ and J are *defined* by the left-hand side of the two equations (2.1).

The reason for such terminology is contained in the following statement, which mirrors the (existence part of) Bonnet's theorem in the Euclidean scenario.

Theorem 2.19 *Let (M, g, h) be an initial data set satisfying the vacuum constraint equations (VCE):*

$$\begin{cases} R_g - \|h\|_g^2 + (tr_g h)^2 = 0 \\ div_g(h - (tr_g h)g) = 0 \end{cases}$$

then, there exists a spacetime $(\mathbb{L}, \gamma)$ and an embedding $\iota : M \to \mathbb{L}$ such that the following assertions are true:

1. *the spacetime $(\mathbb{L}, \gamma)$ is Ricci-flat;*
2. *g is the induced metric by ι, namely $\iota^*(\gamma) = g$;*
3. *h is the second fundamental form of the embedding $\iota : M \to \mathbb{L}$.*

This is a pioneering result of Choquet-Bruhat [39], later improved in joint work with Geroch [22] to prove existence of a maximal, globally hyperbolic extension. In other words, the Einstein constraint equations provide a necessary and sufficient condition for the embeddability of a Riemannian hypersurface inside a spacetime solving the Einstein field equations. It is worth mentioning that in fact, with substantially more effort, one can gain the same conclusion in a more constructive fashion and without invoking Zorn's lemma, which is a recent result of Sbierski [62]. There is also a more general version of the theorem above, which applies to the case with matter (i.e., when the right-hand side of the constraints is not zero).

Two peculiar cases deserve special attention: we shall say that an initial data set (M, g, h) is time-symmetric (resp. maximal) if $h = 0$ (resp. $tr_g h = 0$) identically. Notice that, under these assumptions, the Einstein constraint equations take the special form

$$R_g = 16\pi\mu, \quad \text{or} \quad \begin{cases} R_g - \|h\|_g^2 = 16\pi\mu \\ div_g(h) = 8\pi J \end{cases}$$

respectively.

These cases have been extensively studied, and are a lot better understood than the general one. However, it is a non-trivial issue whether there exist (inside a spacetime $(\mathbb{L}, \gamma)$) spacelike slices that are either time-symmetric (in other words: totally geodesic) or maximal.

We shall present here the proof of Theorem 2.17.

Proof Given $p \in M$, let $\{E_1, \ldots, E_n\}$ be a local orthonormal frame for M which we shall assume (without loss of generality) to be parallel at p. This is completed to a local Lorentzian frame by means of the vector $E_0 = V$ (as in the statement of the theorem).

Let us start by deriving the first constraint: considering the 00-component of the Einstein field equation we get

$$(Ric_\gamma)_{00} + \frac{1}{2} R_\gamma = 8\pi\mu$$

but on the other hand, recalling the definition of scalar curvature,

$$R_\gamma = -\sum_{i=0}^{n}(Riem_\gamma)_{0i0i} - \sum_{i=0}^{n}(Riem_\gamma)_{i0i0} + \sum_{i,j=1}^{n}(Riem_\gamma)_{ijij}$$

$$= -2(Ric_\gamma)_{00} + \sum_{i,j=1}^{n}(Riem_\gamma)_{ijij}$$

hence

$$R_\gamma + 2(Ric_\gamma)_{00} = \sum_{i,j=1}^{n}(Riem_\gamma)_{ijij}$$

so that we can use the (Lorentzian) Gauss equations to get

$$R_\gamma + 2(Ric_\gamma)_{00} = \sum_{i,j=1}^{n}(Riem_g)_{ijij} + \sum_{i=1}^{n}h_{ii}\sum_{j=1}^{n}h_{jj} - \sum_{i,j=1}^{n}h_{ij}^2$$

$$= R_g + (tr_g h)^2 - \|h\|_g^2.$$

Combining the first and the last equation above gives the first constraint.

On the other hand, if we evaluate the Einstein field equation at (V, E_j) for some index $1 \le j \le n$ then we get

$$(Ric_\gamma)_{0j} = 8\pi J(E_j)$$

and one can rewrite the left-hand side using the Codazzi equations as follows:

$$(Ric_\gamma)_{0j} = -(Riem_\gamma)_{000j} + \sum_{i=1}^{n}(Riem_\gamma)_{i0ij} = \sum_{i=1}^{n}(Riem_\gamma)_{iji0}$$

$$= \nabla_i h_{ji} - \nabla_j h_{ii} = (div_g h)_j - (d(tr_g h))_j = (div_g h)_j - div_g((tr_g h)g)_j$$

thus the second constraint follows as well. $\qquad\square$

Now, to focus on the simplest possible instance, let us consider the time-symmetric Einstein constraint equations on a given background manifold such as $\mathbb{R}^3$: like it happens in the familiar case of the Poisson equation, the constraint

$$R_g = 16\pi\mu$$

does not uniquely determine the asymptotic behaviour of the solution, which can instead be rather diverse. In the case of the Einstein equation, this is consistent with the fact that this model is supposed to cover a very broad spectrum of physical phenomena (corresponding to very diverse classes of physical sources). The goal of the next lectures will be the study of asymptotically flat spaces, namely of Riemannian manifolds (M^n, g) such that $M \setminus K$ is diffeomorphic to $\mathbb{R}^n$ and such that, in the induced coordinates one has

$$g - \delta = O(|x|^{-p}), \quad \text{as } |x| \to \infty.$$

For $M \simeq \mathbb{R}^3$ and $p = 1$, the metric g can be regarded as the relativistic counterpart of the Newtonian potential describing an isolated gravitational system, understood as a compactly supported distribution of matter (cf. Sect. 2.1).

If we assume the falloff of the metric to be rapid enough (and that the Einstein constraints are satisfied), then we can define a notion of mass by means of the formula (dating back to [3])

$$m = \frac{1}{2(n-1)\omega_{n-1}} \lim_{r \to \infty} \int_{|x|=r} \sum_{i,j=1}^{n} \left(g_{ij,i} - g_{ii,j}\right) v_0^j \, d\mathcal{H}^{n-1}$$

where ω_{n-1} equals the $(n-1)$-dimensional measure of the unit sphere in flat $\mathbb{R}^n$ and v_0 is the corresponding unit normal. Notice that one can write

$$R_g = g_{ij,ij} - g_{ii,jj} + O((g-\delta)\partial^2 g) + O((\partial g)^2)$$

with sum over repeated indices and where all computations are understood in the Euclidean chart at infinity. If we restrict our analysis, for the sake of consistency, to the case $n = 3$ we see at once that the multiplicative constant in the definition of ADM mass equals $1/16\pi$ and thus by the divergence theorem we have

$$m = \lim_{r \to \infty} \int_{|x| \leq r} (\mu - h.o.t.) \, dx$$

which corresponds to the Newtonian scenario modulo the higher order terms. This formula will be the starting point for our next lecture.

3 Second Lecture: The Riemannian Positive Mass Theorem

The definition, sketched above, of asymptotically flat Riemannian manifold can be formalized and specified in different ways, by requiring that the error terms of the metric belong to suitable weighted Sobolev or Hölder spaces. Our goal here is not to present any results in their maximal generality or under sharp technical assumptions, but rather to quickly introduce the reader to the subject: for this reason we will opt for the simplest possible setup.

3.1 Definitions and Main Statements

One can also use Hölder spaces and prescribe decay in a pointwise sense.

Definition 3.1 Given an integer $n \geq 3$, and numbers $\alpha \in (0, 1)$, $q > \frac{n-2}{2}$, $q_0 > n$, a Riemannian manifold (M, g) is called asymptotically flat if there exists a compact set $K \subset M$ (the *interior* of the manifold) such that $M \setminus K$ consists of a disjoint union of finitely many ends, namely $M \setminus K = \bigsqcup_{\ell=1}^{e} E_\ell$ and for each index ℓ there exists a smooth diffeomorphism $\Phi_\ell : E_\ell \to \mathbb{R}^n \setminus B_\ell$ for some open ball $B_\ell \subset \mathbb{R}^n$ containing the origin so that the pull-back metric $\left(\Phi_\ell^{-1} \right)^{*} g$ satisfies the condition

$$g_{ij} - \delta_{ij} \in C_{-q}^{2,\alpha}(\mathbb{R}^n \setminus B_\ell),$$

andthe scalar curvature satisfies the condition

$$R_g \in C_{-q_0}^{0,\alpha}(\mathbb{R}^n \setminus B_\ell) \text{ for every } \ell = 1, \ldots, e.$$

Roughly speaking, we are simply requiring that the metric g decays along each end to the Euclidean one modulo an error term bounded by $|x|^{-q}$, with its first derivatives bounded by $|x|^{-q-1}$ and so on for all derivatives up to and order two. Furthermore, we require a faster falloff on the scalar curvature, beyond the integrability threshold at infinity: we want $R_g = O(|x|^{-q_0})$ for some $q_0 > n$. Notice that the latter requirement is always satisfied in the important special case of the vacuum (time-symmetric) constraints.

Example For $n = 3$ consider the Riemannian manifold (M, g) where

$$M = \mathbb{R}^3 \setminus \{0\}, \quad g = \left(1 + \frac{m}{2|x|} \right)^4 \delta.$$

This is an asymptotically flat manifold with two ends ($e = 2$) and the set $|x| = m/2$ is a totally geodesic round sphere S^2 corresponding to the set of fixed points for the isometry described in the chart above by $x \mapsto x/|x|^2$.

Remark 3.2 In many applications one also wishes to consider the case when (M, g) is a Riemannian manifold with boundary, in which case one postulates that the boundary consists of finitely many closed minimal hypersurfaces, possibly with the additional requirement that there are no further minimal hypersurfaces in $M \setminus \partial M$ (an assumption often described as *horizon boundary*). For instance, such an example is provided by

$$M = \mathbb{R}^3 \setminus \{|x| \leq m/2\}, \quad g = \left(1 + \frac{m}{2|x|} \right)^4 \delta.$$

A significant problem in the middle of the past century was to define, for asymptotically flat manifolds, a meaningful notion of *mass* and, more generally, an energy-momentum 4-vector for asymptotically flat initial data sets. This was accomplished by Arnowitt–Deser–Misner in [3]. For the sake of simplicity, we shall keep focusing on the purely Riemannian case.

Definition 3.3 Given an asymptotically flat manifold (M, g) *with one end*, the ADM mass m is defined by means of the formula

$$m = \frac{1}{2(n-1)\,\omega_{n-1}} \lim_{r \to \infty} \int_{|x|=r} \sum_{i,j=1}^{n} (g_{ij,i} - g_{ii,j}) \, v_0^j \, d\mathcal{H}^{n-1}$$

where we have set $v_0^j = x^j / |x|$ and ω_{n-1} is the volume of the unit sphere in $\mathbb{R}^n$.

One can check, using the Einstein constraint equations, that this quantity is well-defined in the sense that the limit above exists and is finite. The following step is to show that the quantity m is canonical, meaning that it does not depend on the choice of the charts at infinity: this is a non-trivial fact, for which we refer the reader to Bartnik [5] and Chruściel [23]. We will give this fact for granted in the sequel of these lectures. Instead, our scope now will be to prove the Riemannian positive mass theorem:

Theorem 3.4 *Let (M, g) be an asymptotically flat manifold of non-negative scalar curvature. Then the ADM mass m of (M, g) is non-negative and equals zero if and only if the manifold in question is isometric to the Euclidean space $(\mathbb{R}^n, \delta)$.*

We shall first investigate the *weak* positive mass theorem, namely we shall aim at proving the sole inequality $m \geq 0$. The rigidity part of the statement, namely the full characterization of the case $m = 0$ will be discussed in the next lecture.

The theorem above was proven by Schoen and Yau in 1979 under the technical assumption that the ambient dimension be less than eigth [63]. Such an assumption related to the occurrence of singularities in high-dimensional stable minimal hypersurfaces was removed in 2017 with a rather different approach [64]. We will follow the general scheme given in the latter article, which in particular allows to obtain a remarkably elegant proof in the low-dimensional case.

The first step in our program is to prove the following assertion, which we shall refer to as *compactification theorem*:

Theorem 3.5 *Suppose all closed manifolds of the form $M^n \simeq M_0^n \# T^n$ support no metric of positive scalar curvature. Then the ADM mass is non-negative for all asymptotically flat manifolds having non-negative scalar curvature.*

The idea is to reduce the question of positivity of the ADM mass in general relativity to a question about those manifolds that do/do not carry metrics of positive scalar curvature.

3.2 Proof of the Compactification Theorem

We shall now present the proof of Theorem 3.5. We will employ a simple result about the behaviour of the mass functional with respect to conformal deformations.

Lemma 3.6 *Let (M, g) be asymptotically flat and set $\overline{g} = u^4 g$, for some C^1 function u having the asymptotic expansion*

$$u(x) = 1 + A|x|^{2-n} + o_1(|x|^{2-n}) \tag{3.1}$$

as one lets $|x| \to \infty$ in the asymptotically flat coordinates along the only end of M. Then the limit

$$\lim_{r \to \infty} \int_{|x|=r} \sum_{i,j=1}^{n} (\overline{g}_{ij,i} - \overline{g}_{ii,j}) \, v_0^j \, d\mathcal{H}^{n-1}$$

exists, and attains a finite value such that

$$m(\overline{g}) = m(g) + 2A.$$

Example Let (M, g) be any asymptotically flat manifold which coincides, outside a compact set, with the one-ended Schwarzschild manifold described in Remark 3.2. Then the parameter m equals the ADM mass of such a manifold.

In the scalar curvature deformation arguments we are about to present, we will make repeated use of the following basic result in linear analysis on asymptotically flat manifolds:

Theorem 3.7 *Let (M^n, g) be an asymptotically flat manifold as specified in Definition 3.1. Then there exists a constant $\epsilon_0 = \epsilon_0(n, g)$ such that if $f \in C^{0,\alpha}_{-q'}$ for $q' > n$ and $f_- \in L^{\frac{n}{2}}$ with $\|f_-\|_{L^{n/2}} < \epsilon_0$ then the equation*

$$\Delta_g u - f u = 0$$

has a positive solution, which has near infinity an asymptotic expansion of the form (3.1).

Remark 3.8 With respect to the statement of the previous theorem, let us observe that

$$A = -\frac{1}{(n-2)\omega_{n-1}} \int_M (fu) \, d\mathcal{V}_g^n.$$

To the scope of proving Theorem 3.5 we first need to present two simple scalar curvature deformations theorems. Both of them can be regarded as direct consequences of Theorem 3.7.

Corollary 3.9 *Let (M, g) be an asymptotically flat manifold having non-negative scalar curvature, not identically equal to zero. Then there exists a conformal metric $\overline{g} = u^{\frac{4}{n-2}}g$ on M, for $u \in C_{loc}^{2,\alpha}$ satisfying (3.1), that is scalar flat, and such that $m(\overline{g}) < m(g)$.*

Proof The conformal factor u is obtained by a direct application of Theorem 3.7 with $f = c(n)R_g$ (thus $q' = q_0$), where $c(n) = (n-2)/4(n-1)$. Notice in particular that for such a choice $f_- = 0$. For the second claim, recall Lemma 3.6 and Remark 3.8: it suffices to show that

$$\int_M c(n)R_g u \, d\mathcal{V}_g^n > 0$$

but this follows straight from our assumption and the positivity of u. $\square$

Corollary 3.10 *Let (M, g) be a scalar flat, asymptotically flat manifold. Given any $\epsilon > 0$ there exists an asymptotically flat metric $\overline{g}$ on M that is scalar flat, takes the form $u^{4/(n-2)}\delta$ outside a compact set (for some harmonic u such that $u \to 1$ as one lets $|x| \to \infty$) and $|m(\overline{g}) - m(g)| < \epsilon$.*

Proof Let $\chi : [0, +\infty) \to [0, +\infty)$ a smooth, monotonically decreasing cutoff function that equals 1 for $t \le 1$ and vanishes for $t \ge 2$. For any $k \ge 1$ set $\chi_k := \chi(x/k)$ so that one has that the first (resp. second) derivatives of χ_k are bounded by some uniform constant times k^{-1} (resp. k^{-2}) as one lets $k \to \infty$. Consider the interpolating metric

$$\hat{g}_k = \chi_k g + (1 - \chi_k)\delta.$$

Set $\overline{g}_k = u_k^{4/n-2}\hat{g}_k$: as usual, we want to solve for u_k the equation

$$\Delta_{\hat{g}_k} u_k - c(n)R_{\hat{g}_k} u_k = 0.$$

To that scope, and based on the statement of Theorem 3.7, we need to observe that the sole region where the scalar curvature of $\hat{g}_k$ is not zero is the annulus of radii k and $2k$ so that one easily checks (based on the decay assumptions on the metric g) that

$$\|R_{\hat{g}_k}\|_{L^{n/2}} \to 0, \text{ as one lets } k \to \infty$$

and thus there exists $k_0 = k_0(\epsilon)$ such that for all $k \ge k_0$ the equation above is solvable, in the sense described by Theorem 3.7. What we need to check now is that $\overline{m}_k \to m$ as one lets $k \to \infty$, where we have denoted by m (resp. $\overline{m}_k$) the ADM mass of g (resp. of $\overline{g}_k$). Recall that $\overline{m}_k = 2A_k$ where A_k is the coefficient of $|x|^{2-n}$

in the asymptotic expansion of u_k. On the other hand, we also know that

$$A_k = -\frac{1}{(n-2)\omega_{n-1}} \int_M c(n) R_{\hat{g}_k} u_k \, d\mathcal{V}_{\hat{g}_k}^n.$$

If we set $u_k = 1 + v_k$, one can readily check that

$$\left| \int_M R_{\hat{g}_k} v_k \, d\mathcal{V}_{\hat{g}_k}^n \right| \to 0 \quad \text{as one lets } k \to \infty.$$

Given the pointwise bounds on the terms of the difference

$$R_{\hat{g}_k} - ((\hat{g}_k)_{ij,ij} - (\hat{g}_k)_{ii,jj})$$

we then have

$$A_k = -\frac{1}{(n-2)\omega_{n-1}} \int_M c(n) R_{\hat{g}_k} \, d\mathcal{V}_{\hat{g}_k}^n$$

$$= -\frac{1}{4(n-1)\omega_{n-1}} \int_{k<|x|<2k} \left[((\hat{g}_k)_{ij,ij} - (\hat{g}_k)_{ii,jj}) \right] d\mathcal{V}_{\hat{g}_k}^n + o(1)$$

and so, applying the divergence theorem in the shell, and keeping in mind the definition of $\hat{g}_k$ as interpolation of g and δ one finally gets

$$A_k = \frac{1}{4(n-1)\omega_{n-1}} \int_{|x|=k} (g_{ij,i} - g_{ii,j}) \frac{x^j}{|x|} \, d\mathcal{H}^{n-1} + o(1)$$

so the conclusion $A_k \to m$ for $k \to \infty$ follows directly from the definition of ADM mass.

$\square$

Before proceeding, let us give a simple lemma of independent interest.

Proposition 3.11 *Let M^n be a smooth compact manifold, without boundary. If M supports a C^2 metric of non-negative scalar curvature that is not Ricci flat, then it also supports a C^∞ metric of positive scalar curvature.*

Proof Let g_0 be a Riemannian metric on M that has non-negative scalar curvature and is not Ricci flat. Let us then consider the Ricci flow emanating from the initial datum g_0, namely the evolution problem:

$$\begin{cases} \partial_t g(t) = -2Ric_{g(t)} \\ g(0) = g_0. \end{cases}$$

We know that the problem above has a unique (smooth) solution on a time interval $(0, \tau)$ for some $\tau > 0$. Notice that for the present argument we do not need τ to be the maximal time of existence for the flow above. The scalar curvature of $g(t)$ evolves according to the equation

$$\partial_t R_{g(t)} = \Delta_{g(t)} R_{g(t)} + 2\|Ric_{g(t)}\|^2_{g(t)}.$$

In particular, a standard application of the parabolic maximum principle ensures that, if g_0 is not Ricci flat, then $g_1 := g(\tau/2)$ has positive scalar curvature at all points, as we had claimed.

$\square$

Let us now give the proof of Theorem 3.5.

Proof Arguing by contradiction, let (M^n, g_0) be an asymptotically flat manifold of non-negative scalar curvature and negative mass. Applying, one after the other, Corollaries 3.9 and 3.10 we easily obtain another Riemannian metric g on M that is scalar flat, has strictly negative mass, and takes the form

$$g = u^{\frac{4}{n-2}}\delta$$

at least outside a compact set. In particular, the function u is harmonic with respect to the Euclidean metric so it has an asymptotic expansion of the form

$$u(x) = 1 + \frac{A}{|x|^{n-2}} + O_1(|x|^{1-n})$$

outside of the coordinate ball of radius σ_0. Recall that $m(g) = 2A$ so in particular $A < 0$. Now, due to this fact, one can find $\eta \in (0, 1/2)$ such that the level set defined by $\{x \in M : u(x) = 1 - \eta\}$ is regular and (neglecting all but the outermost connected component) is diffeomorphic to S^{n-1}. In fact, we can (and we shall) assume that such a level set consists of only one connected component: if not, the following truncation is understood to happen only in the outermost component, where the radial derivative of u is strictly positive. The continuous function

$$v(x) := \min\{u(x), 1 - \eta\}$$

is, in the sense of distributions, superharmonic (being the infimum of two harmonic functions) and thus it can be approximated by a smooth function w such that

$$w(x) = \begin{cases} 1 - \eta & |x| \geq \sigma_2 \\ u(x) & |x| \leq \sigma_1 \end{cases}$$

for positive constants $\sigma_1 < \sigma_2$ (both larger than σ_0), in a way that

$$\Delta_g w \leq 0$$

with strict inequality in some subdomain of M. If we now recall the equation describing the change in scalar curvature associated with a conformal deformation of the metric, we must conclude that the metric

$$\overline{g} = \begin{cases} (1 - \eta)^{-\frac{4}{n-2}} g & |x| \leq \sigma_1 \\ ((1 - \eta)^{-1} w)^{\frac{4}{n-2}} \delta & |x| \geq \sigma_0 \end{cases}$$

has non-negative scalar curvature, and in fact positive scalar curvature somewhere. In particular, notice that $\overline{g}$ is not Ricci flat. Now, it is obvious that such a metric is exactly Euclidean outside a compact set and hence we can identify opposite sides of a large coordinate box in M and take the compactified manifold $\overline{M} \simeq M_0 \# T^n$ where M_0 is the one-point compactification of M. This manifold is naturally equipped with the metric $\overline{g}$, which is scalar non-negative but not Ricci flat. In particular, by Proposition 3.11 the manifold $\overline{M}$ would also support a smooth metric of positive scalar curvature, which completes the proof.   □

3.3   Torus Rigidity Theorems

We shall now focus on the question whether n-dimensional tori support Riemannian metrics of positive, or non-negative, scalar curvature apart from the flat one.

To answer this question we will employ minimal hypersurfaces, so let us first gather some basic facts. In what follows, we always let (M, g) be a complete Riemannian manifold of dimension $n + 1$ (for $n \geq 2$) and we let Σ be a smooth, closed embedded hypersurface (so that its dimension equals n). We always assume M to be orientable and Σ to be two-sided, namely we postulate the existence of a (smooth) unit vector field $N : \Sigma \to TM$ that is normal to $T_p \Sigma$ at each point $p \in \Sigma$ (we remind the reader that this assumption is in fact equivalent to the orientability of Σ itself, given the orientability of M). Hence, we let $X : \Sigma \to TM$ be any smooth vector field defined over Σ and in such case we have the natural orthogonal decomposition $X = X^{\perp} + X^{\parallel}$, where $X^{\perp} = uN$ for $u = g(X, N)$, the function describing the normal component of X.

In this setting, we let $F : (-\varepsilon, \varepsilon) \times M \to M$ be any flow of diffeomorphisms of M subject to the initial conditions

$$\begin{cases} F(0, x) = x & \text{for } x \in M \\ \left[\frac{\partial F}{\partial t} \right]_{t=0} (0, x) = X(x) & \text{for } x \in \Sigma. \end{cases}$$

It is then natural to study the rate of change of the area of Σ as we deform it by means of an admissible variation.

Proposition 3.12 (Setting as Above) *Given a smooth vector field $X : \Sigma \to TM$ and $F : (-\varepsilon, \varepsilon) \times M \to M$ be an associated flow of variations, then*

$$\left[\frac{d}{dt}\right]_{t=0} |F(t, \Sigma)| = -\int_{\Sigma} g(X, H) \, d\mathscr{H}_g^n$$

where $|F(t, \Sigma)|$ denotes the n-dimensional volume of $F(t, \Sigma)$ and H denotes the mean-curvature vector of Σ.

Remark 3.13 Throughout these notes, we adopt, for the mean curvature vector, the following convention: if $\{E_1, \ldots, E_n\}$ a local orthonormal frame for $T\Sigma$ then $H = \sum (D_{E_i} E_i)^{\perp}$, where D denotes the Levi-Civita connection of the ambient manifold (M, g).

Corollary 3.14 *Given a complete Riemannian manifold (M, g) a smooth, closed embedded hypersurface Σ is a critical point for the n-dimensional area functional if and only if its mean-curvature vector vanishes identically.*

This simple fact justifies the following definition:

Definition 3.15 Given a complete Riemannian manifold (M, g) we shall say that a smooth, closed embedded hypersurface Σ is minimal if it has vanishing mean curvature.

In particular, let us explicitly observe that a minimal surface does not need to be area-minimizing (the definition just refers to a first derivative test). Then we can proceed and study the second variation.

Proposition 3.16 (Setting as Above) *Given a minimal, closed embedded hypersurface Σ and a smooth vector field $X : \Sigma \to TM$ such that $X = X^{\perp} = uN$ and denoted by $F : (-\varepsilon, \varepsilon) \times M \to M$ an associated flow of variations, then*

$$\left[\frac{d^2}{dt^2}\right]_{t=0} |F(t, \Sigma)| = \int_{\Sigma} |\nabla_{\Sigma} u|^2 - (|A|^2 + Ric_g(N, N))u^2 \, d\mathscr{H}_g^n$$

where A denotes the second fundamental form of Σ in (M, g).

Furthermore, one can integrate by parts the right-hand side of the second variation formula, thereby getting

$$\frac{d^2}{dt^2}_{t=0} |F(t, \Sigma)| = -\int_{\Sigma} u J_{\Sigma} u \, d\mathscr{H}_g^n$$

where J is the Jacobi operator associated with Σ, namely

$$J_{\Sigma} u = \Delta_{\Sigma} u + (|A|^2 + Ric_g(N, N))u.$$

Now, the spectral theorem for bounded operators ensures the existence of an Hilbertian orthonormal basis $\{\phi_q\}_{q=1}^{\infty}$ and associated eigenvalues $\{\lambda_q\}_{q=1}^{\infty}$ with

$$\lambda_1 < \lambda_2 \leq \lambda_3 \ldots, \quad \lambda_q \to +\infty \text{ as } q \to \infty$$

satisfying

$$J_{\Sigma}\phi_q = -\lambda_q\phi_q.$$

Furthermore, the first eigenfunction ϕ_1 can be chosen to be positive (and, as can be inferred from the above inequality $\lambda_1 < \lambda_2$, the corresponding first eigenvalue is simple).

Definition 3.17 Let Σ be a closed, two-sided, embedded minimal hypersurface in a complete Riemannian manifold (M, g). We shall say it is stable if the second variation of the associated area functional is non-negative, namely if $\lambda_1(\Sigma) \geq 0$; if that is not the case we define the Morse index of Σ as $\text{Ind}(\Sigma) = \text{card}\{q \geq 1 : \lambda_q < 0\}$.

Example Let S^3 be the unit sphere with its standard round metric, identified with the set of points in $\mathbb{R}^4$ having unit distance from the origin. One can easily check that:

1. any round equatorial $S^2 \subset S^3$ is a minimal, in fact totally geodesic, surface of Morse index one;
2. any Clifford tours $S^1(1/\sqrt{2}) \times S^1(1/\sqrt{2}) \subset S^3$ is a minimal surface of Morse index five.

The following one is a first basic theorem providing some gist of the interactions that occur between stability and positive curvature conditions.

Theorem 3.18 *Let (M, g) be a complete 3-manifold and let $\Sigma \subset M$ be a closed, two-sided, embedded minimal surface.*

1. *If $Ric_g > 0$ then Σ is unstable;*
2. *If $R_g > 0$ and Σ is stable then it is diffeomorphic to S^2.*

Proof For the first part, observe that if Σ were stable then for every function $f \in C^1(\Sigma; \mathbb{R})$ one would have

$$\int_{\Sigma} |\nabla_{\Sigma}f|^2 \, d\mathcal{H}_g^2 \geq \int_{\Sigma} (|A|^2 + Ric_g(N, N)) f^2 \, d\mathcal{H}_g^2$$

but on the other hand this functional inequality is violated by simply choosing the function $f = 1$ identically on Σ. For the second part, we use the so-called rearrangement trick due to Schoen and Yau. First of all, using the Gauss equations for Σ one checks at once that

$$|A|^2 + Ric_g(N, N) = \frac{1}{2}(|A|^2 + R_g) - K$$

where K stands for the Gauss curvature of Σ. Hence, the stability inequality can be rewritten in the form

$$\int_\Sigma (|\nabla_\Sigma f|^2 + K f^2)\, d\mathcal{H}_g^2 \geq \int_\Sigma (|A|^2 + R_g) f^2\, d\mathcal{H}_g^2$$

again to hold for every differentiable function on Σ. When $f = 1$ this gives, using the Gauss–Bonnet theorem,

$$2\pi \chi(\Sigma) = 4\pi(1 - \mathfrak{g}) \geq \frac{1}{2}\int_\Sigma (|A|^2 + R_g)\, d\mathcal{H}_g^2 > 0$$

where $\mathfrak{g}$ denotes the genus of Σ. This is only possible if $\mathfrak{g} = 0$ and thereby the proof is complete.   □

In fact, with some extra effort one can strengthen the second assertion to analyze the relevant case when the ambient scalar curvature is only assumed to be non-negative: if $R_g \geq 0$ then Σ is diffeomorphic to either the sphere S^2 or the torus T^2 and in the latter case infinitesimal rigidity holds, namely Σ is totally geodesic, intrinsically flat and the ambient scalar curvature vanishes identically on Σ. Furthermore, when Σ is area-minimizing in fact M isometrically splits as a Riemannian product. We shall come back to the proof of this fact, in any dimension, later in this lecture. Instead, let us employ the theorem above to prove that tori cannot support Riemannian metrics of positive scalar curvature.

Theorem 3.19 *Let (M^n, g) be an n-dimensional Riemannian manifold, with $n < 8$, of the form $M_0^n \# T^n$ for some closed n-dimensional orientable manifold M_0^n. Then its scalar curvature must be non-positive somewhere.*

In order to proceed with the proof, we first need a simple lemma characterizing stability. While we will not need its full strength, we provide its proof for the sake of completeness.

Lemma 3.20 *Let (M^{n+1}, g) be a complete Riemannian manifold and let $\Sigma \subset M$ be a closed, embedded minimal hypersurface. Then Σ is stable (i.e., $\lambda_1(\Sigma) \geq 0$) if and only if there exists a smooth, positive function ϕ such that $J_\Sigma \phi \leq 0$.*

Proof That the first eigenfunction of a Schrödinger-type operator (in fact: of much more general elliptic operators) is positive is an easy consequence of the variational characterization of eigenvalues, so that the second assertion is trivially implied by the first by choosing $\phi = \phi_1$. The non-trivial part is the converse, namely gaining stability. This is known as Barta's criterion and here is the argument behind it: set $\psi = \log \phi$ so that

$$\nabla_\Sigma \psi = \frac{\nabla_\Sigma \phi}{\phi}, \text{ and } \Delta_\Sigma \psi = \frac{\Delta_\Sigma \phi}{\phi} - \frac{|\nabla_\Sigma \phi|^2}{\phi^2}$$

and thus the assumption $J_\Sigma \phi \leq 0$ gives

$$\Delta_\Sigma \psi \leq -(|A|^2 + Ric_g(N, N)) - \frac{|\nabla_\Sigma \phi|^2}{\phi^2}$$

hence given any function $f \in C^1(\Sigma)$ multiplying the previous inequality by f^2 one has

$$\int_\Sigma (|A|^2 + Ric_g(N, N)) f^2 \, d\mathcal{H}_g^n + \int_\Sigma \frac{|\nabla_\Sigma \phi|^2}{\phi^2} f^2 \, d\mathcal{H}_g^n$$

$$\leq - \int_\Sigma f^2 \Delta_\Sigma \psi \, d\mathcal{H}_g^n$$

$$= 2 \int_\Sigma g\left(\nabla_\Sigma f, f \frac{\nabla_\Sigma \phi}{\phi}\right) d\mathcal{H}_g^n$$

$$\leq \int_\Sigma |\nabla_\Sigma f|^2 \, d\mathcal{H}_g^n + \int_\Sigma \frac{|\nabla_\Sigma \phi|^2}{\phi^2} f^2 \, d\mathcal{H}_g^n$$

and so, cancelling out equal terms we have obtained that the stability inequality is satisfied, which completes the proof. □

Remark 3.21 All results presented in this section only require the background metric in the ambient manifold M to be of class C^2, and thus apply, as a special case, to asymptotically flat manifolds according to Definition 3.1.

The proof of Theorem 3.19, will be obtained as a special case of the following (more general) assertion:

Theorem 3.22 *Let (M^n, g) be an n-dimensional Riemannian manifold, with $n < 8$, and assume the existence of a smooth map $F : M^n \to T^n$ of degree one. Then its scalar curvature must be non-positive somewhere.*

Notice that such a degree one map certainly exists in the special case when $M^n \simeq M_0^n \# T^n$, for one can construct $F : M^n \to T^n$ by simply collapsing the first summand while keeping the second one fixed except for a small tubular transition region.

Proof Let $F : M^n \to T^n$ be a smooth, degree one map as in the statement of the theorem. For the sake of clarity, take $x^1, \ldots, x^n$ to be the coordinates on each S^1 factor of $T^n = S^1 \times \ldots \times S^1$ and, possibly scaling if necessary, assume that $\int_{S^1} dx^j = 1$. Let then F_j denote the composition of F and the projection π_j onto the j-the copy of S^1, namely

$$F_j : M^n \xrightarrow{F} T^n \xrightarrow{\pi_j} S^1$$

and set $\omega^j = F_j^*(dx^j)$. Since $\deg(F) = 1$ by the standard change of variables formula one has

$$\int_M \omega^1 \wedge \ldots \wedge \omega^n = 1.$$

At this stage, we consider the following class of $(n-1)$-currents:

$$\mathcal{C}_{n-1} = \left\{ \Sigma \text{ is an integral } (n-1) \text{ current in } M : \int_\Sigma \omega^1 \wedge \ldots \wedge \omega^{n-1} = 1 \right\}.$$

Observe that the class $\mathcal{C}_{n-1}$ is not empty: indeed at any regular point $p \in S^1$ for the map $F_n : M \to T^n \to S^1$ one has that $F_n^{-1}(p) \in \mathcal{C}_{n-1}$ because

$$\int_{F_n^{-1}(p)} \omega^1 \wedge \ldots \wedge \omega^{n-1} = \int_{F(F_n^{-1}(p))} dx^1 \wedge \ldots \wedge dx^{n-1} = \int_{T_n^{n-1}} dx^1 \wedge \ldots \wedge dx^{n-1} = 1.$$

Here, the second equality relies on the fact that $F(F_n^{-1}(p)) = \pi_n^{-1}(p)$ is an $(n-1)$-dimensional (unit) torus T_n^{n-1} in T^n, while the first uses the fact that the restriction of F to the pre-image $F_n^{-1}(p)$ is still a degree one map. Incidentally, once this is checked it follows from Sard's lemma that *almost every* $p \in S^1$ has this property, namely $F_n^{-1}(p) \in \mathcal{C}_{n-1}$.

That being said, we next minimize the $(n-1)$-dimensional volume functional in the class $\mathcal{C}_{n-1}$, namely we consider the variational problem

$$\inf \{ |\Sigma| \ : \ \Sigma \in \mathcal{C}_{n-1} \}.$$

Such a problem can be solved by direct methods (cf. [37]): indeed, by the compactness theorem for currents with uniformly bounded mass[2] the infimum is achieved by some integral $(n-1)$ current Σ_{n-1}, but the normalization condition

$$\int_\Sigma \omega^1 \wedge \ldots \wedge \omega^{n-1} = 1$$

is satisfied when taking limits of currents in the sense above[3] so in the end $\Sigma_{n-1} \in \mathcal{C}_{n-1}$. It follows from the regularity theory for minimal hypersurfaces that if $n < 8$, Σ_{n-1} is a *smooth*, embedded minimal hypersurface. Notice also that Σ_{n-1} is stable (in the classical sense) because the normalization condition is preserved by any small perturbation (by homological invariance). Also, Σ_{n-1} comes together with a smooth degree one map to T^{n-1} which is obtained by simply restricting the map F

---

[2]We refer to compactness of currents with respect to the weak* topology, namely with respect to the canonical duality with spaces of differential forms.

[3]By its very definition, considering the special case of $\omega^1 \wedge \ldots \wedge \omega^{n-1}$ as a test form.

and then projecting onto the first $(n-1)$ factors of $T^n = S^1 \times \ldots \times S^1$, namely setting $F^{n-1} = \pi^n(F_{|\Sigma_{n-1}})$ (notice the position of the indices, π^n is not to be confused with the one-dimensional projection $\pi_n : T^n \to S^1$). The fact that this map has degree one follows from the normalization condition:

$$1 = \int_{\Sigma_{n-1}} \omega^1 \wedge \ldots \wedge \omega^{n-1} = \deg(F^{n-1}) \int_{F^{n-1}(\Sigma_{n-1})} dx^1 \wedge \ldots \wedge dx^{n-1}$$

$$= \deg(F^{n-1}) \int_{T_n^{n-1}} dx^1 \wedge \ldots \wedge dx^{n-1} = \deg(F^{n-1})$$

where the second to last step relies on the fact that F^{n-1} must be surjective (or else necessarily $\int_{F^{n-1}(\Sigma_{n-1})} dx^1 \wedge \ldots \wedge dx^{n-1} = 0$). Observe that it also follows that Σ_{n-1} is orientable, hence two-sided because the ambient M^n is itself orientable.

The idea now is to proceed inductively, proving the assertion for increasing dimension, starting with $n = 3$.

$\boxed{\text{Case } n = 3:}$ let g be a positive scalar curvature metric on T^3. Following the argument above, one can find inside (T^3, g) a stable, embedded minimal surface Σ_2 coming endowed with a degree one map to a two-dimensional torus T^2. By virtue of Theorem 3.18 we have that $\Sigma_2 \simeq S^2$ but on the other hand recall that any smooth map $S^2 \to T^2$ must have degree zero,[4] contradiction.

$\boxed{\text{Case } n - 1 \Rightarrow n:}$ suppose that the conclusion of the theorem holds in dimension $n - 1 \geq 3$, and let then g be a positive scalar curvature metric on T^n. Once again, let Σ_{n-1} be a stable, embedded minimal hypersurface in the ambient manifold in question. By the above discussion, one can construct Σ_{n-1} together with a degree one map to the $(n-1)$ dimensional torus T^{n-1}.

Now, we proceed by means of a conformal deformation in the sense we are about to explain.

If we denote by g_Σ the induced Riemannian metric on Σ_{n-1}, one can consider the conformally related metric $\tilde{g}_\Sigma = \phi^{2/(n-2)} h$ (for ϕ a first positive eigenfunction of the stability operator of Σ_{n-1}), whose scalar curvature is given by

$$R_{\tilde{g}_\Sigma} = \phi^{-\frac{n}{n-2}} \left(-2\Delta_\Sigma \phi + R_{g_\Sigma} \phi + \frac{n-1}{n-2} \frac{|\nabla_\Sigma \phi|^2}{\phi} \right).$$

On the other hand, the higher-dimensional counterpart of the rearrangement trick presented above gives the identity

$$|A|^2 + Ric_g(N, N) = \frac{1}{2}(|A|^2 + R_g - R_{g_\Sigma})$$

---

[4] In fact, it is a standard result in geometric topology that any smooth map $F : S_h \to S_k$ has degree zero if $h < k$ where S_h (resp. S_k) is a genus h (resp. k) surface.

and so we can plug it in the above expression for the scalar curvature of the conformally deformed metric to obtain

$$R_{\tilde{g}_\Sigma} = \phi^{-\frac{2}{n-2}} \left(2\lambda_1 + R_g + |A|^2 + \frac{n-1}{n-2} \frac{|\nabla_\Sigma \phi|^2}{\phi^2} \right)$$

where $\lambda_1 = \lambda_1(\Sigma)$ denotes the first eigenvalue of the Jacobi operator of the minimal surface Σ_{n-1}. By assumption, each term of the right-hand side is non-negative and $R_g > 0$, so we must conclude that $R_{\tilde{g}_\Sigma} > 0$ at all points, so Σ_{n-1} would support a smooth metric of positive scalar curvature, contrary to the inductive assumption. Thereby the proof is complete.   □

Now, recalling the compactification theorem we have proven above (Theorem 3.5), we immediately derive the following weak form of the Riemannian positive mass theorem:

Theorem 3.23 ([63]) *The ADM mass is non-negative for all asymptotically flat manifolds of dimension less than eight having non-negative scalar curvature.*

The equality case will instead be discussed in the next lecture, thereby completing the presentation of the positive mass theorem in dimension less than eight. Instead, we conclude this lecture getting back to the aforementioned splitting theorem:

Theorem 3.24 *Let (M^n, g) ($n \geq 3$) be a complete Riemannian manifold of non-negative scalar curvature and let $\Sigma \subset M$ be a closed, embedded area-minimizing[5] (two-sided) hypersurface. If the only metrics of non-negative scalar curvature carried by Σ are scalar flat, then there is a (local) isometric splitting, namely there is a neighborhood U of M such that (U, g_U) is isometric to the Riemannian product $((-\varepsilon, \varepsilon) \times \Sigma, dt^2 + g_\Sigma)$.*

Proof Let us start by checking that Σ is *infinitesimally rigid*. To that aim, if we denote by g_Σ the induced Riemannian metric on Σ, one can again consider the conformally related metric $\tilde{g}_\Sigma = \phi^{2/(n-2)} h$ (for ϕ a first positive eigenfunction of the stability operator of Σ) and recall

$$R_{\tilde{g}_\Sigma} = \phi^{-\frac{2}{n-2}} \left(2\lambda_1 + R_g + |A|^2 + \frac{n-1}{n-2} \frac{|\nabla_\Sigma \phi|^2}{\phi^2} \right).$$

By assumption, each term of the right-hand side is non-negative and so we conclude that *each term* has to vanish and so in particular the Jacobi operator of Σ takes the very simple form $J = \Delta$. That being said, we are in position to proceed.

---

[5]A smooth, embedded, hypersurface is called area-minimizing if it is a local minimum for the area functional under smooth perturbations. In particular, an area-minimizing hypersurface is a stable, minimal hypersurface.

Let us consider for fixed $\alpha \in (0, 1)$ and $u \in C^{2,\alpha}(\Sigma; \mathbb{R})$ (of sufficiently small norm) the hypersurface given by $\Sigma(u) := \{\exp_x(u(x)N) \ x \in \Sigma\}$. Hence, if we let $\mathcal{H} : C^{2,\alpha}(\Sigma; \mathbb{R}) \to C^{0,\alpha}(\Sigma; \mathbb{R})$ denote the mean curvature map we have

$$\mathcal{H}'(0) = -\Delta_\Sigma$$

which has a one dimensional kernel consisting of constants. Thus, a standard application of the implicit function theorem (based on restricting on the closed subspace of the domain consisting of functions having null means) ensures that one can find a local foliation whose leaves have constant mean curvature, and hence one can write on $U = (-\varepsilon_0, \varepsilon_0) \times \Sigma$ the ambient metric in the form

$$\lambda^2 dt^2 + \sigma_t$$

where $\lambda = \lambda(t, x)$ and $\sigma_t = \sum_{i,j} \sigma_{ij}(t, x) dx^i \otimes dx^j$ (for $\{x\}$ local coordinates on Σ) and subject to the constraints that *each* slice Σ_t has constant mean curvature. Now, we want to study for $t \in (-\varepsilon_0, \varepsilon_0)$ the

- area function $S(t)$ of the hypersurface Σ_t;
- mean curvature function $H(t)$ of the hypersurface Σ_t.

The second variation of the area functional (for hypersurfaces that are not necessarily minimal) gives, when considering the variation vector field ∂_t

$$\frac{dH(t)}{dt} = -\Delta_{\Sigma_t}\lambda + \frac{1}{2}(R_{\Sigma_t} - R_g - |A|^2 - H^2)\lambda$$

with initial condition $H(0) = 0$ (by minimality of Σ) and we claim that this forces $H \leq 0$ for $t \in [0, \varepsilon)$. By contradiction: if that were not the case, then one could find $t_0 \in (0, \varepsilon_0)$ where $H'(t_0) > 0$. Consider the scalar curvature of the associated hypersurface Σ_{t_0} in the conformally deformed metric $\tilde{\sigma}_{t_0} = \lambda^{2/(n-2)}\sigma_{t_0}$. Replacing $\Delta\lambda$ by means of the previous equation (involving $H'(t_0)$) we find

$$R_{\tilde{\sigma}_{t_0}} = \lambda^{-\frac{2}{n-2}}\left(2\lambda^{-1}H'(t_0) + R_g + |A|^2 + H^2 + \frac{n-1}{n-2}\frac{|\nabla_\Sigma\lambda|^2}{\lambda^2}\right) > 0$$

which is impossible because Σ_{t_0} (that is diffeomorphic to Σ, by construction) is assumed not to carry any metric of positive scalar curvature. Hence, we gain that $H(t) \leq 0$ for each $t \in (0, \varepsilon_0)$. At this stage, we shall use the assumption that Σ is in fact area-minimizing. Indeed

$$S'(t) = \int_{\Sigma_t} H(t)\lambda(t, x) \, d\mathcal{H}^{n-1} \leq 0 \ \forall \, t \in [0, \varepsilon_0)$$

and thus this leads to the conclusion that necessarily $S'(t) = 0$ for all $t \in (0, \varepsilon_0)$ (because Σ is area-minimizing) and hence (due to the fact that $\lambda(t, x) > 0$ in U)

one has $H(t) = 0$ for all $t \in (0, \varepsilon_0)$ and a similar argument gives that the same conclusion holds true for $t \in (-\varepsilon_0, \varepsilon_0)$. Furthermore, the stability criterion above (Lemma 3.20) applied for $\phi = \lambda$ gives that each hypersurface Σ_t is in fact stable and thus (by the first part of this proof) infinitesimally rigid, in particular totally geodesic. Using this information at the level of the metric on U, namely for $g_U = \lambda^2 dt^2 + \sigma_t$ gives that σ_t is independent of t (so we can denote it simply by σ) because

$$\frac{\partial \sigma_t}{\partial t} = 2A_t$$

and that, on other hand, $\lambda = \lambda(t)$ (namely: a function of t only) because $\Delta \lambda = 0$ for every $t \in (-\varepsilon_0, \varepsilon_0)$. Thus, one has $g = \lambda^2 dt^2 + \sigma$ and, at that stage, a change of variable in the first factor allows to reduce to $\lambda = 1$ identically, which precisely means that U splits as a Riemannian product, as we had claimed.   □

Thereby we can prove the following assertion, which can be regarded as a slight extension of what is known in the literature as *torus rigidity theorem*.

Theorem 3.25 *Let (M^n, g) be an n-dimensional Riemannian manifold, with $n < 8$, admitting a degree one smooth map to the n-dimensional torus T^n. If its scalar curvature is non-negative, then (M^n, g) is flat.*

Proof The idea is, once again, to proceed inductively starting with $n = 3$. We give for granted that one can minimize area subject to a constraint defined by an $(n - 1)$ form as has been described in detail in the proof of Theorem 3.22.

$\boxed{\text{Case } n = 3:}$ a 2-dimensional (orientable) surface of positive genus does not admit a metric of positive scalar (or, equivalently, Gauss) curvature by virtue of the Gauss–Bonnet theorem, hence we are in position to apply Theorem 3.24 which provides a local splitting of (M^3, g) around Σ. But clearly, a standard continuation argument gives that in fact a global splitting holds, namely (M^3, g) is isometric $S^1 \times \Sigma$ where both factors are endowed with the flat metric. Recall that the fact that the metric on Σ is flat follows, once again, by the conformal deformation trick that we presented along the course of the previous proof, namely by the formula

$$R_{\tilde{g}_\Sigma} = \phi^{-\frac{2}{n-2}} \left(2\lambda_1 + R_g + |A|^2 + \frac{n-1}{n-2} \frac{|\nabla_\Sigma \phi|^2}{\phi^2} \right)$$

where $R_{\tilde{g}_\Sigma} = 0$ (consequence of Gauss-Bonnet) implies, among other things, $\phi = $ constant and thus

$$R_{\tilde{g}_\Sigma} = \phi^{-\frac{n}{n-2}} \left(-2\Delta\phi + R_{g_\Sigma}\phi + \frac{n-1}{n-2} \frac{|\nabla_\Sigma \phi|^2}{\phi} \right)$$

gives $R_{g_\Sigma} = 0$ identically, as claimed.

$\boxed{\text{Case } n - 1 \Rightarrow n:}$ suppose that the conclusion of the theorem holds in dimension $n - 1 \geq 3$. In particular, this assumption also says that Σ_{n-1} (coming with a degree one map to T^{n-1}) cannot be endowed with a metric g of non-negative scalar curvature unless it is flat. Thus, applying Theorem 3.24 to (M^n, g) we obtain an isometric splitting as a Riemannian product $S^1 \times \Sigma_{n-1}$. The two formulae above give that Σ_{n-1} is scalar flat and thus the inductive hypothesis (applied for the second time) gives that Σ_{n-1} is in fact flat (and so, incidentally, must be a flat torus), which completes the proof. $\square$

4 Third Lecture: Scalar Curvature Rigidity Phenomena

In this lecture, we start by studying the rigidity part of the positive mass theorem and show, following Schoen and Yau, that the only asymptotically flat manifold of non-negative scalar curvature and zero ADM mass is the Euclidean space. Moving from the ideas and techniques that emerge in such a proof, we then give a brief survey of various scalar curvature rigidity phenomena related to this result (the interested reader may also wish to consult the beautiful article by Brendle [10]).

4.1 Characterizing the Zero Mass Case

We want to present the proof of the following assertion:

Theorem 4.1 *Let the weak positive mass theorem hold in dimension n, then the strong positive mass theorem in dimension n also holds.*

We are adopting here the following terminology:

- the expression *weak positive mass theorem* means that $m \geq 0$ for all asymptotically flat manifolds of non-negative scalar curvature;
- the expression *strong positive mass theorem* means that $m \geq 0$ for all asymptotically flat manifolds of non-negative scalar curvature and $m = 0$ is only satisfied by the Euclidean space.

We refer our readers to Sect. 3.1 for further details and for the precise definitions we are using above, including the specific decay assumption we are making. In particular, based on Theorem 3.23, we immediately derive this statement:

Theorem 4.2 *If (M^n, g) is asymptotically flat of dimension $3 \leq n < 8$, with non-negative scalar curvature and zero ADM mass, then it is (globally) isometric to the flat Euclidean space $\mathbb{R}^n$.*

The 2017 paper by Schoen and Yau [64] is only concerned with the weak positive mass theorem in all dimensions: in particular, its conclusion suffices to also (automatically) imply the characterization of the $m = 0$ case by virtue of the general argument we are about to present, which does not rely on the use of minimal hypersurfaces.

The strategy to handle this borderline case is roughly as follows: we first show that (M^n, g) is Ricci-flat via a variational argument, and then present two different ways of concluding the argument that appeal to different geometric tools. As a first, preliminary step we shall obtain the following:

Lemma 4.3 *Let the weak positive mass theorem hold in dimension n. If (M^n, g) is asymptotically flat of dimension n, with non-negative scalar curvature and zero ADM mass, then it is Ricci-flat.*

Remark 4.4 A straightforward variation of the scalar curvature deformation argument presented in the previous lecture (cf. Corollary 3.9) ensures that any such (M^n, g) with $m = 0$ must be scalar flat. If not, namely if one could find an open region where $R_g > 0$, then one could also obtain a conformal factor u (converging to one at infinity, and with an asymptotic expansion of the form (3.1)) such that $u^{4/(n-2)} g$ is scalar flat and has ADM mass *strictly less* than that of (M^n, g), hence necessarily negative, thereby violating the weak positive mass theorem.

Proof Let k be a compactly supported, symmetric $(0, 2)$ tensor and consider, for $t \in (-\varepsilon, \varepsilon)$ the Riemannian metrics $g_t = g + tk$ (which are well-defined provided we only choose $\varepsilon > 0$ small enough). Furthermore, possibly by choosing ε even smaller, we can ensure that the value of $\int_M |R^-_{g_t}|^{n/2}$ is below the threshold required by Theorem 3.7: hence, we can find $u_t > 0$ such that the conformally deformed metric $u_t^{4/(n-2)} g_t$ has zero scalar curvature and $u_t \to 1$ at infinity. The family (u_t) can be constructed so to be locally continuously differentiable in t, and satisfying $u_0 = 0$.

Now, set

$$\hat{g}_t = u_t^{\frac{4}{n-2}} g_t, \quad m(t) := m(\hat{g}_t) \tag{4.1}$$

and observe that $m(0) = 0$ while necessarily (by the weak positive mass theorem) $m(t) \geq 0$ for all $t \in (-\varepsilon, \varepsilon)$.

Thus $t = 0$ must be a local minimum for the function $t \mapsto m(t)$, so that $m'(0) = 0$. We now need to compute $m'(t)$ to see the geometric content of such a condition. First of all, the fact that the tensor k is compactly supported implies that $m(g_t) = m(g) = 0$ for all $t \in (-\varepsilon, \varepsilon)$ and thus the contribution to the ADM mass is solely due to the conformal factor, and is equal to

$$m(\hat{g}_t) = m(g_t) - \frac{2}{(n-2)\omega_{n-1}} \int_M R_{g_t} u_t \, d\mathcal{V}^n_{g_t}$$

where $\mathcal{V}^n_{g_t}$ is the volume measure associated with the Riemannian metric $g_t = g + tk$.

If we differentiate such an expression in t, the fact that $u_0 = 1$ and $R_{g_0} = 0$ identically on M implies

$$m'(0) = -\frac{2}{(n-2)\omega_{n-1}} \int_M \left(\left[\frac{d}{dt} \right]_{t=0} R_{g_t} \right) d\mathscr{V}_g^n$$

$$= \frac{2}{(n-2)\omega_{n-1}} \int_M g(Ric_g, k) \, d\mathscr{V}_g^n.$$

Here we have exploited the formula for the first variation of the scalar curvature map, which reads (in the special case under consideration)

$$\left[\frac{d}{dt} \right]_{t=0} R_{g_t} = -\Delta_g(tr_g k) + div_g(div_g k) - g(Ric_g, k)$$

and have performed integration by parts, on large spheres enclosing the support of k, to get rid of the contribution of the first two summands.

As a result, the aforementioned first-derivative condition forces

$$\int_M g(Ric_g, k) \, d\mathscr{V}_g^n = 0$$

for all compactly supported tensors k, hence the conclusion is that (M^n, g) is Ricci-flat. $\qquad \square$

We can now proceed and present a proof of the following assertion, of independent interest:

Theorem 4.5 *Let (M^n, g) be an asymptotically flat manifold with non-negative Ricci curvature. Then it is flat.*

Proof Given any $p \in M$ consider for $r > 0$ the mass ratio

$$\Theta(r) := \frac{|B_r(p)|}{v_n r^n}$$

where $|B_r(p)|$ stands for the volume of the metric ball of center p and radius r, while v_n equals the volume of the unit ball in $\mathbb{R}^n$ (so that, with our notation, $\omega_{n-1} = n v_n$). Then, the Bishop–Gromov comparison theorem (cf. [40], 4.19 and 4.20) asserts that whenever the Ricci curvature of the ambient manifold in question is non-negative (which we are assuming), then the function $\Theta(\cdot)$ is non-increasing, and is actually constant if the manifold is flat Euclidean space. In our setting, we have that

$$\lim_{r \to 0^+} \Theta(r) = 1$$

because (M^n, g) is smooth, but also

$$\lim_{r \to +\infty} \Theta(r) = 1$$

because (M^n, g) is asymptotically flat.[6] Hence, it follows at once that the function Θ is constant and equal to one for all values of r. Thus (M^n, g) is isometric to Euclidean space $\mathbb{R}^n$.   □

Remark 4.6 It has been known in the relativity community since at least the work by Ashtekar–Hansen [4], see also Chruściel [23], that (under suitable decay assumptions on the metric) one has

$$m_{ADM} = \lim_{r \to \infty} \frac{1}{(n-1)(2-n)\omega_{n-1}} \int_{|x|=r} \left(Ric_g - \frac{1}{2} R_g g \right) (X, v_g) \, d\mathcal{H}^{n-1}$$

where X denotes the conformal Killing vector field $x^1 \frac{\partial}{\partial x^1} + \ldots + x^n \frac{\partial}{\partial x^n}$.

In particular, if one assumes

$$|g_{ij} - \delta_{ij}| = o_2(|x|^{-\frac{n-2}{2}}) \text{ and } |R_g| = O(|x|^{-q_0}), q_0 > n$$

which is certainly the case if one considers asymptotically flat Riemannian manifolds as per Definition 3.1, then in fact

$$m_{ADM} = \lim_{r \to \infty} \frac{r}{(n-1)(2-n)\omega_{n-1}} \int_{|x|=r} Ric_g(v_g, v_g) \, d\mathcal{H}^{n-1}.$$

This equality is proven either via direct computation (cf. Miao–Tam [56]) or checking it for conformally flat data and then arguing by density and continuity, as suggested by Schoen.

Such a formula immediately implies that an asymptotically flat manifold of non-negative Ricci curvature must have non-positive ADM mass, hence *a posteriori* must be flat.

We will now sketch a different proof of Theorem 4.5 in the special case of vanishing Ricci curvature. This argument, which relies on the use of harmonic coordinates, goes back to [63].

Lemma 4.7 *If (M^n, g) is an asymptotically flat Riemannian manifold of dimension $n \geq 3$ and satisfying, for some $\varepsilon > 0$, $Ric_g = O(|x|^{-n-\varepsilon})$ as one lets $x \to \infty$ then*

---

[6]Recall from Definition 3.3 that we are assuming, for the sake of simplicity, to only deal with asymptotically flat manifolds with one end. More generally, one will have $\lim_{r \to +\infty} \Theta(r) \geq 1$ and thus Theorem 4.5 still holds true, with the very same proof.

there exist asymptotically flat coordinates $\{y\}$ near infinity such that

$$g_{ij} = \delta_{ij} + o_2(|y|^{2-n}).$$

Proof We let $\{y\}$ denote a system of harmonic coordinates for (M^n, g): namely we have $\Delta_g y^i = 0$ for every $i = 1, 2, \ldots, n$ and, furthermore, $\{y^1, \ldots, y^n\}$ provide a system of coordinates outside a compact set (the existence of such coordinate system is a result of Bartnik, see Theorem 3.1 in [5]). Now, recall that in harmonic coordinates the Ricci tensor has a remarkably simple expansion (cf. equation (3.6) in [5]):

$$(Ric_g)_{ij} = -\frac{1}{2}g^{kl}\partial_{kl}(g_{ij}) + O(|\partial g|^2)$$

and so, using the decay assumption on the metric and on the Ricci tensor (which determine for g_{ij} an elliptic equation whose right-hand side is known to decay like $|x|^{-\sigma}$ where $\sigma = \min\{2q + 2, n + \varepsilon\}$), we get by [53] an asymptotic expansion of the form

$$g_{ij} = \delta_{ij} + \frac{c_{ij}}{|y|^{n-2}} + o_2(|y|^{2-n}).$$

By virtue of the symmetry of the metric tensor g we also have $c_{ij} = c_{ji}$ for all couples of indices i, j and hence, diagonalizing, we can always assume that in fact

$$g_{ij} = \delta_{ij} + \frac{c_i \delta_{ij}}{|y|^{n-2}} + o_2(|y|^{2-n})$$

holds. So, all we need now is to show that $c_i = 0$ for every choice of $i = 1, 2, \ldots, n$. This follows by plugging-in the expansions

$$g_{ij} = \delta_{ij} + \frac{c_i \delta_{ij}}{|y|^{n-2}} + o_2(|y|^{2-n})$$

$$g^{ij} = \delta_{ij} - \frac{c_i \delta_{ij}}{|y|^{n-2}} + o_2(|y|^{2-n})$$

$$\sqrt{\det(g_{k\ell})} = 1 + \frac{1}{2}\frac{\sum_{i=1}^n c_i}{|y|^{n-2}} + o_2(|y|^{2-n})$$

in the equation solved by each coordinate functions y^i, thereby getting

$$0 = (\sqrt{\det(g_{k\ell})}g^{ij})_j = \left((n-2)c_i - \frac{n-2}{2}\sum_{k=1}^n c_k\right)\frac{\delta_{ij}y^j}{|y|^n} + o(|y|^{1-n}).$$

For indeed, if we multiply these equations by $|y|^{n-1}$, evaluate along y^j axis, and let $|y^j| \to \infty$ we end up getting an $n \times n$ homogeneous, linear system which has the sole solution $c_1 = \ldots c_n = 0$ and thus the proof is complete.   □

To move further and conclude the (second) proof of the rigidity result, we shall first recall the (scalar) Bochner identity

$$\frac{1}{2}\Delta_g |\nabla_g u|^2 = |\nabla_g^2 u|^2 + g(\nabla_g u, \nabla_g \Delta_g u) + Ric_g(\nabla_g u, \nabla_g u).$$

When u is harmonic and (M^n, g) is Ricci-flat this simplifies to

$$\frac{1}{2}\Delta_g |\nabla_g u|^2 = |\nabla_g^2 u|^2$$

which we are about to use for the coordinate functions constructed in the previous section.

Proof Applying Bochner's formula to each harmonic coordinate function y^i (following the same notation as in the previous proof) and integrating such an identity over any ball of sufficiently large radius (say r) one finds

$$2\sum_{i=1}^{n}\int_{B_r} |\nabla_g^2 y^i|^2 \, d\mathcal{V}_g^n = \sum_{i=1}^{n}\int_{B_r} \Delta_g |\nabla_g y^i|^2 \, d\mathcal{V}_g^n = \int_{S_r} \frac{\partial}{\partial \nu}\left(\sum_{i=1}^{n}|\nabla_g y^i|^2\right) d\mathcal{H}_g^{n-1}$$

where the last equality relies on the divergence theorem (applied with respect to the metric g) and where ν is the (outward-pointing) g-unit normal to the coordinate sphere S_r. Patently $|\nabla y^i|^2 = g^{ii}$ and thus the integrand on the right-hand side satisfies

$$\frac{\partial}{\partial \nu}\left(\sum_{i=1}^{n}|\nabla y^i|^2\right) = \sum_{i=1}^{n}\frac{\partial g^{ii}}{\partial \nu} = o(|y|^{1-n})$$

where $\tau > n - 2$ is as in the statement of Lemma 4.7, so that

$$\lim_{r \to +\infty}\int_{S_r} \frac{\partial}{\partial \nu}\left(\sum_{i=1}^{n}|\nabla y^i|^2\right) d\mathcal{H}_g^{n-1} = 0.$$

Hence, the identity above also implies for each $i = 1, 2, \ldots, n$

$$\int_M |\nabla_g^2 y^i|^2 \, d\mathcal{V}_g^n = 0.$$

The conclusion is that each (globally well-defined) vector field ∇y^i is parallel and thus

$$g^{ij} = g(\nabla_g y^i, \nabla_g y^j)$$

is constant along each curve in M, and thus does not depend on the point. But the asymptotic flatness condition then forces $g_{ij} = \delta_{ij}$ and so, this implies that the map

$$Y : M^n \to R^n, \quad Y(p) = (y^1(p), \ldots, y^n(p))$$

is in fact a global isometry, which completes the proof. □

4.2 A Survey of Related Rigidity Results

The Schoen–Yau compactification trick, which we have presented during the course of the second lecture, has the advantage of allowing a neat recursive argument when dealing with the possible occurrence of singularities of high-dimensional minimal hypersurfaces. On the other hand, the original approach relying on the construction of complete, unbounded, minimal hypersurfaces in asymptotically flat spaces (without any compactification) suggested a number of natural geometric questions which have been thoroughly investigated, and are connected to a variety of results concerning scalar curvature rigidity phenomena.

Our starting point, directly connected to [63], are the following two rigidity theorems which should be read and interpreted in relation to the proof of Theorem 3.19.

Theorem 4.8 (See [14]) *Let (M, g) be an asymptotically flat Riemannian 3-manifold with non-negative scalar curvature. Let $\Sigma \subset M$ be a non-compact properly embedded stable minimal surface. Then Σ is a totally geodesic flat plane and the ambient scalar curvature vanishes along Σ. Such a surface cannot exist under the additional assumption that (M, g) is asymptotic to Schwarzschild with mass $m > 0$.*

Here, we shall say that an asymptotically flat Riemannian manifold (as per Definition 3.1) is asymptotic to Schwarzschild with mass $m > 0$ if one has

$$g_{ij} - \left(1 + \frac{m}{2|x|}\right)^4 \delta_{ij} \in C^{2,\alpha}_{-q}, \quad q > 1.$$

This assumption is by no means sharp, see Appendix A of [17] for the precise asymptotic requirements. Here is instead the second statement:

Theorem 4.9 (See [17]) *The only asymptotically flat Riemannian 3-manifold with non-negative scalar curvature that admits a non-compact area-minimizing boundary is flat $\mathbb{R}^3$.*

Both of these theorems can be regarded as suitable extensions of the classical Bernstein theorem in flat $\mathbb{R}^3$. As the reader shall notice, there is a subtle balance between the stronger variational assumptions (area-minimizing vs. stable) and the asymptotic assumptions concerning the ambient manifold (asymptotically flat vs. Schwarzschildean). In fact, these two results are essentially sharp, meaning that one cannot expect such a global rigidity statement to hold in general asymptotically flat manifolds due to the sole presence of a complete, unbounded stable minimal surface. Indeed, we anticipate the following consequence of the gluing construction we shall present in our fourth lecture:

Theorem 4.10 (See [16]) *There exists an asymptotically flat Riemannian metric g with non-negative scalar curvature and positive mass on $\mathbb{R}^3$ that is Euclidean on a half-space.*

Said $\mathbb{R}^2 \times (0, \infty)$ the half-space in question, the coordinate planes $\mathbb{R}^2 \times \{z\}$ with $z > 0$ in Theorem 4.10 are stable minimal surfaces. In particular, the area-minimizing condition in Theorem 4.9 cannot be relaxed. We also see that the condition that (M, g) be asymptotic to Schwarzschild in Theorem 4.8 is necessary.

Both Theorem 4.8 and Theorem 4.9 are part of the rich theory of scalar curvature rigidity results, and naturally connect to the well-known cylinder theorem by Fischer-Cobrie and Schoen [38], which was recently sharpened by Chodosh et al. in [20]. We further refer the reader to the papers [2, 8, 9, 13, 51, 52, 57, 60] for a selection of rigidity and splitting results in presence of (compact) minimal surfaces in Riemannian 3-manifolds with a lower bound (not necessarily zero) on the scalar curvature. Interesting contributions to the general question whether manifolds of the form $T^k \times \mathbb{R}^{n-k}$ support metrics of positive scalar curvature (possibly under additional requirements) have been obtained by Gromov in [42].

In view of the discussion we are about to present in the next section, we should also mention the following refinement of Theorem 4.8, which is needed to handle the case of (possibly improper) minimal immersions, both in the complete case and in the case of horizon boundary.

Theorem 4.11 (See [17]) *Let (M, g) be an asymptotically flat Riemannian 3-manifold with non-negative scalar curvature. Assume that there is an unbounded complete stable minimal immersion $\varphi : \Sigma \to M$ that does not cross itself. Then (M, g) admits a complete non-compact properly embedded stable minimal surface. If (M, g) is asymptotic to Schwarzschild with mass $m > 0$ and has horizon boundary, then the only non-trivial complete stable minimal immersions $\varphi : \Sigma \to M$ that do not cross themselves are embeddings of components of the horizon.*

A first, significant application of Theorem 4.9 concerns the classification of initial data sets that admit a global static potential. To provide some context and motivation, let us first recall the basic definitions.

We shall say that a (connected) Riemannian manifold (M, g) is static if there exists a non-constant function $f \in C^\infty(M)$ with $L_g^* f = 0$, where

$$L_g^* f = -(\Delta_g f)g + \nabla_g^2 f - f \, Ric_g$$

is the formal adjoint of the linearization L of the scalar curvature operator at g. A static manifold has constant scalar curvature, hence if (M, g) is asymptotically flat, then its scalar curvature vanishes and the condition that $L_g^* f = 0$ is equivalent to the system

$$\begin{cases} \nabla_g^2 f = f \, Ric_g \\ \Delta_g f = 0. \end{cases}$$

In that case, the spacetime

$$(M_* \times \mathbb{R}, g - f^2 dt \otimes dt) \quad \text{where} \quad M_* = \{x \in M : f(x) > 0\}$$

is a static solution of the vacuum Einstein equations. It is well-known that the zero set of a static potential is a totally geodesic hypersurface (cf. [31]) and, more recently, Galloway and Miao have shown that when (M, g) is an asymptotically flat Riemannian 3-manifold—possibly with several ends—such that f vanishes on the boundary of M, then every unbounded component of such a level set is an area-minimizing plane (see [41]). Hence, Theorem 4.9 implies that such unbounded components can only exist when (M, g) is flat $\mathbb{R}^3$ and f is a linear function. Combining this fact with some recent result by Miao and Tam [55], one obtains the following classification result:

Corollary 4.12 (See [17], cf. [12, 47, 58, 61]) *Let (M, g) be an asymptotically flat Riemannian 3-manifold, possibly with several ends, that admits a non-constant function $f \in C^\infty(M)$ with $L_g^* f = 0$ that vanishes on the boundary of M. Then (M, g) is isometric to either flat $\mathbb{R}^3$, or, for some $m > 0$, to the (one-ended) Schwarzschild manifold*

$$\left(\{x \in \mathbb{R}^3 : |x| \geq m/2\}, \left(1 + \frac{m}{2|x|}\right)^4 \sum_{i=1}^{3} dx^i \otimes dx^i \right)$$

or its double

$$\left(\mathbb{R}^3 \setminus \{0\}, \left(1 + \frac{m}{2|x|}\right)^4 \sum_{i=1}^{3} dx^i \otimes dx^i \right).$$

4.3 The Isoperimetric Structure of Asymptotically Flat Spaces

The rigidity theorems presented above, and various improvements thereof, also have important implications concerning the isoperimetric structure of asymptotically flat 3-manifolds. It has been known for more than two decades (cf. Huisken–Yau [46]) that the complement of a certain compact subset C of a Riemannian 3-manifold (M, g) that is asymptotic to Schwarzschild with mass $m > 0$ admits a foliation by closed stable CMC spheres $\{\Sigma_H\}_{H \in (0, H_0]}$ where Σ_H has (outward) mean curvature H. Such a construction can be employed to provide a geometric definition for the center of mass of an isolated gravitational system, and in fact to even provide an (a priori) alternative notion of mass. Therefore, a conceptually important point is to obtain suitable uniqueness theorems (at the level of foliation, or even for single leaves having, say, small mean curvature or large area) to ensure that these sorts of definitions are, in some appropriate sense, canonical. We will not attempt to survey the very many questions one can ask and the progress on each of them, but limit ourselves to mention a corollary of Theorem 4.11 with respect to the uniqueness of leaves of large area:

Corollary 4.13 *Let (M, g) be a Riemannian 3-manifold with non-negative scalar curvature that is asymptotic to Schwarzschild with mass $m > 0$ and which has horizon boundary. Let $p \in M$. Every connected closed volume-preserving stable CMC surface $\Sigma \subset M$ that contains p and which has sufficiently large area is part of the canonical foliation.*

Roughly speaking, we want to think about this theorem in terms of a uniqueness result for *centered* large stable CMC spheres. It is then important, to the scope of answering the unconditional uniqueness question, the behavior of *outlying* large CMC spheres, where we mean that one considers closed surfaces that are allowed to be disjoint from the center of the ambient manifold (i.e., from any pre-assigned compact set whose complement is covered by the Huisken–Yau foliation). This study turns out to be rather subtle, but in recent years we have witnessed some dramatic progress.

First of all, Brendle and Eichmair have constructed in [11] examples of Riemannian 3-manifolds asymptotic to Schwarzschild with positive mass that contain a sequence of larger and larger volume-preserving stable CMC surfaces that diverge to infinity together with the regions they bound: hence, uniqueness fails even under strong asymptotic assumptions if one allows these surfaces to drift-off at infinity. However, these examples do not have non-negative scalar curvature. A recent and somewhat surprising breakthrough by Chodosh and Eichmair ensures the aforementioned uniqueness result holds *unconditionally* if one assumes that the scalar curvature (of the asymptotically Schwarzschildean manifold in question) be identically equal to zero [18, 19]: this beautiful conclusion comes by combining the delicate analysis that is needed to understand the three possible regimes, namely studying sequences of large CMC spheres $\{\Sigma_k\}$ in each of the three cases

$$r_0(\Sigma_k)H(\Sigma_k) \to \begin{cases} \eta > 0 & (\text{cf. } [11]) \\ 0 & (\text{cf. } [18]) \\ \infty & (\text{cf. } [19]) \end{cases}$$

where

$$r_0(\Sigma) = \sup\{r > 1 \ : \ B_r(0) \cap \Sigma = \emptyset\}$$

is a scale measuring the distance of the surface in question from the center of the ambient manifold. We notice that (Theorem 1.3 in in [19]) Chodosh and Eichmair also proved that one cannot expect unconditional uniqueness (even in the class of data that are asymptotic to Schwarzschild) if one allows for non-negative scalar curvature (i.e., for matter sources) opposed to zero (i.e., vacuum).

Unconditional uniqueness is patently false, even in the vacuum case, if one considers the larger class of asymptotically flat Riemannian manifolds (not necessarily asymptotic to Schwarzschild): indeed one just needs to consider the data produced by Theorem 4.10 (which, by the way, also allows to obtain examples with horizon boundary). Now, it turns out one can also produce outer CMC foliations for general asymptotically flat manifolds (cf. [45, 50, 59]) and thus, with these more general ambient manifolds, one should require something more on the surfaces in question to hope for uniqueness: a related, and geometrically natural problem is then whether each leaf of such a canonical foliation has least possible area for the volume it encloses (i.e., if it solves the classical isoperimetric problem in the class of, say, finite perimeter sets) and if each such leaf is then the *unique* solution to the isoperimetric problem for any sufficiently large volume. The first question was answered in the affirmative by Eichmair and Metzger in [33] for 3-manifolds that are asymptotic to Schwarzschild (which is the natural context for the Huisken-Yau foliation we have alluded to above). For general asymptotics, a significant step in this direction was indeed obtained in [17] (see Corollary 1.13 therein) and then, about 2 years ago, the problem was fully solved in [21]: even in the category of asymptotically flat manifolds of non-negative scalar curvature, each leaf of the (unique) CMC foliation is uniquely isoperimetric for the volume it encloses.

As mentioned above, these questions have great importance even at a purely physical level. Over the past two decades considerable effort has been spent on recasting the notion of ADM mass (which one may simply regard as a suitable flux integral at infinity) in terms of geometric properties of (M, g). In particular, an isoperimetric notion of mass was first proposed by G. Huisken

$$m_{ISO} = \lim_{r \to \infty} \frac{2}{\text{area}(S_r)}\left(\text{vol}(B_r) - \frac{\text{area}(S_r)^{3/2}}{6\sqrt{\pi}}\right)$$

which does not involve derivatives of the metric at all. This formula should be read as follows: a small geodesic ball in a Riemannian manifold that is centered at a point of positive scalar curvature bounds more volume than a Euclidean ball of the same surface area, and similarly explicit computations give that large centered coordinate balls in Schwarzschild have the same property, and that the "isoperimetric deficit" encodes the mass.

In fact, Fan, Miao, Shi, and Tam have shown in [36] that

$$m_{ISO} = m_{ADM}$$

for any asymptotically flat Riemannian manifold (M, g) whose scalar curvature is integrable. Together with the positive mass theorem, this result should be interpreted as a remarkable large-scale manifestation of non-negative scalar curvature. We explicitly note that the equality above implies that, in the examples by Schoen and the author that we mentioned above, sufficiently large spheres in the Euclidean half-space, though evidently stable CMC surfaces, are not isoperimetric. The theorem by Chodosh–Eichmair–Shi–Yu significantly deepens our understanding of the global picture, by unambiguously identifying all large isoperimetric domains.

5   Fourth Lecture: Localized Solutions of the Einstein Equations

In this lecture, we describe a gluing construction aimed at obtaining asymptotically flat initial data sets that have positive ADM mass but are exactly trivial outside a cone of given angle. The scheme that we develop allows to produce a new class of N-body solutions for the Einstein equation, which exhibit the phenomenon of gravitational shielding.

5.1   Shielding in Newtonian Gravity

In order to contextualize the construction we are about to present, let us first make a digression on the problem of shielding in Newtonian gravity. Let then μ be a compactly suppported, finite measure representing the sources and let φ be a weak solution of the Poisson equation

$$\Delta \varphi = 4\pi \mu$$

that is uniquely determined once we set the zero level at spatial infinity. A general question one may ask is whether there exist (for non-trivial mass distributions) open regions where no gravitational forces are measured at all. It is not hard to see that this cannot be the case for any finite distribution of point masses, but on the other hand one can definitely construct explicit examples of continuous distributions where this shielding does occur. For instance, let us consider the measure μ that is absolutely continuous with respect to the Lebesgue measure in $\mathbb{R}^3$ and whose density function is given by

$$\varrho = \begin{cases} \varrho_0 & \text{if } a \leq |x| \leq b \\ 0 & \text{if } |x| < a, |x| > b \end{cases}$$

for positive numbers $a < b$, and $\varrho_0 > 0$ constant. The gravitational field is radial and (being attractive) inward-pointing, its magnitude being given by

$$-G(x) \cdot \frac{x}{|x|} = \begin{cases} 0 & \text{if } |x| \leq a \\ \frac{4\pi \varrho_0 (|x|^3 - a^3)}{3|x|^2} & \text{if } a \leq |x| \leq b \\ \frac{m}{|x|^2} & \text{if } |x| \geq b \end{cases}$$

where we have set $m = \frac{4\pi \varrho_0}{3} (b^3 - a^3)$, the total mass of this distribution. Correspondingly, the potential φ is constant in the ball centered at the origin and whose radius equals $|a|$. However, the potential will not be constant in the unbounded domain $|x| > b$: this is in fact a general phenomenon, corresponding to the absence of (long-range) gravitational shielding in Newtonian gravity. Indeed, let μ be the mass distribution describing an isolated gravitational system: we have seen in our first lecture that its potential has an asymptotic expansion of the form

$$\varphi(x) = -\frac{1}{|x|} \int_{\mathbb{R}^3} d\mu(y) - \frac{x}{|x|^3} \cdot \int_{\mathbb{R}^3} y d\mu(y) + O(|x|^{-3})$$

so that, in particular, if φ is constant in an open region that is unbounded then necessarily the leading term in the expansion above must vanish, i.e.,

$$m = \int_{\mathbb{R}^3} d\mu(y) = 0$$

which is only possible if there are no sources at all. We will now present some asymptotically flat manifolds of positive ADM mass that contain large (in fact, scaling-invariant) regions where no gravity is perceived, as encoded by the fact that the metric there is exactly Euclidean.

5.2 The Gluing Theorem

We shall now state a simplified, special version of the gluing theorem proven in [16]. In particular, for the sake of clarity, we shall only present the construction of Riemannian (i.e., time-symmetric) data in dimension three and in the most

convenient functional setup, without aiming at the sharpest possible results in terms
of decay and regularity. However, none of these assumptions is made in the paper
in question. As we shall discuss in Remark 5.8 below, the case of dimension two is,
instead, very peculiar.

Following the notation introduced in our second lecture, we consider an asymptotically flat manifold $(M, \check{g})$ that satisfies, outside of a coordinate ball B, the decay
estimate encoded by the condition

$$g_{ij} - \delta_{ij} \in C^{4,\alpha}_{-\check{p}}(\mathbb{R}^3 \setminus B)$$

for some $\check{p} \in (1/2, 1]$ (cf. Sect. 4.1) and solves the vacuum Einstein constraint
equations (i.e., the metric g is scalar flat). We are tacitly assuming that the manifold
M has only end, which we can do with no loss of generality as our construction is
local to a given end.

Given an angle $0 < \vartheta < \pi$ and a point $a \in \mathbb{R}^3$ sufficiently far from the origin we
denote by $C_\vartheta(a)$ the region of M consisting of the compact part together with the
set of points p in the exterior region such that $p - a$ makes an angle less than ϑ with
the vector $-a$. If we are given two angles $0 < \vartheta_1 < \vartheta_2 < \pi$ we consider the region
between the cones $C_{\vartheta_1}(a)$ and $C_{\vartheta_2}(a)$. To the scope of simplifying our discussion,
so to avoid the analysis at the tip of the cone, we consider a regularization of
this domain near the vertex a, and introduce three regions Ω_I, Ω, and Ω_O whose
(unbounded) boundary components are two smooth surfaces that coincide with the
boundaries of the two cones above outside of the unit ball centered at a. We shall
refer to Ω_I as the inner region, Ω the transition region, and Ω_O the outer region.
The specific shape of the regularization plays no role in the construction (nor in the
proof), and we only require the region Ω to be scaling-invariant outside a compact
set.

Here is the most basic version of our gluing theorem:

Theorem 5.1 *Assume that we are given an asymptotically flat manifold $(M, \check{g})$ as
above together with angles ϑ_1, ϑ_2 less than π. Furthermore, let $1/2 < p < \check{p}$. Then
there exists a_∞ so that for any $a \in \mathbb{R}^3$ satisfying $|a| \geq a_\infty$ we can find a metric $\hat{g}$
so that $(M, \hat{g})$ is scalar flat, $\hat{g}_{ij} - \delta_{ij} \in C^{2,\alpha}_{-p}$ and*

$$\hat{g} = \begin{cases} \check{g}, & in\ \Omega_I(a) \\ \delta & in\ \Omega_O(a). \end{cases}$$

Remark 5.2 From the viewpoint of the regularity of our data $(M, \hat{g})$ we have two
versions of the gluing theorem in [16], which will be for brevity referred to as *finite
regularity version* and *infinite regularity version*.

In the finite regularity version, we are given $\check{g} \in C^{\ell,\alpha}_{loc}$ and produce $\hat{g} \in C^{\ell-2,\alpha}_{loc}$.
Therefore we face the well-known phenomenon of derivative loss that has been
described both in [31] and [32]. With more work it is possible to improve the
theorem to remove this derivative loss, but we will not discuss this aspect here.

In the infinite regularity version, we start with a smooth metric $\check{g}$ and obtain a new metric $\hat{g}$ that is smooth as well.

Among possible applications, these gluing methods allow to construct a new class of N-body initial data sets for the Einstein constraint equations. In the context of Newtonian gravity, a set of initial data for the evolution of N massive bodies can be obtained by solving a single Poisson equation in the complement of a finite number of compact domains (say balls) in $\mathbb{R}^3$, at least if the interior structure and dynamics of the bodies in question is neglected. The nonlinearity of the Einstein constraint equations makes such a task a lot harder; in fact it was only in the last decade that sufficiently general results in this direction, without restricting to rather symmetric configurations [26] or exploiting the presence of multiple ends [28], have been obtained (cf. [29, 30]).

Let us suppose that a finite collection of N asymptotically flat manifolds $(M_1, g_1), \ldots, (M_N, g_N)$ is assigned and let U_i denote a compact, regular subdomain of M_i for each value of the index i. Moreover, let $x_1, \ldots, x_N \in \mathbb{R}^3$ be N vectors which are supposed to prescribe the location of the regions $U_1, \ldots, U_N$ with respect to a flat background. Then, one can use Theorem 5.1 to construct an asymptotically flat manifold (M, g) which is scalar flat and contains N regions that are isometric to the given bodies, with the centers of such bodies in a configuration which is a scaled version of the chosen configuration.

Remark 5.3 An important, and somewhat surprising feature of our construction is that the ADM mass of (M, g) converges to the ADM mass of $(M, \check{g})$ as one lets $a \to \infty$. In particular, the N-body solutions described above can be engineered so that their total mass is arbitrarily close to the sum of the ADM masses of the glued data $(M_1, g_1,), \ldots, (M_N, g_N)$.

Remark 5.4 The gluing theorem in [16] only concerns initial data sets (i.e., solutions of the Einstein constraints), however invoking e.g. [7] one can construct globally hyperbolic spacetimes $(\mathbb{L}, \gamma)$ that contain the exotic data (M, g) as initial data (or, equivalently, as a time-symmetric slice). The evolution of the massive regions has been described in terms of *focussed gravitational waves* (cf. [24]). Thereby, the scheme that we developed allows to produce N-body solutions for the Einstein equation, which patently exhibit the phenomenon of *(dynamical) gravitational shielding*: for any large T we can engineer solutions where any two massive bodies do not interact at all for any time $t \in (0, T)$, in striking contrast with the Newtonian gravity scenario.

Remark 5.5 The following extensions of the construction, listed in the beautiful *exposé* [24], are straightforward:

(a) one can glue two given scalar-flat, asymptotically flat Riemannian metrics to obtain a new one (in other words, there is no need for one of the two background metrics to be Euclidean);

(b) rather than aiming at scalar-flat metrics, one can simply obtain a combination of any two given asymptotically flat metrics g_1 and g_2, where the scalar curvature function is prescribed to be (for instance) the interpolating function $\chi R(g_1) + (1-\chi)R(g_2)$ where χ is a smooth cutoff function only depending on the angular variable;

(c) the gluing region can be a suitable deformation of the region between two coaxial cones.

Over the last few years, some interesting variations on the theme of [16] have appeared. Among them, we shall mention the construction of exotic hyperbolic gluings [27], the construction of localized solutions of the *linearized* Einstein constraints [6] (where no constraints on the gluing regions need to be assumed), and more generally the extension of these gluing schemes to other physical fields satisfying suitable axioms (in particular the so-called *higher spin fields*) [48]. It is still unclear whether the results in [27] are optimal from the perspective of the decay of the data on approach to the conformal boundary, or whether the same technique can be employed to obtain even more general combination of hyperbolic metrics.

5.3 Geometric Implications

We have already described in our previous lecture (cf. Sects. 4.2 and 4.3) some geometric implications of the above construction: one can engineer asymptotically flat solutions of positive mass that are Euclidean on a half-space, and thus contain plenty of stable minimal surfaces and outlying, arbitrarily large, constant mean curvature spheres which are not isoperimetric for the volume they enclose. Thereby, one gets an understanding of how special asymptotically Schwarzschildean data are in the broad class of general asymptotically flat data.

Instead, we would like to focus here on a somewhat different aspect, which is also connected to the positive mass theorem. First of all, let us observe that it follows from the strong form presented in Sect. 4.1 that there cannot exists on $\mathbb{R}^3$ Riemannian metrics of non-negative scalar curvature and positive ADM mass which are exactly flat outside a compact set. In other words, there must be an unbounded region where the metric in question is not Euclidean and one may wonder, roughly speaking, how large such a region should be. Very concretely, one may wonder whether it is possible to construct asymptotically flat metrics of positive mass being flat outside a cylinder or a slab of given height.

In order to give a quantitative description of this problem, we need to introduce a convenient definition.

Definition 5.6 Given an asymptotically flat metric g on $\mathbb{R}^3$ and considered the maximal open domain Ω where it is not Ricci-flat, we define the content at infinity of g to be

$$\kappa(g) = \liminf_{\varrho \to \infty} \left(\frac{\text{area}\,(\Omega \cap \{|x| = \varrho\})}{4\pi \varrho^2} \right).$$

Combining the rigidity part of the positive mass theorem with the formula expressing the ADM in terms of the Ricci curvature (cf. Remark 4.6) one easily proves the following assertion, which one may regard as a *positive content theorem*.

Proposition 5.7 *Let (M, g) be an asymptotically flat manifold with non-negative scalar curvature. Then* either *g is flat or* $\kappa(g) > 0$.

In particular, this proposition does indeed rule out any chance of constructing non-trivial data localize in a cylinder or a slab, and roughly asserts states that if an asymptotically flat metric g of non-negative scalar curvature is not trivial then the region where it is not (Ricci-) flat must contain a cone of positive aperture. Our gluing scheme provides a sort of converse to the previous statement, by ensuring that for any cone we can construct non-trivial data localized inside that cone.

Remark 5.8 One can also consider two-dimensional asymptotically flat Riemannian manifolds, and there is a suitable notion of mass in that context as well. The reader is referred to [15] and references therein for the relevant, elementary results, and to the monograph by Chruściel [25] for a much broader contextualization. Differently from the higher-dimensional counterpart, there exist plenty of (non-trivial) asymptotically flat manifolds that have non-negative scalar (equivalently: Gauss) curvature, but are flat outside a compact region. Indeed, for any $h_0 < 1$ one can simply smoothen the singular, conical metric $dr^2 + (h_0 r)^2 d\vartheta^2$. Such a smoothing can be obtained, for instance, within the class of rotationally symmetric metrics on $\mathbb{R}^2$, by suitably interpolating between the metric in question and a spherical cap. One way to do that is to regularize a C^1 metric following the methods presented in Section 3 of [54], which rely on a localized fiberwise convolution.

5.4 Sketch of the Proof

In the rest of this lecture we will outline the proof of Theorem 5.1, to be regarded as a toy model for the general gluing theorem obtained in [16]. In particular we are dealing here with the purely Riemannian case, which turns out to be much simpler than the full system of constraints, both at the linearized and at the nonlinear level.

We introduce coordinates $\{x\}$ centered at a and let $r : \mathbb{R}^3 \to \mathbb{R}$ be any positive function which equals the usual Euclidean distance $|\cdot|$ outside of the unit ball (for this given set of coordinates). For angles $0 < \vartheta_1 < \vartheta_2 < \pi$ we consider the region Ω described in the Sect. 5.2 (notice that the cones have vertex at $x = 0$). We let g be a Riemannian metric gotten by making a *rough patch* of $\check{g}$ and δ in the region Ω: therefore g agrees with $\check{g}$ in the region Ω_I, and agrees with δ in Ω_O. Here we are making use of an *angular* cutoff function χ (namely a function only depending on the angle between a given point and the vector $-a$) with rapid decay at $\partial\Omega_O$ and such that $1 - \chi$ rapidly decays at $\partial\Omega_I$. We observe that the metric g satisfies the same falloff properties at infinity as $\check{g}$.

The general strategy of the proof is to obtain the manifold $(M, \hat{g})$ starting from the rough patch (M, g) by solving the Einstein constraint equations iteratively, and more specifically by means of a Picard scheme. To that aim, we will need to solve a sequence of linearized problems of the form $L_g h_i = f_i$ and show that indeed the tensors $g + h_i$ converge, in suitable functional spaces, to a solution of the nonlinear constraint system. While this conceptual scheme is similar to that presented in [31] and [32] (see also [26]), we need to face here some peculiar technical obstacles. In order to solve the linear problems we will show that a suitable functional F (whose critical points solve the linearized problem) is coercive. This fact follows from certain Poincaré type estimates, that we called *basic estimates*, which may be of independent interest. The whole construction is performed in doubly weighted functional spaces, since we will have to keep control, at the same time, of both the decay at infinity and of the regularity at the boundary of the gluing region $\partial\Omega$.

For $q \in \mathbb{R}$ we consider the weighted L^2 Sobolev space $H_{k,-q}(\Omega)$ of functions (or tensors) defined as the completion of the smooth functions with bounded support (no condition on $\partial\Omega$) with respect to the norm $\| \cdot \|_{H_{k,-q}}$ defined letting

$$\|f\|^2_{H_{k,-q}} = \sum_{d=0}^{k} \int_{\Omega} |\nabla^d f|^2 r(x)^{-3+2(d+q)} \, dx.$$

The reader shall notice that due to the decay properties of the Riemannian metrics we consider here, such Sobolev spaces could have been equivalently defined by means of covariant derivatives with respect to g and by the volume measure associated with g. The space $H_{k,-q}(\Omega)$ consists of those functions f which roughly decay like $|x|^{-q}$ along the cone and for which a derivative of order $d \leq k$ decays like $|x|^{-q-d}$. Whenever no ambiguity is likely to arise, we will simply adopt the symbol $\|\cdot\|_{k,-q}$ rather than $\| \cdot \|_{H_{k,-q}}$.

We use the L^2 pairing to identify the dual space $H^*_{0,-q}$ with $H_{0,q-3}$ since

$$\left| \int_{\Omega} f_1 f_2 \, dx \right| = \left| \int_{\Omega} (f_1 r^{(-3/2+q)})(f_2 r^{(3/2-q)}) \, dx \right| \leq \|f_1\|_{0,-q} \|f_2\|_{0,q-3}.$$

We need to work in doubly weighted Sobolev spaces, namely we also need to introduce an angular weight which is a certain (large) power of the angular distance of a point from $\partial\Omega$. This is necessary in order to prove that the gluing we perform is smooth (more generally: regular enough) up to the boundary of the gluing domain. More precisely we take $\varrho = \varphi^{2N}$ where N will be chosen to be a large integer and φ is a positive weight function which is equal to $\vartheta - \vartheta_1$ near $\partial\Omega_I \setminus B_1(0)$ and equal to $\vartheta_2 - \vartheta$ near $\partial\Omega_O \setminus B_1(0)$. Near $\Omega \cap B_1(0)$ we further require that φ vanishes with nonzero gradient along $\partial\Omega$. We let φ_0 denote the maximum value of φ and we assume that each set $\Omega_t = \{\varphi \geq t\}$ for $0 \leq t \leq \varphi_0$ is (the closure of) a smooth domain which is a cone outside $B_1(0)$. We define the doubly weighted spaces $H_{k,-q,\varrho}(\Omega)$ and $H_{k,-q,\varrho^{-1}}(\Omega)$ using the norms

$$\|f\|^2_{H_{k,-q,\varrho}} = \sum_{d=0}^{k} \int_{\Omega} |\nabla^d f|^2 r(x)^{-3+2(d+q)} \varrho(x)\, dx,$$

$$\|f\|^2_{H_{k,-q,\varrho^{-1}}} = \sum_{d=0}^{k} \int_{\Omega} |\nabla^d f|^2 r(x)^{-3+2(d+q)} \varrho^{-1}(x)\, dx$$

As above, we will often use the notation $\|\cdot\|_{k,-q,\varrho}$ (respectively $\|\cdot\|_{k,-q,\varrho^{-1}}$) in lieu of $\|\cdot\|_{H_{k,-q,\varrho}}$ (respectively $\|\cdot\|_{H_{k,-q,\varrho^{-1}}}$).

Let us now move on to discuss the solvability of the linearized constraint equation

$$L_g h = f$$

where $L_g h = -\Delta_g(tr_g(h)) + div_g(div_g(h)) - g(h, Ric_g)$ is the linear operator we introduced in our third lecture when talking about the classification of static manifolds.

It is easily checked that for any $p \in (0, 1)$ the linearized map $L_g : H_{2,-p} \to H_{0,-p-2}$ is continuous; correspondingly, the same conclusion holds true for the adjoint operator $L_g^* : H_{2,-1+p} \to H_{0,-3+p}$. Similar mapping properties are true in doubly weighted functional spaces, and we shall make use of the continuity of $L_g^* : H_{2,-1+p,\varrho} \to H_{0,-3+p,\varrho}$.

At this stage we introduce the functional $F : H_{2,-1+p,\varrho} \to \mathbb{R}$ defined by

$$F(u) = \int_{\Omega} \left\{ \frac{1}{2} \left| L_g^* u \right|^2 r^{n-2p} \varrho - fu \right\} d\mathcal{V}_g^n$$

where we are taking $f \in H_{0,-p-2,\varrho^{-1}}$. If we let

$$h = r^{3-2p} \varrho L_g^* u$$

then the Euler–Lagrange equation for the functional F takes the form

$$L_g h - f = 0$$

and h decays as prescribed by the statement of Theorem 5.1.

We shall now prove existence of critical points of the functional F: one can initially assume to work at the trivial data, and specifically at the flat metric $g = \delta$, and then get the general case by a perturbation argument. Moreover, we observe that it is enough to prove coercivity estimates in singly weighted Sobolev spaces, i.e., with only radial but no angular weights: this follows from the following coarea type lemma.

Lemma 5.9 *Let $\zeta \in C_c^\infty(\mathbb{R}^3)$ and let, in the setting above, $\tilde{\varrho} : (0, \varphi_0) \to \mathbb{R}$ be the smooth, monotone increasing function characterized by the identity $\tilde{\varrho}(\varphi(x)) = \varrho(x)$ for all $x \in \Omega$. One has*

$$\int_\Omega \zeta \varrho \, d\mathcal{V}_g^n(x) = \int_0^{\varphi_0} \tilde{\varrho}'(t) \int_{\Omega_t} \zeta \, d\mathcal{V}_g^n(x) dt.$$

Proof We use the following level set formula that holds true for any smooth function ζ on $\bar{\Omega}$ with bounded support in $\mathbb{R}^3$

$$\int_\Omega \zeta \varrho \, d\mathcal{V}_g^3 = -\int_0^{\varphi_0} \frac{d}{dt} \int_{\Omega_t} \zeta \varrho \, d\mathcal{V}_g^3 \, dt + \int_{\Omega_{\varphi_0}} \zeta \tilde{\varrho}(\varphi_0) \, d\mathcal{V}_g^3.$$

Now we have, because of the standard coarea formula and integration by parts

$$-\int_0^{\varphi_0} \frac{d}{dt} \int_{\Omega_t} \zeta \varrho \, d\mathcal{V}_g^3 dt = -\int_0^{\varphi_0} \tilde{\varrho}(t) \frac{d}{dt} \int_{\Omega_t} \zeta \, d\mathcal{V}_g^3 dt$$

$$= -\tilde{\varrho}(\varphi_0) \int_{\Omega_{\varphi_0}} \zeta \, d\mathcal{V}_g^3 + \int_0^{\varphi_0} \tilde{\varrho}'(t) \int_{\Omega_t} \zeta \, d\mathcal{V}_g^3 dt.$$

Combining the previous two equations we obtain the desired conclusion. □

Once it is proven that a linear bounded operator $T : H_{k_1, -q_1} \to H_{k_2, -q_2}$ satisfies a functional inequality of the form

$$\|f\|_{k_1, -q_1} \le C \|Tf\|_{k_2, -q_2}$$

one can employ the previous lemma to obtain that in fact

$$\|f\|_{k_1, -r_1, \varrho} \le C \|Tf\|_{k_2, -r_2, \varrho}.$$

As a result, the key step in order to gain the coercivity of F is proving the following Poincaré inequalities.

Lemma 5.10 *Let $u \in C_c^\infty(\mathbb{R}^3)$. For any real number q with $0 < q < 1/2$ we have*

$$\|u\|_{1, -q} \le C \|\nabla u\|_{0, -q-1}$$

as well as

$$\|u\|_{2, -q} \le C \|\nabla^2 u\|_{0, -q-2}.$$

Proof We recall that we prove the inequality for the Euclidean metric, denoting by ∇u the Euclidean gradient, and by $\nabla^2 u$ the Hessian. We consider the function

$v = |x|^{-1+2q}$ and observe that for $|x| \neq 0$ we have

$$\Delta v = 2q(-1 + 2q)|x|^{-3+2q}$$

where Δ is the Euclidean–Laplace operator. When u has bounded support, as we are assuming, one can integrate by parts to obtain

$$\int_{\Omega \backslash B_1(0)} u^2 \Delta v \, dx = (1 - 2q) \int_{\partial B_1(0) \cap \Omega} u^2 |x|^{-2+2q} \, d\sigma - 2 \int_{\Omega \backslash B_1(0)} u \nabla u \cdot \nabla v \, dx$$

where we have used that fact that Ω is a cone outside of $B_1(0)$, so that the other boundary terms there vanish. It follows from the sign of the boundary term and the Schwarz inequality that

$$\delta \|u\|_{0,-q,\Omega \backslash B_1(0)}^2 = \delta \int_{\Omega \backslash B_1(0)} u^2 |x|^{-3+2q} \, dx \leq 2 \|u\|_{0,-q,\Omega \backslash B_1(0)} \|\nabla u\|_{0,-q-1,\Omega \backslash B_1(0)},$$

for $\delta = 2q(1 - 2q)$. Therefore, we have

$$\int_{\Omega \backslash B_1(0)} u^2 r^{-3+2q} \, dx \leq C \int_{\Omega \backslash B_1(0)} |\nabla u|^2 r^{-1+2q} \, dx.$$

Let now ζ be a smooth cutoff function with support in $B_2(0)$ and with $\zeta = 1$ on $B_1(0)$. A standard Poincaré inequality gives

$$\int_\Omega (\zeta u)^2 \, dx \leq C \int_\Omega |\nabla(\zeta u)|^2 \, dx,$$

which patently implies

$$\int_{\Omega \cap B_1(0)} u^2 r^{-3+2q} \, dx \leq C \int_{\Omega \cap B_2(0)} |\nabla u|^2 r^{-1+2q} \, dx + C \int_{\Omega \backslash B_1(0)} u^2 r^{-3+2q} \, dx$$

since r is bounded above and below by positive constants on $B_2(0)$. Combining this inequality with the previous one we obtain

$$\int_\Omega u^2 r^{-3+2q} \, dx \leq C \int_\Omega |\nabla u|^2 r^{-1+2q} \, dx,$$

thus the first estimate in the statement follows at once.

To justify the second one, we now apply the previous argument to each partial derivative of u. We may use essentially the same argument with the function $v = |x|^{1+2q}$. The boundary term one gets integrating by parts can be thrown away and we end up showing

$$\int_\Omega |\nabla u|^2 r^{-1+2q}\, dx \leq C \int_\Omega |\nabla^2 u|^2 r^{1+2q}\, dx$$

which allows to complete the proof.   □

Hence, the coercivity estimates for the adjoint of the linearized scalar curvature operator easily follow.

Proposition 5.11 *For any real number p with $1/2 < p < 1$ we have*

$$\|u\|_{2,-1+p} \leq C \|L^*_\delta u\|_{0,-3+p} \text{ for all } u \in H_{2,-1+p}.$$

Hence, thanks to Lemma 5.9 we deduce that

$$\|u\|_{2,-1+p,\varrho} \leq C \|L^*_\delta u\|_{0,-3+p,\varrho} \text{ for all } u \in H_{2,-1+p,\varrho}.$$

Proof Since $q = 1 - p$ satisfies $0 < q < 1/2$, we may apply Lemma 5.10 to obtain

$$\|u\|_{2,-1+p} \leq C \|\nabla^2 u\|_{0,-3+p}.$$

Thus to complete the proof it suffices to show

$$\|\nabla^2 u\|_{0,-3+p} \leq C \|L^*_\delta u\|_{0,-3+p}.$$

Recalling that we are working at the Euclidean metric (and will then deduce a general coercivity result by perturbation) we can simply take the trace in the definition of the operator L^*_δ thereby obtaining $-2\Delta u$, from which it follows that

$$\nabla^2 u = L^*_\delta u - \frac{1}{2}\mathrm{tr}(L^*_\delta u)\delta$$

and the conclusion follows.   □

Proposition 5.12 *In the setting above, there exist constants $a_{\infty,L}$ and C (depending only on g, p, ϑ_1, ϑ_2) such that uniformly for $|a| > a_{\infty,L}$*

$$\|u\|_{2,-1+p,\varrho} \leq C \left\|L^*_g u\right\|_{0,-3+p,\varrho}$$

for all $u \in H_{2,-1+p,\varrho}$.

Proof The assertion above follows by simply combining the conclusion of Proposition 5.11 with a standard perturbation argument: given $\varepsilon > 0$ we can find $a_{\infty,L}$ such that

$$|a| > a_{\infty,L} \implies \left\|L^*_g u - L^*_\delta u\right\|_{0,-3+p} \leq \varepsilon \|u\|_{2,-1+p}$$

hence, by virtue of Lemma 5.9

$$\left\| L_g^* u - L_\delta^* u \right\|_{0,-3+p,\varrho} \leq \varepsilon \, \|u\|_{2,-1+p,\varrho} \, .$$

Thus, picking $\varepsilon_0 = 1/(3C)$ we have that for all $|a| > a_{\infty,L}$ the conclusion comes from the triangle inequality. $\qquad\square$

At this stage, we can use a direct method to solve our linearized problem.

Proposition 5.13 *In the setting above, for any $f \in H_{0,-p-2,\varrho^{-1}}$ there exists a unique $\tilde{u} \in H_{2,-1+p,\varrho}$ which minimizes the functional F on the Hilbert space $H_{2,-1+p,\varrho}$.*

Proof The argument follows the direct method of the Calculus of Variations. Indeed, the functional F is bounded from below on $H_{2,-1+p,\varrho}$ since its very definition implies

$$F(u) \geq C_1 \left\| L_g^* u \right\|_{0,-3+p,\varrho}^2 - C_2 \, \|f\|_{0,-p-2,\varrho^{-1}} \, \|u\|_{0,-1+p,\varrho}$$

and hence, thanks to the basic estimate (in the form of Proposition 5.12)

$$F(u) \geq C_1 \, \|u\|_{2,-1+p,\varrho}^2 - C_2 \, \|f\|_{0,-p-2,\varrho^{-1}} \, \|u\|_{2,-1+p,\varrho}$$

which immediately implies that F is coercive. As a result, we can pick a minimizing sequence (u_i) which is bounded in $H_{2,-1+p,\varrho}$ (in fact, the previous estimate shows that *any* minimizing sequence has to be bounded): by Banach–Alaoglu there will be a subsequence which weakly converges to a limit point $\tilde{u}$ and by (weak) lower semicontinuity of the functional we conclude that

$$F(\tilde{u}) \leq \liminf_{i \to \infty} F(u_i).$$

This precisely means that $\tilde{u}$ minimizes the value of F. Finally, the uniqueness statement follows by strict convexity of the functional: indeed, if we had two minima u_1 and u_2 then because of the identity

$$F\left(\frac{u_1 + u_2}{2} \right) = \frac{1}{2} F(u_1) + \frac{1}{2} F(u_2) - \frac{1}{8} \left\| L_g^* [u_2 - u_1] \right\|_{0,-1+p,\varrho}^2$$

we would reach a contradiction unless $L_g^* [u_2 - u_1] = 0$ and by Proposition 5.12 this forces $u_1 = u_2$, which is what we had to prove. $\qquad\square$

Let us now describe how one can proceed to solve the nonlinear equation (which is a scalar curvature prescription problem). Roughly speaking, one would like to set up a sequence of linearized problem whose solutions converge to an actual solution of the scalar curvature equation with datum f. If we want to solve (for

h) the equation

$$R_{g_0+h} = f$$

then we can consider a suitable expansion

$$R_{g_0} + L_{g_0}h + Q_{g_0}(h) = f$$

where $Q_{g_0}(h)$ collects all the terms of R_{g_0+h} that are at least quadratic in h.

We then set up an iteration scheme. As a first step, one can neglect the quadratic term and solve the equation

$$L_{g_0}h = f - R_{g_0}$$

so if we let h_1 to be the unique solution of this equation (in the appropriate functional spaces, based on Proposition 5.13) we can then set $g_1 = g_0 + h_1$ and proceed. In practice, one could be tempted to set up a Newton iterative scheme, where the linearization for the second step happens at g_1 and keeps changing step after step. Unfortunately, this poses a serious technical problem as one faces a well-known loss of regularity phenomenon: if $g_0 \in C^{k+4,\alpha}$ then $g_1 \in C^{k+2,\alpha}$ (here one needs to observe that the equation for the minimizer u_1 is elliptic of order four, with the bilaplacian operator as leading term, and we then set $h_1 = r^{n-2p}\varrho L_g^* u_1$ which involves two derivatives of u_1) so we lose two orders of differentiability and thus we cannot possibly iterate this procedure as long as we wish. For this reason, we proceed differently and rather pick h_2 solving

$$L_{g_0}h = f - R_{g_0} - Q_{g_0}(h_1).$$

The general iteration scheme is defined by setting $f_0 = 0, h_0 = 0$ and then requiring, for $i \geq 1$, that

$$\begin{cases} f_i = (f - R_{g_0}) - Q_{g_0}(h_{i-1}) \\ L_{g_0}h_i = f_i. \end{cases}$$

In order to actually define the iteration one needs to introduce two Banach spaces

$$(X_1, \| \cdot \|_1), \quad (X_2, \| \cdot \|_2)$$

which, in the specific case, are taken to be the closure of the set of smooth functions and symmetric tensors with respect to mixed, doubly-weighted, Sobolev-Hölder norms. The following proposition, concerning the solution operator S for the linearized problem that can be defined relying on Proposition 5.13, ensures that the quadratic error terms get smaller and smaller along the course of the iteration.

Proposition 5.14 *Given any* $\lambda > 0$, *there exists* $r_0 > 0$ *sufficiently small so that if* $\|f_1\|_1 < r_0$ *and* $\|f_2\|_1 < r_0$ *and we let* $h_1 = Sf_1$, $h_2 = Sf_2$ *then we have*

$$\|Q_g(h_1) - Q_g(h_2)\|_1 \le \lambda \|h_1 - h_2\|_2.$$

The proof of this fact, although based on Schauder estimates, is rather lengthy and technical and we skip it entirely. Yet, once this is proven, the conclusion (in terms of existence and boundary regularity of the gluing) follows at once from a simple iteration lemma:

Theorem 5.15 *Given* $f \in X_1$ *sufficiently small, there is a small* $h \in X_2$ *satisfying*

$$L_{g_0} h + Q_{g_0}(h) = f.$$

The interested reader may find all details, in much greater generality, in Section 5 of [16]. As we described in the last paragraph of Sect. 5.2, this construction has been employed in recent years to obtain a number of interesting results and to advance on significant geometric problems. We believe the range of applicability of this technique to be quite broader than has been realized so far, and we expect further developments in the near future.

References

1. M. Abate, F. Tovena, *Curves and Surfaces* (Springer, Milan, 2012)
2. L. Ambrozio, Rigidity of area-minimizing free boundary surfaces in mean convex three-manifolds. J. Geom. Anal. **25**(2), 1001–1017 (2015)
3. R. Arnowitt, S. Deser, C.W. Misner, Dynamical structure and definition of energy in general relativity. Phys. Rev. **2**(116), 1322–1330 (1959)
4. A. Ashtekar, R. Hansen, A unified treatment of null and spatial infinity in general relativity. I. Universal structure, asymptotic symmetries, and conserved quantities at spatial infinity. J. Math. Phys. **19**(7), 1542–1566 (1978)
5. R. Bartnik, The mass of an asymptotically flat manifold. Commun. Pure Appl. Math. **39**(5), 661–693 (1986)
6. R. Beig, P.T. Chruściel, Shielding linearized gravity. Phys. Rev. D **95**(6), 064063, 9pp. (2017)
7. L. Bieri, An extension of the stability theorem of the Minkowski space in general relativity. J. Differ. Geom. **86**(1), 17–70 (2010)
8. H. Bray, S. Brendle, M. Eichmair, A. Neves, Area-minimizing projective planes in 3-manifolds. Commun. Pure Appl. Math. **63**(9), 1237–1247 (2010)
9. H. Bray, S. Brendle, A. Neves, Rigidity of area-minimizing two-spheres in three-manifolds. Commun. Anal. Geom. **18**(4), 821–830 (2010)
10. S. Brendle, *Rigidity Phenomena Involving Scalar Curvature*. Surveys in Differential Geometry, vol. 17 (International Press, Boston, 2012), pp. 179–202
11. S. Brendle, M. Eichmair, Large outlying stable constant mean curvature spheres in initial data sets. Invent. Math. **197**(3), 663–682 (2014)
12. G. Bunting, A. Masood-ul-Alam, Nonexistence of multiple black holes in asymptotically Euclidean static vacuum space-time. Gen. Relativ. Gravit. **19**(2), 147–154 (1987)
13. M. Cai, G. Galloway, Rigidity of area minimizing tori in 3-manifolds of nonnegative scalar curvature. Commun. Anal. Geom. **8**(3), 565–573 (2000)

14. A. Carlotto, Rigidity of stable minimal hypersurfaces in asymptotically flat spaces. Calc. Var. Partial Differ. Equ. **55**(3), 1–20 (2016)
15. A. Carlotto, C. De Lellis, Min-max embedded geodesics lines in asymptotically conical surfaces. J. Differ. Geom. **112**(3), 411–445 (2019)
16. A. Carlotto, R. Schoen, Localizing solutions of the Einstein constraint equations. Invent. Math. **205**(3), 559–615 (2016)
17. A. Carlotto, O. Chodosh, M. Eichmair, Effective versions of the positive mass theorem.Invent. Math. **206**(3), 975–1016 (2016)
18. O. Chodosh, M. Eichmair, Global uniqueness of large stable CMC surfaces in asymptotically flat 3-manifolds (arXiv:1703.02494, preprint)
19. O. Chodosh, M. Eichmair, On far-outlying CMC spheres in asymptotically flat Riemannian 3-manifolds (arXiv:1703.09557, preprint)
20. O. Chodosh, M. Eichmair, V. Moraru, A splitting theorem for scalar curvature. Commun. Pure Appl. Math. **72**(6), 1231–1242 (2019)
21. O. Chodosh, M. Eichmair, Y. Shi, H. Yu, Isoperimetry, scalar curvature, and mass in asymptotically flat Riemannian 3-manifolds (arXiv:1606.04626, preprint)
22. Y. Choquet-Bruhat, R. Geroch, Global aspects of the Cauchy problem in general relativity. Commun. Math. Phys. **14**, 329–335 (1969)
23. P. Chruściel, *Boundary Conditions at Spatial Infinity from a Hamiltonian Point of View. Topological Properties and Global Structure of Space-Time (Erice, 1985)*. NATO Advanced Science Institutes Series B: Physics, vol. 138 (Plenum, New York, 1986)
24. P.T. Chruściel, Anti-gravity à la Carlotto-Schoen. Séminaire Bourbaki **1120**, 1–24 (2016)
25. P.T. Chruściel, *Lectures on Energy in General Relativity* (preprint)
26. P.T. Chruściel, E. Delay, On mapping properties of the general relativistic constraints operator in weighted function spaces, with applications. Mém. Soc. Math. Fr. (N.S.) (94), vi+103pp. (2003)
27. P.T. Chruściel, E. Delay, Exotic hyperbolic gluings. J. Differ. Geom. **108**(2), 243–293 (2018)
28. P.T. Chruściel, J. Isenberg, D. Pollack, Initial data engineering. Commun. Math. Phys. **257**(1), 29–42 (2005)
29. P.T. Chruściel, J. Corvino, J. Isenberg, Construction of N-body initial data sets in general relativity. Commun. Math. Phys. **304**(3), 637–647 (2011)
30. P.T. Chruściel, J. Corvino, J. Isenberg, *Construction of N-Body Time-Symmetric Initial Data Sets in General Relativity*. Complex Analysis and Dynamical Systems IV, Part 2, Contemporary Mathematics, vol. 554 (American Mathematical Society, Providence, 2011), pp. 83–92
31. J. Corvino, Scalar curvature deformation and a gluing construction for the Einstein constraint equations. Commun. Math. Phys. **214**(1), 137–189 (2000)
32. J. Corvino, R. Schoen, On the asymptotics for the vacuum Einstein constraint equations. J. Differ. Geom. **73**(2), 185–217 (2006)
33. M. Eichmair, J. Metzger, Large isoperimetric surfaces in initial data sets. J. Differ. Geom. **94**(1), 159–186 (2013)
34. A. Einstein, Die Feldgleichungen der Gravitation. Sitzungsberichte der Preussischen Akademie der Wissenschaften (1915), pp. 844–847
35. A. Einstein, Die Grundlage der allgemeinen Relativitätstheorie. Ann. Phys. **49**, 769–822 (1916)
36. Y.-S. Fan, Y. Shi, L.-F. Tam, Large-sphere and small-sphere limits of the Brown-York mass. Commun. Anal. Geom. **17**(1), 37–72 (2009)
37. H. Federer, *Geometric Measure Theory. Die Grundlehren der mathematischen Wissenschaften*, Band 153 (Springer, New York 1969), xiv+676pp.
38. D. Fischer-Colbrie, R. Schoen, The structure of complete stable minimal surfaces in 3-manifolds of nonnegative scalar curvature. Commun. Pure Appl. Math. **33**(2), 199–211 (1980)
39. Y. Fourès-Bruhat, Théorème d'existence pour certains systèmes d'équations aux dérivées partielles non linéaires. Acta Math. **88**, 141–225 (1952)
40. S. Gallot, D. Hulin, J. Lafontaine, *Riemannian Geometry*. Universitext, 2nd edn. (Springer, Berlin, 1990), xvi+322pp.

41. G. Galloway, P. Miao, Variational and rigidity properties of static potentials. Commun. Anal. Geom. **25**(1), 163–183 (2017)
42. M. Gromov, Metric inequalities with scalar curvature. Geom. Funct. Anal. **28**(3), 645–726 (2018)
43. V. Guillemin, A. Pollack, *Differential Topology* (Prentice-Hall, Englewood Cliffs, 1974), xvi+222pp.
44. L. Hörmander, *The Analysis of Linear Partial Differential Operators. I. Distribution Theory and Fourier Analysis*, 2nd edn. Springer Study Edition (Springer, Berlin, 1990), xii+440pp.
45. L.-H. Huang, Foliations by stable spheres with constant mean curvature for isolated systems with general asymptotics. Commun. Math. Phys. **300**(2), 331–373 (2010)
46. G. Huisken, S.-T. Yau, Definition of center of mass for isolated physical systems and unique foliations by stable spheres with constant mean curvature. Invent. Math. **124**(1–3), 281–311 (1996)
47. W. Israel, *Black Hole Uniqueness and the Inner Horizon Stability Problem. The Future of the Theoretical Physics and Cosmology (Cambridge, 2002)* (Cambridge University Press, Cambridge, 2003), pp. 205–216
48. J. Joudioux, Gluing for the constraints for higher spin fields. J. Math. Phys. **58**(11), 111513, 10pp. (2017)
49. S. Lang, *Algebra. Revised Third Edition*. Graduate Texts in Mathematics, vol. 211 (Springer, New York, 2002), xvi+914pp.
50. S. Ma, Uniqueness of the foliation of constant mean curvature spheres in asymptotically flat 3-manifolds. Pac. J. Math. **252**(1), 145–179 (2011)
51. F. Marques, A. Neves, Rigidity of min-max minimal spheres in three-manifolds. Duke Math. J. **161**(14), 2725–2752 (2012)
52. D. Máximo, I. Nunes, Hawking mass and local rigidity of minimal two-spheres in three-manifolds. Commun. Anal. Geom. **21**(2), 409–432 (2013)
53. N. Meyers, An expansion about infinity for solutions of linear elliptic equations. J. Math. Mech. **12**(2), 247–264 (1963)
54. P. Miao, Positive mass theorem on manifolds admitting corners along a hypersurface. Adv. Theor. Math. Phys. **6**(6), 1163–1182 (2002)
55. P. Miao, L.-F. Tam, Static potentials on asymptotically flat manifolds. Ann. Henri Poincaré **16**(10), 2239–2264 (2015)
56. P. Miao, L.-F. Tam, Evaluation of the ADM mass and center of mass via the Ricci tensor. Proc. Am. Math. Soc. **144**(2), 753–761 (2016)
57. M. Micallef, V. Moraru, Splitting of 3-manifolds and rigidity of area-minimising surfaces. Proc. Am. Math. Soc. **143**(7), 2865–2872 (2015)
58. H. Müller zum Hagen, D. Robinson, H. Seifert, Black holes in static vacuum space-times. Gen. Relativ. Gravit. **4**(8), 53–78 (1973)
59. C. Nerz, Foliations by stable spheres with constant mean curvature for isolated systems without asymptotic symmetry. Calc. Var. Partial Differ. Equ. **54**(2), 1911–1946 (2015)
60. I. Nunes, Rigidity of area-minimizing hyperbolic surfaces in three-manifolds. J. Geom. Anal. **23**(3), 1290–1302 (2013)
61. D. Robinson, A simple proof of the generalization of Israel's theorem. Gen. Relativ. Gravit. **8**(8), 695–698 (1977)
62. J. Sbierski, On the existence of a maximal Cauchy development for the Einstein equations: a dezornification. Ann. Henri Poincaré **17**(2), 301–329 (2016)
63. R. Schoen, S.T. Yau, On the proof of the positive mass conjecture in general relativity. Commun. Math. Phys. **65**(1), 45–76 (1979)
64. R. Schoen, S.T. Yau, Positive scalar curvature and minimal hypersurface singularities (arXiv: 1704.05490, preprint)
65. R. Wald, *General Relativity* (University of Chicago Press, Chicago, 1984), xiii+491pp.

Lectures on Linear Stability of Rotating Black Holes

Felix Finster

Abstract These lecture notes are concerned with linear stability of the non-extreme Kerr geometry under perturbations of general spin. After a brief review of the Kerr black hole and its symmetries, we describe these symmetries by Killing fields and work out the connection to conservation laws. The Penrose process and superradiance effects are discussed. Decay results on the long-time behavior of Dirac waves are outlined. It is explained schematically how the Maxwell equations and the equations for linearized gravitational waves can be decoupled to obtain the Teukolsky equation. It is shown how the Teukolsky equation can be fully separated to a system of coupled ordinary differential equations. Linear stability of the non-extreme Kerr black hole is stated as a pointwise decay result for solutions of the Cauchy problem for the Teukolsky equation. The stability proof is outlined, with an emphasis on the underlying ideas and methods.

Keywords Black holes · Linear stability · Kerr geometry · Gravitational waves · Linear hyperbolic PDEs · Teukolsky Equation

Mathematics Subject Classification (2000) Primary 83C57, 83C35; Secondary 58J45, 83C20, 83C60, 35Q75, 35L15, 35L52

1 Introduction

These lectures are concerned with the black hole stability problem. Since this is a broad topic which many people have been working on, we shall restrict attention to specific aspects of this problem: First, we will be concerned only with *linear* stability. Indeed, the problem of nonlinear stability is much harder, and at present there are only few rigorous results. Second, we will concentrate on *rotating* black

F. Finster (✉)
Fakultät für Mathematik, Universität Regensburg, Regensburg, Germany
e-mail: finster@ur.de

© Springer Nature Switzerland AG 2019

S. Cacciatori et al. (eds.), *Einstein Equations: Physical and Mathematical Aspects of General Relativity*, Tutorials, Schools, and Workshops in the Mathematical Sciences, https://doi.org/10.1007/978-3-030-18061-4_2

holes. This is because the angular momentum leads to effects (Penrose process, superradiance) which make the rotating case particularly interesting. Moreover, the focus on rotating black holes gives a better connection to my own research, which was carried out in collaboration with Niky Kamran (McGill), Joel Smoller (University of Michigan), and Shing-Tung Yau (Harvard). The linear stability result for general spin was obtained together with Joel Smoller (see [23] and the survey article [22]). Before beginning, I would like to remember Joel Smoller, who sadly passed away in September 2017. These notes are dedicated to his memory.

2 The Kerr Black Hole

In general relativity, space and time are combined to a four-dimensional space-time, which is modelled mathematically by a Lorentzian manifold $(\mathcal{M}, g)$ of signature $(+ - - -)$ (for more elementary or more detailed introductions to general relativity see the textbooks [1, 36, 42, 44]). The gravitational field is described geometrically in terms of the curvature of space-time. Newton's gravitational law is replaced by the Einstein equations

$$R_{jk} - \frac{1}{2} R g_{jk} = 8\pi\kappa T_{jk},\tag{1}$$

where R_{jk} is the Ricci tensor, R is scalar curvature, and κ denotes the gravitational constant. Here T_{jk} is the energy-momentum tensor which describes the distribution of matter in space-time.

A rotating black hole is described by the *Kerr geometry*. It is a solution of the vacuum Einstein equations discovered in 1963 by Roy Kerr. In the so-called Boyer–Lindquist coordinates, the Kerr metric takes the form (see [7, 37])

$$ds^2 = \frac{\Delta}{U} \left(dt - a \sin^2 \vartheta \, d\varphi\right)^2 - U \left(\frac{dr^2}{\Delta} + d\vartheta^2\right) - \frac{\sin^2 \vartheta}{U} \left(a \, dt - (r^2 + a^2)d\varphi\right)^2,\tag{2}$$

where

$$U = r^2 + a^2 \cos^2 \vartheta, \qquad \Delta = r^2 - 2Mr + a^2,\tag{3}$$

and the coordinates $(t, r, \vartheta, \varphi)$ are in the range

$$-\infty < t < \infty, \quad M + \sqrt{M^2 - a^2} < r < \infty, \quad 0 < \vartheta < \pi, \quad 0 < \varphi < 2\pi.$$

The parameters M and aM describe the mass and the angular momentum of the black hole.

In the case $a = 0$, one recovers the Schwarzschild metric

$$ds^2 = \left(1 - \frac{2M}{r}\right) dt^2 - \left(1 - \frac{2M}{r}\right)^{-1} dr^2 - r^2(d\theta^2 + \sin^2\theta \, d\varphi^2) \,.$$

In this case, the function Δ has two roots

$$r = 2M \qquad \text{event horizon}$$

$$r = 0 \qquad \text{curvature singularity} \,.$$

In the region $r > 2M$, the so-called *exterior region*, t is a time coordinate, whereas r, ϑ, and φ are spatial coordinates. More precisely, (ϑ, φ) are polar coordinates, whereas the radial coordinate r is determined by the fact that the two-surface $S = \{t = t_0, r = r_0\}$ has area $4\pi r_0^2$. The region $r < 2M$, on the other hand, is the *interior region*. In this region, the radial coordinate r is time, whereas t is a spatial coordinate. Since time always propagates to the future, the event horizon can be regarded as the "boundary of no escape." The surface $r = 2M$ merely is a coordinate singularity of our metric. This becomes apparent by transforming to Eddington–Finkelstein or Kruskal coordinates. For brevity, we shall not enter the details here.

In the case $a \neq 0$, the singularity structure is more involved. The function U is always strictly positive. The function Δ has the two roots

$$r_0 := M + \sqrt{M^2 - a^2} \qquad \text{event horizon} \tag{4}$$

$$r_1 := M - \sqrt{M^2 - a^2} \qquad \text{Cauchy horizon} \,. \tag{5}$$

If $a^2 > M^2$, these roots are complex. This corresponds to the unphysical situation of a naked singularity. We shall not discuss this case here, but only consider the so-called

$$\textit{non-extreme case} \qquad M^2 < a^2 \,.$$

In this case, the hypersurface

$$r = r_1 := M + \sqrt{M^2 - a^2}$$

again defines the *event horizon* of the black hole. In what follows, we shall restrict attention to the *exterior region* $r > r_1$. This is because classically, no information can be transmitted from the interior of the black hole to its exterior. Therefore, it is impossible for principal reasons to know what happens inside the black hole. With this in mind, it seems pointless to study the black hole inside the event horizon, because this study will never be tested or verified by experiments.

We finally remark that in *quantum gravity*, the situation is quite different because it is conceivable that a black hole might "evaporate," in which case the interior of the black hole might become accessible to observations. In physics, such questions are often discussed in connection with the so-called information paradox, which states that the loss of information at the event horizon is not compatible with the unitary time evolution in quantum theory. I find such questions related to quantum effects of a black hole quite interesting, and indeed most of my recent research is devoted to quantum gravity (in an approach called causal fermion systems; see, for example, the textbook [13] or the survey paper [14]). But since this summer school is devoted to classical gravity, I shall not enter this topic here.

3 Symmetries and Killing Fields

The Kerr geometry is stationary and axisymmetric. This is apparent in Boyer–Lindquist coordinates (2) because the metric coefficients are

$$\text{independent of } t : \text{stationary}$$

$$\text{independent of } \varphi : \text{axisymmetric} .$$

These symmetries can be described more abstractly using the notion of *Killing fields*. We recall how this works because we need it later for the description of the Penrose process and superradiance. We restrict attention to the time translation symmetry, because for the axisymmetry or other symmetries, the argument is similar. Given $\tau \in \mathbb{R}$, we consider the mapping

$$\Phi_\tau \; : \; \mathcal{M} \to \mathcal{M} , \qquad (t, x) \mapsto (t + \tau, x)$$

(where x stands for the spatial coordinates (r, ϑ, φ)). The fact that the metric coefficients are time independent means that Φ_τ is an *isometry*, defined as follows. The derivative of Φ_τ (i.e., the linearization; it is sometimes also denoted by $(\Phi_\tau)_*$) is a mapping between the corresponding tangent spaces,

$$D\Phi_\tau|_p \; : \; T_p\mathcal{M} \to T_{\Phi_\tau(x)}\mathcal{M} .$$

Being an isometry means that

$$g_p(u, v) = g_{\Phi_\tau(p)}\big(D\Phi_\tau|_p u, D\Phi_\tau|_p v\big) \qquad \text{for all } u, v \in T_p\mathcal{M} .$$

Let us evaluate this equation infinitesimally in τ. We first introduce the vector field K by

$$K := \frac{d}{d\tau}\Phi_\tau\Big|_{\tau=0} .$$

Choosing local coordinates, we obtain in components

$$\left(D\Phi_\tau|_p u\right)^a = \frac{\partial \Phi_\tau^a(p)}{\partial x^i} \, u^i \, ,$$

where for clarity we denote the tensor indices at the point $\Phi_\tau(x)$ by a and b. We then obtain

$$
\begin{aligned}
0 &= \frac{d}{d\tau} \, g_{\Phi_\tau(p)}\left(D\Phi_\tau|_p u, D\Phi_\tau|_p v\right)\Big|_{\tau=0} \\
&= \frac{d}{d\tau} \left(g_{ab}\left(\Phi_\tau(p)\right) \frac{\partial \Phi_\tau^a(p)}{\partial x^i} \, u^i \, \frac{\partial \Phi_\tau^b(p)}{\partial x^j} \, v^j \right)\Big|_{\tau=0} \\
&= \partial_k g(u,v) \, K^k + g\left(u^i \partial_i K, v\right) + g\left(u, v^j \partial_j K\right).
\end{aligned}
$$

Choosing Gaussian coordinates at p, one sees that this equation can be written covariantly as

$$0 = g\left(\nabla_u K, v\right) + g\left(u, \nabla_v K\right),$$

where ∇ is the Levi-Civita connection. This is the *Killing equation*, which can also be written in the shorter form

$$0 = \nabla_{(i} K_{j)} := \frac{1}{2} \left(\nabla_i K_j + \nabla_j K_i\right). \tag{6}$$

A vector field which satisfies the Killing equation is referred to as a *Killing field*. We remark that if the flow lines exist on an interval containing zero and τ, then the resulting diffeomorphism Φ_τ is indeed an isometry of $\mathcal{M}$.

A variant of Noether's theorem states that Killing symmetries, which describe infinitesimal symmetries of space-time, give rise to corresponding conservation laws. For *geodesics*, these conservation laws are obtained simply by taking the Lorentzian inner product of the Killing vector field and the velocity vector of the geodesic. Indeed, let $\gamma(\tau)$ be a parameterized geodesic, i.e.,

$$\nabla_\tau \dot{\gamma}(\tau) = 0 \, .$$

Then, denoting the metric for simplicity by $\langle ., . \rangle_p := g_p(., .)$, we obtain

$$
\begin{aligned}
\frac{d}{d\tau} \langle K(\gamma(\tau)), \dot{\gamma}(\tau) \rangle_{\gamma(\tau)} &= \langle \nabla_\tau K(\gamma(\tau)), \dot{\gamma}(\tau) \rangle_{\gamma(\tau)} + \langle K(\gamma(\tau)), \nabla_\tau \dot{\gamma}(\tau) \rangle_{\gamma(\tau)} \\
&= \langle \nabla_\tau K(\gamma(\tau)), \dot{\gamma}(\tau) \rangle_{\gamma(\tau)} = \nabla_i K_j \big|_{\gamma(\tau)} \, \dot{\gamma}^i(\tau) \, \dot{\gamma}^j(\tau) = 0 \, ,
\end{aligned}
$$

where in the last step we used the Killing equation (6). We thus obtain the
conservation law

$$\left\langle K(\gamma(\tau)), \dot\gamma(\tau)\right\rangle_{\gamma(\tau)} = \text{const},$$

which holds for any parameterized geodesic $\gamma(\tau)$ and any Killing field K.

4   The Penrose Process and Superradiance

In the Kerr geometry, the two vector fields ∂_t and ∂_φ are Killing fields. The
corresponding conserved quantities are

$$E := \left\langle \frac{\partial}{\partial t}, \dot\gamma(\tau)\right\rangle_{\gamma(\tau)} \qquad \text{energy} \tag{7}$$

$$A := \left\langle \frac{\partial}{\partial \varphi}, \dot\gamma(\tau)\right\rangle_{\gamma(\tau)} \qquad \text{angular momentum}. \tag{8}$$

Let us consider the energy in more detail for a test particle moving along the
geodesic γ. In this case, $\gamma(\tau)$ is a causal curve (i.e., $\dot\gamma(\tau)$ is timelike or null
everywhere), and we always choose the parameterization such that γ is future-
directed (i.e., the time coordinate $\gamma^0(\tau)$ is monotone increasing in τ). In the
asymptotic end (i.e., for large r), the Killing field ∂_t is timelike and future-directed.
As a consequence, the inner product in (7) is strictly positive. This corresponds to
the usual concept of the energy being a non-negative quantity. We point out that this
result relies on the assumption that the Killing field ∂_t is timelike. However, if this
Killing field is spacelike, then the inner product in (7) could very well be negative.
In order to verify if this case occurs, we compute

$$\left\langle \frac{\partial}{\partial t}, \frac{\partial}{\partial t}\right\rangle = g_{00} = \frac{\Delta}{U} - \frac{a^2 \sin^2 \vartheta}{U} = \frac{1}{U}\left(r^2 - 2Mr + a^2 \cos^2 \vartheta\right),$$

where we read off the corresponding metric coefficient in (2) and simplified it
using (3). Computing the roots, one sees that the Killing field ∂_t indeed becomes
null on the surface

$$r = r_{\text{es}} := M + \sqrt{M^2 - a^2 \cos^2 \vartheta}, \tag{9}$$

the so-called *ergosphere*. Comparing with the formula for the event horizon (4), one
sees that the ergosphere is outside the event horizon and intersects the event horizon
at the poles $\vartheta = 0$, π (see the left of Fig. 1). The region $r_1 < r < r_{\text{es}}$ is the so-called
ergoregion.

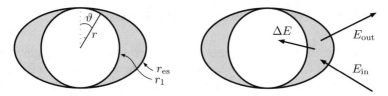

Fig. 1 Schematic picture of the ergosphere (left) and the Penrose process (right)

The ergosphere causes major difficulties in the proof of linear stability of the Kerr geometry. These difficulties are not merely technical, but they are related to physical phenomena, as we now explain step by step. The name ergosphere is motivated from the fact that it gives rise to a mechanism for extracting energy from a rotating black hole. This effect was first observed by Penrose [38] and is therefore referred to as the *Penrose process*. In order to explain this effect, we consider a spaceship of energy E_{in} which flies into the ergoregion (see the right of Fig. 1), where it ejects a projectile of energy ΔE which falls into the black hole. After that, the spaceship flies out of the ergoregion with energy E_{out}. Due to energy conservation, we know that $E_{in} = E_{out} + \Delta E$. By choosing the momentum of the projectile appropriately, one can arrange that the energy ΔE is negative. Then the final energy E_{out} is larger than the initial energy E_{in}, which means that we gained energy. This energy gain does not contradict total energy conservation, because one should think of the energy as being extracted from the black hole (this could indeed be made precise by taking into account the back reaction of the space ship onto the black hole, but we do not have time for entering such computations). Therefore, the Penrose process is similar to the so-called "swing-by" or "gravitational slingshot," where a satellite flies close to a planet of our solar system and uses the kinetic energy of the planet for its own acceleration. The surprising effect is that in the Penrose process, one can extract energy from the black hole, although the matter of the black hole is trapped behind the event horizon.

The wave analogue of the Penrose process is called *superradiance*. Instead of the spaceship one considers a wave packet flying in the direction of the black hole. The wave propagates as described by a corresponding wave equation (we will see such wave equations in more detail later). As a consequence, part of the wave will "fall into" the black hole, whereas the remainder will pass the black hole and will eventually leave the black hole region. If the energy of the outgoing wave is larger than the energy of the oncoming wave, then one speaks of superradiant scattering. This effect is quite similar to the Penrose process. However, one major difference is that, in contrast to the Penrose process, there is no freedom in choosing the momentum of the projectile. Instead, the dynamics is determined completely by the initial data, so that the only freedom is to prepare the incoming wave packet. As we shall see later in this lecture, superradiance indeed occurs for scalar waves in the Kerr geometry.

5   The Scalar Wave Equation in the Kerr Geometry

In preparation of the analysis of general linear wave equations, we begin with the simplest example: the scalar wave equation. It has the useful property that it is of variational form, meaning that it can be derived from an action principle. Indeed, choosing the Dirichlet action

$$S = \int_{\mathcal{M}} g^{ij}\, (\partial_i \phi)\, (\partial_j \phi)\, d\mu_{\mathcal{M}} \,,$$

(where $d\mu_{\mathcal{M}} = \sqrt{|\det g|}\, d^4x$ is the volume measure induced by the Lorentzian metric), demanding criticality for first variations gives the scalar wave equation

$$0 = \Box \phi := \nabla_i \nabla^i \phi \,.$$

The main advantage of the variational formulation is that Noether's theorem relates symmetries to conservation laws. Another method for getting these conservation laws, which is preferable to us because it is closely related to the notion of Killing fields, is to work directly with the energy-momentum tensor of the field. Recall that in the Einstein equations (1), the Einstein tensor on the left is divergence-free as a consequence of the second Bianchi identities. Therefore, the energy-momentum tensor is also divergence-free,

$$\nabla^i T_{ij} = 0 \,. \tag{10}$$

Now let K be a Killing field. Contracting the energy-momentum tensor with the Killing field gives a vector field,

$$u^i := T^{ij}\, K_j \,.$$

The calculation

$$\nabla_i u^i = \left(\nabla_i T^{ij}\right) K_j + T^{ij}\, \nabla_i K_j = 0$$

(where the first summand vanishes according to the conservation law (10), whereas the second summand is zero in view of the Killing equation (6) and the symmetry of the energy-momentum tensor) shows that this vector field is divergence-free. Therefore, integrating the divergence of u over a space-time region Ω and using the Gauß divergence theorem, we conclude that the flux integral of u through the surface $\partial\Omega$ vanishes. The situation we have in mind is that the set Ω is the region between two spacelike hypersurfaces $\mathcal{N}_1$ and $\mathcal{N}_2$ (see Fig. 2). Assuming that the vector field u has suitable decay properties at spatial infinity (in the simplest case that it has spatially compact support), we obtain the conservation law

$$0 = \int_{\Omega} \nabla_i u^i \, d\mu_{\mathcal{M}} = \int_{\mathcal{N}_1} T_{ij}\, \nu^i\, K^j\, d\mu_{\mathcal{N}_1} - \int_{\mathcal{N}_1} T_{ij}\, \nu^i\, K^j\, d\mu_{\mathcal{N}_2} \,, \tag{11}$$

Fig. 2 Conservation law corresponding to a Killing symmetry

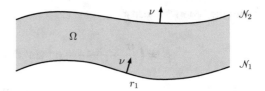

where ν is the future-directed normal on $\mathcal{N}_{1/2}$ and $d\mu_{\mathcal{N}_{1/2}}$ is the volume measure corresponding to the induced Riemannian metric.

In the Kerr geometry in Boyer–Lindquist coordinates, the Dirichlet action takes the explicit form

$$S = \int_{-\infty}^{\infty} dt \int_{r_1}^{\infty} dr \int_{-1}^{1} d(\cos \vartheta) \int_{0}^{\pi} d\varphi \, \mathcal{L}(\phi, \nabla\phi)$$

with

$$\mathcal{L}(\phi, \nabla\phi) = -\Delta |\partial_r \phi|^2 + \frac{1}{\Delta} \left| ((r^2 + a^2)\partial_t + a\partial_\varphi)\phi \right|^2$$
$$- \sin^2 \vartheta \, |\partial_{\cos \vartheta} \phi|^2 - \frac{1}{\sin^2 \vartheta} \left| (a \sin^2 \vartheta \partial_t + \partial_\varphi)\phi \right|^2 .$$

Considering first variations, the scalar wave equation becomes

$$\left[\frac{\partial}{\partial r} \Delta \frac{\partial}{\partial r} - \frac{1}{\Delta} \left\{ (r^2 + a^2) \frac{\partial}{\partial t} + a \frac{\partial}{\partial \varphi} \right\}^2 \right.$$
$$\left. + \frac{\partial}{\partial \cos \vartheta} \sin^2 \vartheta \frac{\partial}{\partial \cos \vartheta} + \frac{1}{\sin^2 \vartheta} \left\{ a \sin^2 \vartheta \frac{\partial}{\partial t} + \frac{\partial}{\partial \varphi} \right\}^2 \right] \phi = 0 . \tag{12}$$

Using the formula for the energy-momentum tensor

$$T_{ij} = (\partial_i \phi)(\partial_j \phi) - \frac{1}{2} (\partial_k \phi) (\partial^k \phi) \, g_{ij} ,$$

the conserved energy becomes

$$E := \int_{\mathcal{N}_t} T_{ij} \, \nu^j \, (\partial_t)^j \, d\mu_{\mathcal{N}_t} = \int_{\mathcal{N}_t} T_{i0} \, (\partial_t)^j \, d\mu_{\mathcal{N}_t} \tag{13}$$
$$= \int_{r_1}^{\infty} dr \int_{-1}^{1} d(\cos \vartheta) \int_{0}^{2\pi} d\varphi \, \mathcal{E} \tag{14}$$

with the "energy density"

$$\mathcal{E} = \left(\frac{(r^2 + a^2)^2}{\Delta} - a^2 \sin^2 \vartheta \right) |\partial_t \phi|^2 + \Delta \, |\partial_r \phi|^2$$

$$+ \sin^2 \vartheta \, |\partial_{\cos \vartheta} \phi|^2 + \left(\frac{1}{\sin^2 \vartheta} - \frac{a^2}{\Delta} \right) |\partial_\varphi \phi|^2 \, .$$

Using (3), one sees that the factor in front of the term $|\partial_\varphi \phi|$ is everywhere positive. However, the factor in front of the term $|\partial_t \phi|^2$ is negative precisely inside the ergosphere (9). This consideration shows that, exactly as for point particles (7), the energy of scalar waves may again be negative inside the ergosphere.

What does the indefiniteness of the energy tell us? We first point out that it does *not* imply that superradiance really occurs, because in order to analyze superradiance, one must study the dynamics of waves. Instead, it only means that there is a possibility for superradiance to occur. In technical terms, the indefiniteness of the energy leads to the difficulty that energy conservation does not give us control of the Sobolev norm of the wave. A possible scenario, which does not contradict energy conservation, is that the amplitude of the wave grows in time both inside and outside the ergosphere. It is a major task in proving linear stability to rule out this scenario.

The basic difficulty can be understood qualitatively in more detail in the scenario of the so-called *black hole bomb* as introduced by Press and Teukolsky [39] and studied by Cardoso et al. [5]. In this gedanken experiment, one puts a metal sphere around a Kerr black hole (as shown schematically on the left of Fig. 3. We consider a wave packet of energy E_{in} inside the sphere flying towards the black hole. Part of the wave will cross the event horizon, while the remainder will pass the black hole. As in the above description of superradiance, we assume that the energy ΔE of the wave crossing the event horizon is negative. Then the energy E_{out} of the outgoing wave is larger than the energy E_{in} of the incoming wave. The outgoing wave is

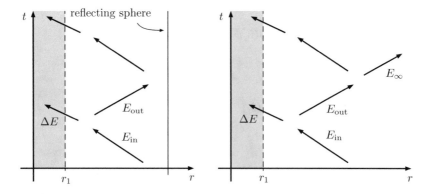

Fig. 3 The black hole bomb (left) and wave propagation in the Kerr geometry (right)

reflected on the metal sphere, becoming a new wave which again flies towards the black hole. If it can be arranged that the new incoming wave has the same shape as the original wave, this process repeats itself, generating in each step a certain positive energy. In this scenario, the energy density inside the metal sphere would grow exponentially fast in time. When the energy density gets too large, the metal sphere would explode, explaining the name "black hole bomb." For clarity, we point out that in this mechanism one always assumes that the total energy extracted from the black hole is much smaller than the total rotational energy of the black hole, so that the back reaction on the black hole need not be taken into account.

The black hole bomb suggests that putting a metal sphere around the black hole could lead to an instability, which would become manifest in an explosion of the metal sphere. The point of interest in connection to the stability problem for rotating black holes is that a very similar scenario might occur even without the metal sphere: We again consider a wave packet flying towards the black hole. Again, part of the wave with energy ΔE crosses the event horizon, whereas the remainder of energy E_{out} passes the black hole. The point is that only part of this wave will reach null infinity. Another part will be backscattered by the gravitational field and will again fly towards the black hole. Therefore, except for the "energy loss" E_{∞} by the part of the wave propagating to null infinity, we are again in the scenario of the black hole bomb where the process repeats itself, potentially leading to an exponential increase in time of the amplitude of the wave.

Clearly, this picture is oversimplified because, instead of wave packets, one must consider waves which are spread out in space, leading to a nonlocal problem. Nevertheless, in this picture it becomes clear why the problem of linear stability of rotating black holes amounts to a quantitative question: Can the initial wave packet be arranged such that the "energy gain" $-\Delta E$ is larger than the "energy loss" E_{∞}? If the answer is yes, a rotating black hole should be unstable, and it should decay by radiation of gravitational waves to a Schwarzschild black hole. It is the main goal of these lectures to explain why this does *not* happen, i.e., why rotating black holes are linearly stable. Before we can enter this problem in mathematical detail, we need to introduce linear wave equations and review a few structural results.

6   An Overview of Linear Wave Equations in the Kerr Geometry

6.1   *The Dirac Equation*

In these lectures I shall not enter the details of the Dirac equation, although most of my work has been concerned with or related to the Dirac equation. I only want to explain why the analysis for the Dirac equation is much *easier* than for other wave equations.

The Dirac equation describes a relativistic quantum mechanical particle with spin. In order to keep the setting as simple as possible, we work in coordinates and local trivializations of the spinor bundle (which has the advantage that we do not need to even define what the spinor bundle is). Then the Dirac wave function $\psi(x) \in \mathbb{C}^4$ has four complex components, which describe the spinorial degrees of freedom of the wave function. The Dirac equation reads

$$\left(i\gamma^j(x)\,\partial_j + \mathcal{B} - m\right)\psi = 0\,.$$

Here m is the rest mass of the Dirac particles, and the four matrices γ^j encode the Lorentzian metric via the anti-commutation relations

$$\left\{\gamma^j(x), \gamma^k(x)\right\} = 2g^{jk}(x)\,\mathbb{1}_{\mathbb{C}^4}\,,$$

where the anti-commutator is defined by

$$\left\{\gamma^j, \gamma^k\right\} := \gamma^j\gamma^k + \gamma^k\gamma^j\,.$$

The multiplication operator $\mathcal{B}$ involves the so-called spin coefficients, which, in analogy to the Christoffel symbols of the Levi-Civita connection, are formed of first partial derivatives of the Dirac matrices. We do not need the details here and refer instead to the explicit formulas in [12] or [30, Chapter 3].

Coming from quantum mechanics, the Dirac equation has additional structures which allow for the probabilistic interpretation of the wave function. In particular, there is a quantity which can be interpreted as the probability density as seen by an observer, and the spatial integral of this probability density is equal to one, for any fixed time of the observer. This probability integral is described mathematically as follows. The spinors at a space-time point $x \in \mathcal{M}$ are endowed with an indefinite inner product of signature $(2, 2)$, which we denote by $\prec\psi|\psi\succ_x$. For any solution ψ of the Dirac equation, the pointwise expectation value of the Dirac matrices with respect to this inner product defines a vector field

$$J^k(x) := \prec\psi(x)\,|\,\gamma^k\,\psi(x)\succ_x\,.$$

This vector field is the so-called *Dirac current*. The structure of the Dirac equation ensures that this vector field is always non-spacelike and future-directed. Moreover, as a consequence of the Dirac equation, this vector field is divergence-free, i.e.,

$$\nabla_k J^k(x) = 0$$

(where ∇ is again the Levi-Civita connection); this is referred to as *current conservation*. Integrating this equation over a region Ω between two spacelike hypersurfaces (as shown in Fig. 2), one obtains the conservation law

$$\int_{\mathcal{N}_2} J_i\,\nu^i\,d\mu_{\mathcal{N}_2} = \int_{\mathcal{N}_1} J_i\,\nu^i\,d\mu_{\mathcal{N}_1} \tag{15}$$

(here we again assume that the Dirac wave function has suitable decay properties at spatial infinity). In view of this conservation law and the linearity of the Dirac equation, one can normalize the Dirac solutions such that the integral in (15) equals one. Then the integrand in (15) has the interpretation as the probability density for an observer for whom the spacelike hypersurface $\mathcal{N}_1$ (or $\mathcal{N}_2$) describes space. We point out that the probability density is non-negative simply as a consequence of the fact that the current vector is non-spacelike and future-directed, and that the normal is timelike and future-directed. In particular, the probability density is non-negative even inside the ergosphere; note also that, in contrast to (11), the integrand in (15) does not involve a Killing field.

These structures coming from the probabilistic interpretation of the Dirac equation are a major simplification for analyzing the long-time dynamics of Dirac waves in the Kerr geometry. Namely, the current integral (15) can be used to define a scalar product on the solutions of the Dirac equation by

$$(\psi|\phi)_t := \int_{N_t} \prec\psi|\gamma^j\phi\succ_x v_j \, d\mu_{\mathcal{N}_t} \,,$$

where $\mathcal{N}_t$ is the surface of constant time in Boyer–Lindquist coordinates outside the event horizon, and where we restrict attention to Dirac solutions with suitable decay properties on the event horizon and near spatial infinity (for example, wave functions with spatially compact support outside the event horizon). Taking the completion, we obtain the Hilbert space $(\mathcal{H}_t, (.|.)_t)$ of Dirac solutions. The conservation law (15) means that the time evolution operator $U_{t,t_0} : \mathcal{H}_{t_0} \to \mathcal{H}_t$ from time t_0 to time t is a unitary operator. Since the Kerr geometry is stationary, we can canonically identify the Hilbert space $\mathcal{H}_t$ with $\mathcal{H}_{t_0}$ by time translation of the wave functions. Moreover, the unitary time evolution can be written as

$$U_{t,t_0} = e^{-i(t-t_0)H} \,, \tag{16}$$

where the so-called Dirac Hamiltonian H is a self-adjoint operator on the Hilbert space (the self-adjointness extension can be constructed in general using Stone's theorem or Chernoff's method [8]). In this way, the long-time dynamics can be related to spectral properties of a *self-adjoint operator on a Hilbert space*. No superradiance phenomena occur.

More details on the Dirac equation and the above method can be found in my joint papers with Niky Kamran, Joel Smoller, and Shing-Tung Yau [24–26]. For the general method of constructing self-adjoint extensions, the more recent paper [15] may be useful.

6.2 Massless Equations of General Spin, the Teukolsky Equation

After the short excursion to quantum mechanics, we now return to *classical waves*. The waves of interest are

$$\left\{ \begin{array}{l} \text{scalar waves} \\ \text{electromagnetic waves} \\ \text{gravitational waves .} \end{array} \right.$$

Scalar waves were already considered in Sect. 5; they are studied mainly because of their mathematical simplicity. The waves of physical interest are electromagnetic and gravitational waves. Note that all these waves are *massless*.

All the above wave equations can be described in a unified framework due to Teukolsky [43]. We now explain schematically how the Teukolsky formulation works. Since the involved computations are quite lengthy, we cannot enter the details but refer instead to the textbook [7]. The Teukolsky equation is derived in the *Newman–Penrose formalism*, which we now briefly introduce. In this formalism, one works with a *double null frame*, i.e., with a set of vectors $(l, n, m, \overline{m})$ of the complexified tangent space with inner products

$$\langle l, n \rangle = 1, \qquad \langle m, \overline{m} \rangle = -1,$$

whereas all other inner products vanish,

$$\langle l, l \rangle = \langle n, n \rangle = \langle m, m \rangle = \langle \overline{m}, \overline{m} \rangle = 0.$$

In the example of Minkowski space in Cartesian coordinates (t, x, y, z), one can choose l and n as two null vectors, for example,

$$l = \frac{1}{\sqrt{2}} \left(\frac{\partial}{\partial t} + \frac{\partial}{\partial x} \right) \qquad \text{and} \qquad n = \frac{1}{\sqrt{2}} \left(\frac{\partial}{\partial t} - \frac{\partial}{\partial x} \right).$$

The orthogonal complement of these two vectors is the two-dimensional spacelike plane spanned by the vectors ∂_y and ∂_z. Therefore, the only way to obtain additional null vectors is to *complexify* by choosing, for example,

$$m = \frac{1}{\sqrt{2}} \left(\frac{\partial}{\partial y} + i \frac{\partial}{\partial z} \right) \qquad \text{and} \qquad \overline{m} = \frac{1}{\sqrt{2}} \left(\frac{\partial}{\partial y} - i \frac{\partial}{\partial z} \right).$$

Likewise, on a Lorentzian manifold, the vectors $(l, n, m, \overline{m})$ form a basis of the complexified tangent space. The Lorentzian inner product $\langle ., . \rangle$ is extended to the complexified tangent space as a bilinear form (not sesquilinear; thus no complex conjugation is involved). The double null frame is well-suited for the analysis of the vacuum Einstein equations (indeed, the Kerr solution was discovered in the Newman–Penrose formalism).

Next, one combines the tensor components in the double null frame in complex-valued functions. In the example of the *Maxwell field*, this works as follows. The electromagnetic field tensor F_{ij} has six real components (three electric and three magnetic field components). One combines these six real components to the three complex functions

$$\Psi_0 = F_{lm} , \qquad \Psi_1 = \frac{1}{2}\left(F_{ln} + F_{m\bar{m}}\right) , \qquad \Psi_2 = F_{n\bar{m}} . \tag{17}$$

Then the homogeneous Maxwell equations

$$dF = 0 \qquad \text{and} \qquad \nabla^k F_{jk} = 0$$

give rise to a first-order system of partial differential equations for $\Psi = (\Psi_0, \Psi_1, \Psi_2)$. For the *gravitational field*, one considers similarly the Weyl tensor C_{ijkl}. Linearized gravitational waves are described by linear perturbations of the Weyl tensor in the Newman–Penrose frame. Denoting the linear perturbation of the Weyl tensor by W, its ten real components are combined to the five complex functions

$$\Psi_0 = -W_{lmlm} , \qquad \Psi_1 = -W_{lnlm} , \qquad \Psi_2 = -W_{lm\bar{m}n}$$

$$\Psi_3 = -W_{ln\bar{m}n} , \qquad \Psi_4 = -W_{l\bar{m}n\bar{m}} .$$

Linearizing the second Bianchi identities $R_{ij(kl;m)} = 0$ gives a first-order system of partial differential equations for $\Psi = (\Psi_0, \ldots, \Psi_4)$. In this formulation, the connection to the *spin* can be obtained simply by counting the number of degrees of freedom. In quantum mechanics, a wave function of spin s has $2s + 1$ complex components. Therefore, we obtain the correct number of degrees of freedom if we set

$$\text{electromagnetic waves:} \quad \text{spin } s = 1$$
$$\text{gravitational waves:} \quad \text{spin } s = 2 .$$

We remark that the connection to the spin is more profound than merely counting the number of degrees of freedom, but we have no time to explain how this works. For our purposes, it suffices to take the spin s as a parameter which characterizes the massless wave equation by counting the number of components of the Newman–Penrose wave function $\Psi = (\Psi_0, \ldots, \Psi_{2s})$.

We write the first-order system of partial differential equations for electromagnetic waves or linearized gravitational waves symbolically as

$$\mathcal{D}\begin{pmatrix} \Psi_0 \\ \vdots \\ \Psi_{2s} \end{pmatrix} = 0 . \tag{18}$$

Working with this first-order system is not convenient for larger spin because the number of equations gets large, and the equations are coupled in a complicated way. But, as discovered by Teukolsky, the system of equations can be decoupled such as to obtain a second-order partial differential equation for one complex-valued function. This *decoupling* works schematically as follows: One chooses a Newman–Penrose null frame where l and n are aligned with the repeated principal null directions of the Weyl tensor (in this frame, the Newman–Penrose components of the Weyl tensor satisfy the equations $\psi_0 = \psi_1 = \psi_3 = \psi_4 = 0$, with ψ_2 being the only non-zero component). Multiplying the linear first-order system (18) in this frame by a suitable first-order differential operator D, we obtain the equation

$$
0 = D\mathcal{D} \begin{pmatrix} \Psi_0 \\ \vdots \\ \Psi_{2s} \end{pmatrix} = \begin{pmatrix} T_0 & 0 & \cdots & 0 & 0 \\ * & * & \cdots & * & * \\ \vdots & \vdots & \ddots & \vdots & \vdots \\ * & * & \cdots & * & * \\ 0 & 0 & \cdots & 0 & T_{2s} \end{pmatrix} \begin{pmatrix} \Psi_0 \\ \Psi_1 \\ \vdots \\ \Psi_{2s-1} \\ \Psi_{2s} \end{pmatrix} ,
$$

where the stars stand for differential operators which we do not need to specify here. The point is that this procedure generates zeros in the first and last row of the matrix, giving rise to decoupled equations for the first and last components of the Newman–Penrose wave function Ψ,

$$
T_0 \, \Psi_0 = 0 = T_{2s} \, \Psi_{2s} . \tag{19}
$$

Once the solution Ψ_0 or Ψ_{2s} is known, all the other components of Ψ can be obtained by employing the so-called *Teukolsky–Starobinsky identities*, which have similarities to the "ladder operator" for the harmonic oscillator used for obtaining the excited states from the ground state. With this in mind, in what follows we restrict attention to the equations for Ψ_0 or Ψ_{2s} in (19). After detailed computations for the electromagnetic field and for linearized gravitational fields, in both cases one ends up with the same equation, except for a parameter s describing the spin. We thus obtain the *Teukolsky equation* (sometimes called Teukolsky master equation; we use the form of the equation as given in [45])

$$
\left(\frac{\partial}{\partial r} \Delta \frac{\partial}{\partial r} - \frac{1}{\Delta} \left\{ (r^2 + a^2) \frac{\partial}{\partial t} + a \frac{\partial}{\partial \varphi} - (r - M) s \right\}^2 - 4s \, (r + ia \cos \vartheta) \frac{\partial}{\partial t} \right.
$$

$$
\left. + \frac{\partial}{\partial \cos \vartheta} \sin^2 \vartheta \frac{\partial}{\partial \cos \vartheta} + \frac{1}{\sin^2 \vartheta} \left\{ a \sin^2 \vartheta \frac{\partial}{\partial t} + \frac{\partial}{\partial \varphi} + is \cos \vartheta \right\}^2 \right) \phi = 0 . \tag{20}
$$

For $s = 1$, this equation describes the first component Ψ_0 of the Newman–Penrose wave function (Ψ_0, Ψ_1, Ψ_2) for electromagnetic waves, whereas the parameter value $s = -1$ gives the equation for Ψ_2. Likewise, setting $s = 2$ gives the first component Ψ_0 of the Newman–Penrose wave function $(\Psi_0, \ldots, \Psi_4)$ for gravitational waves, whereas $s = -2$ gives the equation for Ψ_4. By direct inspection, one sees that setting $s = 0$ gives back the scalar wave equation (12). We remark that setting $s = \frac{1}{2}$ gives the massless Dirac equation [43], and $s = \frac{3}{2}$ gives the massless Rarita–Schwinger equation [32].

We close with a remark on gravitational perturbations. As outlined above, our method is to consider perturbations of the Weyl tensor. Alternatively, one could consider perturbations of the metric (indeed, this was historically the first approach, going back to the stability analysis by Regge and Wheeler [40]). Working with *metric perturbations* has the disadvantage that infinitesimal coordinate transformations also lead to perturbations of the metric, which, however, have no geometric significance. In other words, when working with metric perturbations, the diffeomorphism invariance leads to a gauge freedom which is not easy to handle. This was the original motivation for Teukolsky and Press to consider instead perturbations of geometric quantities like the Newman–Penrose components of the Weyl tensor, leading to the Teukolsky framework. However, for some applications (for example, in order to include matter models or to describe nonlinear waves) it is necessary to work with metric perturbations. Therefore, working in the Teukolsky formulation, the following question remains: Given a linear perturbation of the Weyl tensor, how can it be realized by a metric perturbation? This is an interesting and in general difficult question which we cannot analyze here (see, however, [46] and the references therein).

7 Separation of the Teukolsky Equation

The Teukolsky equation (20) has the remarkable property that it can be completely separated into a system of ordinary differential equations (ODEs): Due to the stationarity and axisymmetry, we can separate the t- and φ-dependence with the usual plane-wave ansatz

$$\phi(t, r, \vartheta, \varphi) = e^{-i\omega t - ik\varphi} \phi(r, \vartheta), \tag{21}$$

where ω is a quantum number which could be real or complex and which corresponds to the "energy," and $k \in \mathbb{Z}/2$ is a quantum number corresponding to the projection of angular momentum onto the axis of symmetry of the black hole (if s is half an odd integer, then so is k). Substituting (21) into (20), we see that the Teukolsky operator splits into the sum of radial and angular parts, giving rise to the equation

$$(\mathcal{R}_{\omega,k} + \mathcal{A}_{\omega,k}) \phi = 0,$$

where $\mathcal{R}_{\omega,k}$ and $\mathcal{A}_{\omega,k}$ are given by (for details see [23, Section 6])

$$\mathcal{R}_\omega = -\frac{\partial}{\partial r}\Delta\frac{\partial}{\partial r} - \frac{1}{\Delta}\Big(\omega\,(r^2 + a^2) + ak - i\,(r - M)\,s\Big)^2 - 4isr\omega + 4k\,a\omega$$

$$\tag{22}$$

$$\mathcal{A}_\omega = -\frac{\partial}{\partial \cos\vartheta}\,\sin^2\vartheta\,\frac{\partial}{\partial \cos\vartheta} + \frac{1}{\sin^2\vartheta}\Big(-a\omega\sin^2\vartheta + k - s\cos\vartheta\Big)^2. \tag{23}$$

We can therefore separate the variables r and ϑ with the multiplicative ansatz

$$\phi(r, \vartheta) \;=\; R(r)\,\Theta(\vartheta)\,, \tag{24}$$

to obtain for given ω and k the system of ODEs

$$\mathcal{R}_{\omega,k}\,R_\lambda \;=\; -\lambda\,R_\lambda, \qquad \mathcal{A}_{\omega,k}\,\Theta_\lambda \;=\; \lambda\,\Theta_\lambda\,. \tag{25}$$

Solutions of the coupled system (25) are referred to as *mode solutions*.

We point out that the last separation (24) is not obvious because it does not correspond to an underlying space-time symmetry. Instead, as discovered by Carter for the scalar wave equation [6], it corresponds to the fact that in the Kerr geometry there exists an irreducible quadratic Killing tensor (i.e., a Killing tensor which is not a symmetrized tensor product of Killing vectors). The separation constant λ is an eigenvalue of the angular operator $\mathcal{A}_{\omega,k}$ and can thus be thought of as an angular quantum number. In the spherically symmetric case $a = 0$, this separation constant goes over to the usual eigenvalues $\lambda = l(l + 1)$ of the total angular momentum operator.

8   Results on Linear Stability and Superradiance

Being familiar with the structure of the different linear wave equations, we can now state our results on stability and superradiance. The problem of *linear stability* of black holes amounts to the question whether solutions of the corresponding linear wave equations decay for large times. In order to put our results into context, we point out that the problem of linear stability of black holes has a long history. It goes back to the study of the Schwarzschild black hole by Regge and Wheeler [40] who showed that an integral norm of the perturbation of each angular mode is bounded uniformly in time. Decay of these perturbations was first proved in [31]. More detailed estimates of metric perturbations in Schwarzschild were obtained in [9, 34]. For the Kerr black hole, linear stability under perturbations of general spin has been an open problem for many years, which was solved in the dynamical setting in [23] (for related results obtained with different methods see [2, 3, 10, 35] and the references in these papers). A key ingredient to our proof is the so-called *mode stability* result obtained by Whiting [45], who proved that the Teukolsky equation does not admit solutions which decay both at spatial infinity and at the event horizon and increase exponentially in time.

We consider the Cauchy problem for the Teukolsky equation. Thus we seek a solution ϕ of the Teukolsky equation (20) for given initial data

$$\phi|_{t=0} = \phi_0 \quad \text{and} \quad \partial_t \phi|_{t=0} = \phi_1 .$$

Being a linear hyperbolic PDE, the Cauchy problem for the Teukolsky equation has unique global solutions. Also, taking smooth initial data, the solution is smooth for all times. Our task is to show that solutions decay for large times. In order to avoid specifying decay assumptions at the event horizon and at spatial infinity, we restrict attention to compactly supported initial data outside the event horizon,

$$\phi_0, \phi_1 \in C_0^\infty \big((r_1, \infty) \times S^2 \big) . \tag{26}$$

Since the Kerr geometry is axisymmetric, the Teukolsky equation decouples into separate equations for each azimuthal mode. Therefore, the solution of the Cauchy problem is obtained by solving the Cauchy problem for each azimuthal mode and taking the sum of the resulting solutions. With this in mind, we restrict attention to the Cauchy problem for a single azimuthal mode, i.e.,

$$\phi_0(r, \vartheta, \varphi) = e^{-ik\varphi} \, \phi_0^{(k)}(r, \vartheta) , \qquad \phi_1(r, \vartheta, \varphi) = e^{-ik\varphi} \, \phi_1^{(k)}(r, \vartheta) \tag{27}$$

for given $k \in \mathbb{Z}/2$. The main result of [23] is stated as follows:

Theorem 8.1 *Consider a non-extreme Kerr black hole of mass M and angular momentum aM with $M^2 > a^2 > 0$. Then for any $s \geq \frac{1}{2}$ and any $k \in \mathbb{Z}/2$, the solution of the Teukolsky equation with initial data of the form (26) and (27) decays to zero in $L_{loc}^\infty((r_1, \infty) \times S^2)$.*

This theorem establishes in the dynamical setting that the non-extreme Kerr black hole is linearly stable.

Our method of proof uses an integral representation of the time evolution operator involving the radial and angular solutions of the separated system of ODEs (25). Such an integral representation was derived earlier for the *scalar wave equation* in [27], and it was used for proving decay in time [28]. Moreover, in [29] it was proven in the dynamical setting that *superradiance* occurs for scalar waves. We now explain this result. Superradiance for scalar waves in the Kerr geometry was first studied by Zel'dovich [47] and Starobinsky [41] on the level of modes. More precisely, they computed the transmission and reflection coefficients for the radial ODE in (25). The absolute value squared of these coefficients can be interpreted as the energy flux of the incoming and outgoing waves, respectively. Comparing these fluxes, one obtains the relative energy gain. Starobinsky computed the relative gain of energy to about 5% for $k = 1$ and less than 1% for $k \geq 2$.

Unfortunately, this mode analysis does not give information on the dynamics. Thus for a rigorous treatment of energy extraction one needs to consider the time-dependent situation. In [29], this is accomplished by constructing initial data of the form of wave packets, in such a way that the energy gain agrees with the results of the mode analysis up to an arbitrarily small error. The crucial analytical ingredient to the proof is the time-independent energy estimate for the outgoing wave as derived in [17].

The remainder of these lectures is devoted to giving an outline of the proof of Theorem 8.1. Before entering the constructions, we point out the main difficulties:

► The Teukolsky equation for $s \neq 0$ is *not of variational form*, i.e., it cannot be obtained as the Euler–Lagrange equation of an action.
► As a consequence, we cannot apply Noether's theorem to obtain conserved quantities. In particular, there is *no conserved energy*, being an integral of an energy density. This means that, in contrast to the situation described for the Dirac equation in Sect. 6.1, the time evolution cannot be described by a unitary operator on a Hilbert space. As a consequence, we cannot use the spectral theorem for self-adjoint or unitary operators on Hilbert spaces.
► A related difficulty is that the *coefficients* of the first derivative terms in the Teukolsky equation for $s \neq 0$ are *complex*. Such complex potentials in a wave equation usually describe dissipation, implying that (depending of the sign of the dissipation terms) the solutions typically decay or increase exponentially in time. This means that, in order to show that the solution of the Teukolsky equation decays for large times, one must carefully control the signs and the size of the complex coefficients by quantitative estimates.
► In the separation of variables (25), both the radial and angular differential operators $\mathcal{R}_{\omega,k}$ and $\mathcal{A}_{\omega,k}$ depend on the separation constants k and ω. As a consequence, it is not at all obvious if and how for given initial data one can decompose the corresponding solution of the Cauchy problem into a superposition of mode solutions. An obvious difficulty is that, for such a *mode decomposition*, one would have to know the separation constant ω, which in turn can be specified only if we already know the full dynamics of the wave.

9   Hamiltonian Formulation and Integral Representations

In order to analyze the dynamics of the Teukolsky wave, it is useful to work with contour integrals of the resolvent of the Hamiltonian, as we now outline. In preparation, we must rewrite the Teukolsky equation in Hamiltonian form. To this end, we introduce the two-component wave function

$$\Psi = \sqrt{r^2 + a^2} \begin{pmatrix} \phi \\ i\,\partial_t \phi \end{pmatrix}$$

and write the Teukolsky equation as

$$i\partial_t \Psi = H\Psi \,, \tag{28}$$

where H is a second-order spatial differential operator. We consider H as an operator on a Hilbert space $\mathcal{H}$ with the domain

$$\mathcal{D}(H) = C_0^\infty\big((r_1, \infty) \times S^2, \mathbb{C}^4\big) \,.$$

It would be desirable to represent H as a self-adjoint operator on a Hilbert space $\mathcal{H}$, because it would then be possible to apply the spectral calculus and write the time evolution operator similar as for the Dirac equation in the form (16). Unfortunately, this procedure does *not* work here, as can be understood as follows. As already mentioned at the end of the previous section, the Teukolsky equation is not of variational form, implying that there is no conserved energy. If there were a conserved bilinear form $\langle \Psi | \Phi \rangle$ on the solutions, then the calculation

$$0 = \partial_t \langle \Psi | \Phi \rangle = \langle \dot\Psi | \Phi \rangle + \langle \Psi | \dot\Phi \rangle = i\big(\langle H\Psi | \Phi \rangle - \langle \Psi | H\Phi \rangle\big)$$

would imply that the Hamiltonian were symmetric with respect to this bilinear form. But, having no conserved energy, there is also no bilinear form with respect to which the Hamiltonian is symmetric. In order to avoid confusion, we remark that there is a conserved physical energy, which in the example of a Maxwell field could be written in the form (13) with T_{ij} the energy-momentum tensor of the Maxwell field. However, this energy involves all the components of the field tensor or, in other words, all the components of the Newman–Penrose wave function in (17). Since the Teukolsky equation only gives Ψ_0 or Ψ_2, we would have to compute the other components using the Teukolsky–Starobinsky identities. As a consequence, the resulting formula for the Maxwell energy would involve higher derivatives of the Teukolsky wave function, making the situation very complicated. This is why we decided not to use the physical energy in our construction.

We conclude that we shall treat the operator H as a non-symmetric operator on a Hilbert space. In order to get an idea for how to work with non-symmetric operators, it is helpful get a motivation from the finite-dimensional setting. Thus let A be a linear operator on a finite-dimensional Hilbert space $\mathcal{H}$. Clearly, this operator need not be diagonalizable, because Jordan chains may form. Nevertheless, one can get a spectral calculus by working with contour integrals.

Lemma 9.1 *Let A be a linear operator A on a Hilbert space $\mathcal{H}$ of dimension $n < \infty$. Then*

$$e^{-itA} = -\frac{1}{2\pi i} \oint_\Gamma e^{-i\omega t} \big(A - \omega\big)^{-1} d\omega \,, \tag{29}$$

where Γ is a contour which encloses the whole spectrum of A with winding number one.

Proof If A is diagonalizable, we can choose a basis where A is diagonal,

$$A = \operatorname{diag}(\lambda_1, \ldots, \lambda_n) \, .$$

In this case, (29) is obtained immediately by carrying out the contour integral for each matrix entry with the help of the Cauchy integral formula.

The case that A is not diagonalizable can be obtained by approximation, noting that the diagonalizable matrices are dense and that both sides of (29) are continuous on the space of matrices (endowed with the topology of $\mathbb{C}^{n \cdot n}$). $\square$

Motivated by this formula for matrices, we can hope that the Cauchy problem for Eq. (28) with initial data Ψ_0 could be solved with the Cauchy integral formula by

$$\Psi(t) = -\frac{1}{2\pi i} \oint_{\Gamma} e^{-i\omega t} \left(H - \omega \right)^{-1} \Psi_0 \, d\omega \, , \tag{30}$$

where Γ is a contour which encloses all eigenvalues of H (note that this formula holds for any matrix H, even if it is not diagonalizable). It turns out that in our infinite-dimensional setting, this formula indeed holds. The first step in making sense of this formula is to localize the spectrum of H and to make sure that the resolvent exists along the integration contour. To this end, we choose the scalar product on $\mathcal{H}$ as a suitable weighted Sobolev scalar product in such a way that that the operator $H - H^*$ is bounded, i.e.,

$$\| H - H^* \| \leq \frac{c}{2}$$

with a suitable constant $c > 0$. Then we prove that the resolvent $R_\omega := (H - \omega)^{-1}$ exists if ω lies outside a strip enclosing the real axis (see [23, Lemma 4.1]).

Lemma 9.2 *For every ω with*

$$| \operatorname{Im} \omega | > c \, ,$$

the resolvent $R_\omega = (H - \omega)^{-1}$ exists and is bounded by

$$\| R_\omega \| \leq \frac{1}{| \operatorname{Im} \omega | - c} \, .$$

When forming contour integrals, one must always make sure to stay outside the strip $| \operatorname{Im} \omega | \leq c$, making it impossible to work with closed contours enclosing the spectrum. But we can work with unbounded contours as follows (see [23, Corollary 5.3]):

Proposition 9.3 *For any integer $p \geq 1$, the solution of the Cauchy problem for the Teukolsky equation with initial data $\Psi|_{t=0} = \Psi_0 \in \mathcal{D}(H)$ has the representation*

$$\Psi(t) = -\frac{1}{2\pi i} \int_C e^{-i\omega t} \frac{1}{(\omega + 3ic)^p} \left(R_\omega \left(H + 3ic \right)^p \Psi_0 \right) d\omega , \qquad (31)$$

where C is the contour

$$C = \{\omega \mid \operatorname{Im}\omega = 2c\} \cup \{\omega \mid \operatorname{Im}\omega = -2c\} \qquad (32)$$

with counter-clockwise orientation.

Here the factor $(\omega + 3ic)^{-p}$ gives suitable decay for large $|\omega|$ and ensures that the integral converges in the Hilbert space $\mathcal{H}$.

The representation (31) gives an explicit solution of the Cauchy problem in terms of a Cauchy integral of the resolvent. Unfortunately, this representation does not immediately give information on the long-time dynamics of the Teukolsky wave. This shortcoming can be understood immediately from the fact that the factor $e^{-i\omega t}$ in the integrand increases exponentially for large times because $|e^{-i\omega t}| = e^{\operatorname{Im}\omega t} = e^{\pm 2ct}$. In order to bypass this shortcoming, our strategy is to move the contour onto the real axis. Once this has been accomplished, the integral representation (31) simplifies to a Fourier transform,

$$\Psi(t) = \int_{-\infty}^{\infty} e^{-i\omega t} \, \hat{\Psi}(\omega) \, d\omega .$$

The decay of such a Fourier transform can be obtained from the *Riemann–Lebesgue lemma*, stating that

$$\hat{\Psi} \in L^1(\mathbb{R}, d\omega) \quad \Longrightarrow \quad \lim_{t \to \pm\infty} \Psi(t) = 0$$

(where the wave functions are evaluated pointwise in space). One of the difficulties in making this strategy work is to prove that the contour can indeed be moved onto the real axis. This makes it necessary to show that the Hamiltonian has no spectrum away from the real axis. We did not succeed in proving this result using operator theoretic methods. Instead, our method is to first make use of the separation of variables, making it possible rule out the spectrum in the complex plane using Whiting's mode stability result [45].

10 A Spectral Decomposition of the Angular Teukolsky Operator

Following the strategy we just outlined, our next task is to employ the separation of variables in the integrand of the integral representation (31). Regarding the angular equation (25) as an eigenvalue equation, we are led to considering the angular operator $\mathcal{A}_\omega$ in (23) as an operator on the Hilbert space

$$\mathcal{H}_k := L^2(S^2) \cap \{e^{-ik\varphi} \Theta(\vartheta) \mid \Theta : (0, \pi) \to \mathbb{C}\}$$

with dense domain $\mathcal{D}(\mathcal{A}_\omega) = C^\infty(S^2) \cap \mathcal{H}_k$. Unfortunately, the parameter ω is not real but lies on the contour (32). As a consequence, the operator $\mathcal{A}_\omega$ is not symmetric, because its adjoint is given by

$$\mathcal{A}_\omega^* = \mathcal{A}_{\overline{\omega}} \neq \mathcal{A}_\omega .$$

The operator $\mathcal{A}_\omega$ is not even a normal operator, making it impossible to apply the spectral theorem in Hilbert spaces. Indeed, $\mathcal{A}_\omega$ does not need to be diagonalizable, because there might be Jordan chains. On the other hand, in order to make use of the separation of variables, we must decompose the initial data into angular modes. This can be achieved by decomposing the angular operator into invariant subspaces of bounded dimension, as is made precise in the following theorem (see [21, Theorem 1.1]):

Theorem 10.1 *Let $U \subset \mathbb{C}$ be the strip*

$$|Im\, \omega| < 3c .$$

Then there is a positive integer N and a family of bounded linear operators Q_n^ω on $\mathcal{H}_k$ defined for all $n \in \mathbb{N} \cup \{0\}$ and $\omega \in U$ with the following properties:

(i) *The image of the operator Q_0^ω is an N-dimensional invariant subspace of $\mathcal{A}_k$.*
(ii) *For every $n \geq 1$, the image of the operator Q_n^ω is an at most two-dimensional invariant subspace of $\mathcal{A}_k$.*
(iii) *The Q_n^ω are uniformly bounded in $L(\mathcal{H}_k)$, i.e., for all $n \in \mathbb{N} \cup \{0\}$ and $\omega \in U$,*

$$\|Q_n^\omega\| \leq c_2$$

for a suitable constant $c_2 = c_2(s, k, c)$ (here $\|\cdot\|$ denotes the sup-norm on $\mathcal{H}_k$).
(iv) *The Q_n^ω are idempotent and mutually orthogonal in the sense that*

$$Q_n^\omega Q_{n'}^\omega = \delta_{n,n'}\, Q_n^\omega \qquad \text{for all } n, n' \in \mathbb{N} \cup \{0\} .$$

(v) *The Q_n^ω are complete in the sense that for every $\omega \in U$,*

$$\sum_{n=0}^{\infty} Q_n^\omega = \mathbb{1} \tag{33}$$

with strong convergence of the series.

11 Invariant Disk Estimates for the Complex Riccati Equation

In order to locate the spectrum of $\mathcal{A}_\omega$, we use detailed ODE estimates. The operators Q_n^ω are then obtained similar to (30) as Cauchy integrals,

$$Q_n^\omega := -\frac{1}{2\pi i} \oint_{\Gamma_n} s_\lambda \, d\lambda \,, \qquad n \in \mathbb{N}_0 \,,$$

where the contour Γ_n encloses the corresponding spectral points, and $s_\lambda = (\mathcal{A}_\omega - \lambda)^{-1}$ is the resolvent of the angular operator. What makes the analysis doable is the fact that $\mathcal{A}_\omega$ is an ordinary differential operator. Transforming the angular equation in (25) into Sturm–Liouville form

$$\left(-\frac{d^2}{du^2} + V(u) \right) \phi = 0 \,, \tag{34}$$

(where $u = \vartheta$ and $V \in C^\infty((0, \pi), \mathbb{C})$ is a complex potential), the resolvent s_λ can be represented as an integral operator whose kernel is given explicitly in terms of suitable fundamental solutions ϕ_L^D and ϕ_R^D,

$$s_\lambda(u, u') = \frac{1}{w(\phi_L^D, \phi_R^D)} \times \begin{cases} \phi_L^D(u) \, \phi_R^D(u') & \text{if } u \le u' \\ \phi_L^D(u') \, \phi_R^D(u) & \text{if } u' < u \,, \end{cases} \tag{35}$$

where $w(\phi_L^D, \phi_R^D)$ denotes the Wronskian.

The main task is to find good approximations for the solutions of the Sturm–Liouville equation (34) with rigorous error bounds which must be uniform in the parameters ω and λ. These approximations are obtained by "glueing together" suitable WKB, Airy and parabolic cylinder functions. The needed properties of these special functions are derived in [19]. In order to obtain error estimates, we combine several methods:

(a) Osculating circle estimates (see [21, Section 6])
(b) The T-method (see [20, Section 3.2])
(c) The κ-method (see [20, Section 3.3])

The method (a) is needed in order to separate the spectral points of $\mathcal{A}_\omega$ (gap estimates). The methods (b) and (c) are particular versions of *invariant disk* estimates as derived for complex potentials in [18] (based on previous estimates for real potentials in [16] and [28]). These estimates are also needed for the analysis of the radial equation, see Sect. 12 below. We now explain the basic idea behind the invariant disk estimates.

Let ϕ be a solution of the Sturm–Liouville equation (34) with a complex potential V. Then the function y defined by

$$y = \frac{\phi'}{\phi}$$

is a solution of the Riccati equation

$$y' = V - y^2 . \tag{36}$$

Conversely, given a solution y of the Riccati equation, a corresponding fundamental system for the Sturm–Liouville equation is obtained by integration. With this in mind, it suffices to construct a particular approximate solution $\tilde{y}$ and to derive rigorous error estimates. The invariant disk estimates are based on the observation that the Riccati flow maps disks to disks (see [18, Sections 2 and 3]). In fact, denoting the center of the disk by $m \in \mathbb{C}$ and its radius by $R > 0$, we get the flow equations

$$R' = -2R \operatorname{Re} m$$
$$m' = V - m^2 - R^2 .$$

Clearly, this system of equations is as difficult to solve as the original Riccati equation (36). But suppose that m is an approximate solution in the sense that

$$R' = -2R \operatorname{Re} m + \delta R$$
$$m' = V - m^2 - R^2 + \delta m ,$$

with suitable error terms δm and δR, then the Riccati flow will remain inside the disk provided that its radius grows sufficiently fast, i.e., (see [18, Lemma 3.1])

$$\delta R \geq |\delta m| .$$

This is the starting point for the invariant disk method. In order to reduce the number of free functions, it is useful to solve the linear equations in the above system of ODEs by integration. For more details we refer the reader to [18, 20].

12 Separation of the Resolvent and Contour Deformations

The next step is to use the spectral decomposition of the angular operator in Theorem 10.1 in the integral representation of the solution of the Cauchy problem. More specifically, inserting (33) into (31) gives

$$\Psi(t) = -\frac{1}{2\pi i} \int_C \sum_{n=0}^{\infty} e^{-i\omega t} \frac{1}{(\omega + 3ic)^p} \left(R_\omega \, Q_n^\omega \, (H + 3ic)^p \, \Psi_0 \right) d\omega \,. \qquad (37)$$

At this point, the operator product $R_\omega Q_n^\omega$ can be expressed in terms of solutions of the radial and angular ODEs (25) which arise in the separation of variables (see [23, Theorem 7.1]). Namely, the operator Q_n^ω maps onto an invariant subspace of $\mathcal{A}_\omega$ of dimension at most N, and it turns out that the operator product $R_\omega \, Q_n^\omega$ leaves this subspace invariant. Therefore, choosing a basis of this invariant subspace, the PDE $(H - \omega) R_\omega Q_\omega^n = Q_\omega^n$ can be rewritten as a radial ODE involving matrices of rank at most N. The solution of this ODE can be expressed explicitly in terms of the resolvent of the radial ODE. In order to compute this resolvent, it is useful to also transform the radial ODE into Sturm–Liouville form (34). To this end, we introduce the Regge-Wheeler coordinate $u \in \mathbb{R}$ by

$$\frac{du}{dr} = \frac{r^2 + a^2}{\Delta} \,,$$

mapping the event horizon to $u = -\infty$. Then the radial ODE takes again the form (34), but now with u defined on the whole real axis. Thus the resolvent can be written as an integral operator with kernel given in analogy to (35) by

$$s_\omega(u, v) = \frac{1}{w(\acute{\phi}, \grave{\phi})} \times \begin{cases} \acute{\phi}(u) \, \grave{\phi}(v) \ \text{if } v \geq u \\ \grave{\phi}(u) \, \acute{\phi}(v) \ \text{if } v < u \,, \end{cases}$$

where $\acute{\phi}$ and $\grave{\phi}$ form a specific fundamental system for the radial ODE. The solutions $\acute{\phi}$ and $\grave{\phi}$ are constructed as Jost solutions, using methods of one-dimensional scattering theory (see [11] and [23, Section 6], [28, Section 3]).

The next step is to deform the contour in the integral representation (37). Standard arguments show that the integrand in (37) is holomorphic on the resolvent set (i.e., for all ω for which the resolvent R_ω in (31) exists). Thus the contour may be deformed as long as it does not cross singularities of the resolvent. Therefore, it is crucial to show that the integrand in (37) is meromorphic and to determine its pole structure. Here we make essential use of Whiting's mode stability result [45] which states, in our context, that every summand in (37) is holomorphic off the real axis. In order to make use of this mode stability, we need to interchange the integral in (37) with the infinite sum. To this end, we derive estimates which show that the summands in (37) decay for large n uniformly in ω. Here we again use ODE techniques, in the same spirit as described above for the angular equation (see [23, Section 10]). In this way, we can move the contour in the lower half plane arbitrarily close to the real axis. Moreover, the contour in the upper half plane may be moved to infinity. We thus obtain the integral representation (see [23, Corollary 10.4])

$$\Psi(t) = -\frac{1}{2\pi i} \sum_{n=0}^{\infty} \lim_{\varepsilon \searrow 0} \int_{\mathbb{R}-i\varepsilon} \frac{e^{-i\omega t}}{(\omega + 3ic)^p} \left(R_{\omega,n} \, Q_n^\omega \, (H + 3ic)^p \, \Psi_0 \right) d\omega \,.$$

The remaining issue is that the integrands in this representation might have poles on the real axis. These so-called *radiant modes* are ruled out by a causality argument (see [23, Section 11]); for an alternative proof see [4]. We thus obtain the following result (see [23, Theorem 12.1].

Theorem 12.1 *For any* $k \in \mathbb{Z}/2$, *there is a parameter* $p > 0$ *such that for any* $t < 0$, *the solution of the Cauchy problem for the Teukolsky equation with initial data*

$$\Psi|_{t=0} = e^{-ik\varphi} \, \Psi_0^{(k)}(r, \vartheta) \qquad \text{with} \qquad \Psi_0^{(k)} \in C^\infty(\mathbb{R} \times S^2, \mathbb{C}^2)$$

has the integral representation

$$\Psi(t, u, \vartheta, \varphi)$$

$$= -\frac{1}{2\pi i} e^{-ik\varphi} \sum_{n=0}^{\infty} \int_{-\infty}^{\infty} \frac{e^{-i\omega t}}{(\omega + 3ic)^p} \left(R_{\omega,n}^- \, Q_n^\omega (H + 3ic)^p \, \Psi_0^{(k)} \right)(u, \vartheta) \, d\omega \,,$$

$$(38)$$

where $R_{\omega,n}^- \Psi := \lim_{\varepsilon \searrow 0} \left(R_{\omega-i\varepsilon,n} \Psi \right)$. *Moreover, the integrals in* (38) *all exist in the Lebesgue sense. Furthermore, for every* $\varepsilon > 0$ *and* $u_\infty \in \mathbb{R}$, *there is* N *such that for all* $u < u_\infty$,

$$\sum_{n=N}^{\infty} \int_{-\infty}^{\infty} \left\| \frac{1}{(\omega + 3ic)^p} \left(R_{\omega,n}^- \, Q_n^\omega \, (H + 3ic)^p \, \Psi_0^{(k)} \right)(u) \right\|_{L^2(S^2)} d\omega < \varepsilon \,. \quad (39)$$

13   Proof of Pointwise Decay

Theorem 8.1 is a direct consequence of the integral representation (38) in Theorem 12.1. Namely, combining the estimate (39) with Sobolev methods, one can make the contributions for large n pointwise arbitrarily small. On the other hand, for each of the angular modes $n = 0, \ldots, N - 1$, the desired pointwise decay as $t \to -\infty$ follows from the Riemann–Lebesgue lemma. For details we refer to [23, Section 12].

14 Concluding Remarks

We first point out that the integral representation of Theorem 12.1 is a suitable starting point for a detailed analysis for the dynamics of the solutions of the Teukolsky equation. In particular, one can study decay rates (similar as worked out for massive Dirac waves in [28]) and derive uniform energy estimates outside the ergosphere (similar as for scalar waves in [17]). Moreover, using the methods in [29], one could analyze superradiance phenomena for wave packets in the time-dependent setting.

Clearly, the next challenge is to prove *nonlinear stability* of the Kerr geometry. This will make it necessary to refine our results on the linear problem, for example, by deriving weighted Sobolev estimates and by analyzing the k-dependence of our estimates. Moreover, it might be useful to combine our methods and results with microlocal techniques (as used, for example, in the proof of nonlinear stability results in the related Kerr-De Sitter geometry [33]).

Acknowledgements I would like to thank the organizers of the first "Domoschool—International Alpine School of Mathematics and Physics" held in Domodossola, 16–20 July 2018, for the kind invitation. This article is based on my lectures delivered at this summer school. I am grateful to Niky Kamran and Igor Khavkine for helpful comments on the manuscript.

References

1. R. Adler, M. Bazin, M. Schiffer, *Introduction to General Relativity* (McGraw-Hill Book, New York, 1965)
2. L. Andersson, P. Blue, Uniform energy bound and asymptotics for the Maxwell field on a slowly rotating Kerr black hole exterior. J. Hyperbolic Differ. Equ. **12**(4), 689–743 (2015). arXiv:1310.2664 [math.AP]
3. L. Andersson, T. Bäckdahl, P. Blue, A new tensorial conservation law for Maxwell fields on the Kerr background. J. Differ. Geom. **105**(2), 163–176 (2017). arXiv:1412.2960 [gr-qc]
4. L. Andersson, S. Ma, C. Paganini, B.F. Whiting, Mode stability on the real axis. J. Math. Phys. **58**(7), 072501, 19pp. (2017). arXiv:1607.02759 [gr-qc]
5. V. Cardoso, Ó.J.C. Dias, J.P.S. Lemos, S. Yoshida, Black-hole bomb and superradiant instabilities. Phys. Rev. D **70**(4), 044039, 9pp. (2004). hep-th/0404096
6. B. Carter, Republication of: black hole equilibrium states. Part I. Analytic and geometric properties of the Kerr solutions. Gen. Relativ. Gravit. **41**(12), 2873–2938 (2009)
7. S. Chandrasekhar, *The Mathematical Theory of Black Holes*. Oxford Classic Texts in the Physical Sciences (The Clarendon Press/Oxford University Press, New York, 1998)
8. P.R. Chernoff, Essential self-adjointness of powers of generators of hyperbolic equations. J. Funct. Anal. **12**, 401–414 (1973)
9. M. Dafermos, G. Holzegel, I. Rodnianski, The linear stability of the Schwarzschild solution to gravitational perturbations (2016). arXiv:1601.06467 [gr-qc]
10. M. Dafermos, G. Holzegel, I. Rodnianski, Boundedness and decay for the Teukolsky equation on Kerr spacetimes I: the case $|a| \ll m$ (2017). arXiv:1711.07944 [gr-qc]
11. V. de Alfaro, T. Regge, *Potential Scattering* (North-Holland, Amsterdam, 1965)
12. F. Finster, Local U(2, 2) symmetry in relativistic quantum mechanics. J. Math. Phys. **39**(12), 6276–6290 (1998). arXiv:hep-th/9703083

13. F. Finster, *The Continuum Limit of Causal Fermion Systems*. Fundamental Theories of Physics, vol. 186 (Springer, Berlin, 2016), arXiv:1605.04742 [math-ph]
14. F. Finster, J. Kleiner, Causal fermion systems as a candidate for a unified physical theory. J. Phys. Conf. Ser. **626**, 012020 (2015). arXiv:1502.03587 [math-ph]
15. F. Finster, C. Röken, Self-adjointness of the Dirac Hamiltonian for a class of non-uniformly elliptic boundary value problems. Ann. Math. Sci. Appl. **1**(2), 301–320 (2016). arXiv:1512.00761 [math-ph]
16. F. Finster, H. Schmid, Spectral estimates and non-selfadjoint perturbations of spheroidal wave operators. J. Reine Angew. Math. **601**, 71–107 (2006)
17. F. Finster, J. Smoller, A time-independent energy estimate for outgoing scalar waves in the Kerr geometry. J. Hyperbolic Differ. Equ. **5**(1), 221–255 (2008). arXiv:0707.2290 [math-ph]
18. F. Finster, J. Smoller, Error estimates for approximate solutions of the Riccati equation with real or complex potentials. Arch. Ration. Mech. Anal. **197**(3), 985–1009 (2010). arXiv:0807.4406 [math-ph]
19. F. Finster, J. Smoller, Absence of zeros and asymptotic error estimates for Airy and parabolic cylinder functions. Commun. Math. Sci. **12**(1), 175–200 (2014). arXiv:1207.6861 [math.CA]
20. F. Finster, J. Smoller, Refined error estimates for the Riccati equation with applications to the angular Teukolsky equation. Methods Appl. Anal. **22**(1), 67–100 (2015). arXiv:1307.6470 [math.CA]
21. F. Finster, J. Smoller, A spectral representation for spin-weighted spheroidal wave operators with complex aspherical parameter. Methods Appl. Anal. **23**(1), 35–118 (2016). arXiv:1507.05756 [math-ph]
22. F. Finster, J. Smoller, Linear stability of rotating black holes: outline of the proof, in *Nonlinear Analysis in Geometry and Applied Mathematics* ed. by L. Bieri, P.T. Chruściel, S.-T. Yau. Harvard University Center of Mathematical Sciences and Applications (CMSA) Series in Mathematics, vol. 1 (International Press, Somerville, 2017), pp. 77–90. arXiv:1609.03171 [math-ph]
23. F. Finster, J. Smoller, Linear stability of the non-extreme Kerr black hole. Adv. Theor. Math. Phys. **21**(8), 1991–2085 (2017). arXiv:1606.08005 [math-ph]
24. F. Finster, N. Kamran, J. Smoller, S.-T. Yau, Nonexistence of time-periodic solutions of the Dirac equation in an axisymmetric black hole geometry. Commun. Pure Appl. Math. **53**7, 902–929 (2000). gr-qc/9905047
25. F. Finster, N. Kamran, J. Smoller, S.-T. Yau, Decay rates and probability estimates for massive Dirac particles in the Kerr-Newman black hole geometry. Commun. Math. Phys. **230**(2), 201–244 (2002). arXiv:gr-qc/0107094
26. F. Finster, N. Kamran, J. Smoller, S.-T. Yau, The long-time dynamics of Dirac particles in the Kerr-Newman black hole geometry. Adv. Theor. Math. Phys. **7**(1), 25–52 (2003). arXiv:gr-qc/0005088
27. F. Finster, N. Kamran, J. Smoller, S.-T. Yau, An integral spectral representation of the propagator for the wave equation in the Kerr geometry. Commun. Math. Phys. **260**(2), 257–298 (2005). gr-qc/0310024
28. F. Finster, N. Kamran, J. Smoller, S.-T. Yau, Decay of solutions of the wave equation in the Kerr geometry. Commun. Math. Phys. **264**(2), 465–503 (2006). gr-qc/0504047
29. F. Finster, N. Kamran, J. Smoller, S.-T. Yau, A rigorous treatment of energy extraction from a rotating black hole. Commun. Math. Phys. **287**(3), 829–847 (2009). arXiv:gr-qc/0701018
30. F. Finster, J. Kleiner, J.-H. Treude, An introduction to the fermionic projector and causal fermion systems (in preparation)
31. J.L. Friedman, M.S. Morris, Schwarzschild perturbations die in time. J. Math. Phys. **41**(11), 7529–7534 (2000)
32. R. Güven, Black holes have no superhair. Phys. Rev. D **22**, 2327–2330 (1980)
33. P. Hintz, A. Vasy, The global non-linear stability of the Kerr–de Sitter family of black holes. Acta Math. **220**(1), 1–206 (2018). arXiv:1606.04014 [math.DG]
34. P.-K. Hung, J. Keller, M.-T. Wang, Linear stability of Schwarzschild spacetime: the Cauchy problem of metric coefficients (2017). arXiv:1702.02843 [gr-qc]

35. J. Metcalfe, D. Tataru, M. Tohaneanu, *Pointwise decay for the Maxwell field on black hole space–times*. Adv. Math. **316**, 53–93 (2017). arXiv:1411.3693 [math.AP]
36. C.W. Misner, K.S. Thorne, J.A. Wheeler, *Gravitation* (W.H. Freeman, San Francisco, 1973)
37. B. O'Neill, *The Geometry of Kerr Black Holes* (A K Peters, Wellesley, 1995)
38. R. Penrose, Gravitational collapse: The role of general relativity. Rev. del Nuovo Cimento **1**, 252–276 (1969)
39. W.H. Press, S.A. Teukolsky, Floating orbits, superradiant scattering and the black-hole bomb. Nature **238**, 211–212 (1972)
40. T. Regge, J.A. Wheeler, Stability of a Schwarzschild singularity. Phys. Rev. **108**, 1063–1069 (1957)
41. A.A. Starobinsky, Amplification of waves during reflection from a black hole. Soviet Phys. JETP **37**, 28–32 (1973)
42. N. Straumann, *General Relativity*. Graduate Texts in Physics, 2nd edn. (Springer, Dordrecht, 2013)
43. S.A. Teukolsky, Perturbations of a rotating black hole I. Fundamental equations for gravitational, electromagnetic, and neutrino-field perturbations. Astrophys. J. **185**, 635–647 (1973)
44. R.M. Wald, *General Relativity* (University of Chicago Press, Chicago, 1984)
45. B.F. Whiting, Mode stability of the Kerr black hole. J. Math. Phys. **30**(6), 1301–1305 (1989)
46. B.F. Whiting, L.R. Price, Metric reconstruction from Weyl scalars. Classical Quant. Gravity **22**(15), S589–S604 (2005)
47. Y.B. Zel'dovich, Amplification of cylindrical electromagnetic waves from a rotating body. Soviet Phys. JETP **35**, 1085–1087 (1972)

The Bianchi Classification of the Three-Dimensional Lie Algebras and Homogeneous Cosmologies and the Mixmaster Universe

Alexander Yu. Kamenshchik

Abstract The Bianchi classification of the three-dimensional Lie algebras and of the spatially homogeneous universes is presented in a rather pedagogical manner. The dynamics of the Bianchi-I, Bianchi-II and Bianchi-IX universes are discussed in detail. A special attention is paid to the phenomenon of the oscillatory approach to the cosmological singularity (BKL) known also as the Mixmaster universe. The stochasticity in cosmology and the connection between cosmological billiards and infinite-dimensional Lie algebras are briefly illustrated.

Keywords Cosmology · Symmetries · Singularity

Mathematics Subject Classification (2000) Primary 83F05; Secondary 83C20

1 Introduction

Almost all modern cosmology is based on the Friedmann–Lemaître spatially homogeneous and isotropic cosmological models [1–4]. Indeed, the Friedmann–Lemaître cosmology

$$ds^2 = dt^2 - a^2(t) \left(\frac{dr^2}{1 - kr^2} + r^2(d\theta^2 + \sin^2\theta d\phi^2) \right), \tag{1}$$

presenting an expanding or contracting spatially homogeneous and isotropic universe is very successful in the description of its global evolution from the times of inflation [5, 6] until the present epoch of the cosmic acceleration [7, 8]. Let us

A. Yu. Kamenshchik (✉)
Department of Physics and Astronomy of the University of Bologna and INFN, Bologna, Italy

L.D. Landau Institute for Theoretical Physics, Moscow, Russia
e-mail: kamenshchik@bo.infn.it

© Springer Nature Switzerland AG 2019
S. Cacciatori et al. (eds.), *Einstein Equations: Physical and Mathematical Aspects of General Relativity*, Tutorials, Schools, and Workshops in the Mathematical Sciences, https://doi.org/10.1007/978-3-030-18061-4_3

remind that in the formula (1) the numbers $k = 1$, $k = 0$ and $k = -1$ correspond to the spatially closed, spatially flat and spatially open universes, respectively.

For the description of the inhomogeneities of the universe one uses the theory of the cosmological perturbations on the Friedmann background [9]:

$$g_{ij} = g_{ij}^{(0)} + h_{ij}, \tag{2}$$

where $g_{ij}^{(0)}$ is the Friedmann metric, while h_{ij} is a small perturbation. Moreover, one can explain the origin of the large-scale structure of the contemporary universe starting from the quantum fluctuations in the very early universe [10].

However, there are very interesting things in gravity and cosmology beyond Friedmann models and the perturbations on their background. The study of spatially homogeneous, but anisotropic models takes its origin in the work by Luigi Bianchi, written as early as in the year 1898 [11]. In this work the complete classification of the three-dimensional homogeneous Riemann spaces and the three-dimensional Lie groups was presented long before the creation of the General Relativity by Einstein in 1916 [12]. Later, in 50th and 60th the Bianchi classification was modernized, simplified and applied to the cosmology [13–17].

In 1963 Khalatnikov and Lifshitz have begun applying the Bianchi universes for the study of the problem of the singularity in cosmology [18]. At the end of sixties Belinski, Khalatnikov and Lifshitz have discovered the phenomenon of the oscillatory approach to the cosmological singularity [19–23]. Using the Hamiltonian formalism Misner has described this phenomenon as the Mixmaster Universe [24].

Later it was understood that when universe tends to the singularity its dynamics becomes chaotic [25, 26]. Finally, at the beginning of the new millennium the connection between the chaotic behaviour in the cosmological models and the infinite-dimensional Lie algebras [27] was discovered [28–30].

The study of the oscillatory approach to the cosmological singularity and of its relation with some advanced mathematical structures has become an important branch of mathematical physics. It can become useful for the description of some physical effects in the very early universe. Moreover, it is not excluded that this connection is important for quantum gravity [31].

The structure of these lecture notes is the following: in Sect. 2 we introduce briefly such basic notions of the differential geometry as manifolds, vector fields, affine connection, metric and curvature; in the third section we discuss symmetries and Killing vector fields; the fourth section is devoted to the presentation of the Bianchi classification of the three-dimensional Lie algebras and of the homogeneous cosmologies; in the fifth section we present the solutions of the Einstein equations for the Bianchi-I universes; Sect. 6 is devoted to the solutions of the Einstein equations for the Bianchi-II universes; in the seventh section we present the geometry of the Bianchi-IX universe and introduce the oscillatory approach to the cosmological singularity (BKL) and the model of the Mixmaster universe; in Sect. 8 we discuss the appearance of chaos in cosmology; the ninth section is devoted to connections between the Mixmaster universe and infinite-dimensional Lie algebras; the last section contains some concluding remarks.

2 Manifolds, Vectors Fields, Affine Connection, Metric and Curvature

2.1 Manifolds, Vectors Fields, One-Forms

Let us recall briefly some basic notions of the differential geometry (see e.g. [32–34]). We shall consider the differentiable manifolds $\mathcal{M}$ of dimensionality n.

A real function $f(P)$, $P \in \mathcal{M}$, defined on the manifold $\mathcal{M}$, is a map from the manifold to the field of real numbers $\mathbb{R}$:

$$f : \mathcal{M} \to \mathbb{R}. \tag{3}$$

A curve $l = P(\lambda)$ is a map from an interval on the real line to the manifold:

$$(a, b) \subset \mathbb{R}; l : (a, b) \to \mathcal{M}. \tag{4}$$

In a point P of the manifold $\mathcal{M}$ we can define a vector X:

$$X \equiv \frac{d}{d\lambda}. \tag{5}$$

We can say that the vector X is a linear operator, acting on the space of all differentiable functions and satisfying the Leibniz rule:

$$X(fg) = (Xf)g + f(Xg). \tag{6}$$

The vectors attached to some point P constitute a tangent space V_P, which is a real linear space of the dimensionality n.

Choosing in any point P a vector $X(P)$, we define a vector field.

The one-form ω defined on the space V is a linear functional:

$$\omega : V \to \mathbb{R}. \tag{7}$$

One-forms constitute a cotangent space V^*.

Defining a one-form in any point, we obtain a differential form $\omega(P)$.

A special differential form called "gradient" df is defined by its action on the vector field X:

$$df(X) = Xf. \tag{8}$$

If in a neighbourhood of a point P we have a local coordinate system: $x^1, x^2, \ldots, x^n$, then we can introduce the coordinate basis of vector fields:

$$e_i \equiv \frac{\partial}{\partial x^i}, \quad [e_i, e_j] = 0. \tag{9}$$

Obviously, these vector fields e_i defined in Eq. (9) commute and this basis is called also holonomic.

Generally, vector fields do not commute and

$$[e_i, e_j] = C_{ij}^k e_k, \tag{10}$$

where C_{ij}^k are the unholonomity coefficients.

The one-forms $\omega^i = dx^i$ constitute a dual basis with respect to the coordinate basis of vector fields $e_i = \frac{\partial}{\partial x^i}$:

$$\omega^i(e_j) = \delta^i_j. \tag{11}$$

2.2 Tensor Fields

A polylinear map

$$T : \underbrace{V^* \times \cdots \times V^*}_{p \text{ times}} \times \underbrace{V \times \cdots \times V}_{q \text{ times}} \to \mathbb{R} \tag{12}$$

is called p-contravariant and q-covariant tensor. A tensor defined on the manifold is a tensor field. The components of the tensor are defined as

$$T^{i_1 \cdots i_p}_{j_1 \cdots j_q} \equiv T(\omega^{i_1}, \ldots, \omega^{i_p}; e_{j_1}, \ldots, e_{j_q}). \tag{13}$$

The tensor product of two tensors of types (p_1, q_1) and (p_2, q_2) gives the new tensor of the type $(p_1 + p_2, q_1 + q_2)$.

The contraction of a tensor of the type (p, q) with respect to one covariant and one contravariant indices gives the tensor of the type $(p - 1, q - 1)$.

2.3 Parallel Transport and Affine Connection

To define a parallel transport of the vectors on a manifold we introduce an affine connection by means of the covariant derivative:

$$\nabla_{e_i} e_j = \Gamma^k_{ji} e_k. \tag{14}$$

The symbols Γ_{ij}^k are called Christoffel symbols.

The Christoffel symbols are not tensor components because

$$\nabla_X(fY) \neq f\nabla_X Y. \tag{15}$$

However, the map

$$T(U, V) = \nabla_U V - \nabla_V U - [U, V] \tag{16}$$

$$T_{ij}^k = \Gamma_{ij}^k - \Gamma_{ji}^k - C_{ij}^k \tag{17}$$

is a two times covariant and one time contravariant tensor. This tensor is antisymmetric in lower indices and it is called torsion.

The map

$$R(U, V)W = \nabla_U \nabla_V W - \nabla_V \nabla_U W - \nabla_{[U,V]} W \tag{18}$$

with the components

$$R_{kij}^l = e_i \Gamma_{kj}^l - e_j \Gamma_{ki}^l + \Gamma_{kj}^m \Gamma_{mi}^l - \Gamma_{ki}^m \Gamma_{mj}^l - C_{ij}^m \Gamma_{km}^l \tag{19}$$

defines a curvature tensor, or the Riemann or the Riemann-Christoffel tensor, which is three times covariant and one time contravariant.

Its contraction gives the Ricci tensor:

$$R_{kj} \equiv R_{kij}^i, \tag{20}$$

which enters into the Einstein equations.

2.4 Metric and Affine Connection

The two times covariant, symmetric, non-degenerate tensor field g, called metric, defines the scalar product between the vectors:

$$X \cdot Y \equiv g(X, Y), \quad g_{ij} = g(e_i, e_j). \tag{21}$$

We wish to define the affine connection in such a way that the scalar product is not changed as a result of the parallel transport. That means that

$$\nabla_X g = 0. \tag{22}$$

This requirement implies the following definition of the Christoffel symbols:

$$\Gamma^m_{ij} = \frac{1}{2} g^{mk} (g_{kj,i} + g_{ik,j} - g_{ij,k})$$

$$+ \frac{1}{2} (g^{mk} g_{nj} C^n_{ki} + g^{mk} g_{in} C^n_{kj} + C^m_{ji})$$

$$+ \frac{1}{2} (g^{mk} g_{nj} T^n_{ik} + g^{mk} g_{in} T^n_{jk} + T^m_{ij}). \tag{23}$$

In the case when the torsion is absent, the third line vanishes. In the case when the basis is holonomic, the second line vanishes. When the basis is chosen in such a way that the metric components do not depend on coordinates (for example, the tetrad basis, $g_{ij} = \eta_{ij}$, where η_{ij} is the metric of the Minkowski spacetime), the first line vanishes.

It would be interesting to consider the case without torsion and using the basis, where both the metric components and the unholonomity coefficients are constant. Then

$$\Gamma^m_{ij} = \frac{1}{2} (g^{mk} g_{nj} C^n_{ki} + g^{mk} g_{in} C^n_{kj} + C^m_{ji}). \tag{24}$$

$$R^l_{kij} = \frac{1}{4} [C^m_{ij} C^l_{km} - g_{kp} g^{lq} C^m_{ij} C^p_{qm}$$

$$+ g_{kp} g^{mq} (c^p_{qj} C^l_{im} - C^p_{qi} C^l_{jm}) + g_{jp} g^{ql} C^m_{kq} C^p_{im}$$

$$+ g_{mp} g^{lq} (C^p_{qi} C^m_{jk} - C^p_{qj} C^m_{ik}) + g_{ip} g^{lq} C^m_{qk} C^p_{jm}$$

$$+ g_{jp} g^{mq} C^p_{qk} C^l_{im} - g_{ip} g^{mq} C^p_{qk} C^l_{jm} - 2 g_{pm} g^{ql} C^p_{qk} C^m_{ij}$$

$$+ g^{mr} g^{ql} (g_{kp} g_{is} C^p_{rj} + g_{jp} g_{is} C^p_{rk} - g_{kp} g_{js} C^p_{ri} - g_{ip} g_{js} C^p_{rk}) C^s_{qm}]. \tag{25}$$

$$R_{kj} = \frac{1}{2} (C^m_{ij} C^i_{km} + g_{kp} g^{mq} C^p_{qj} C^i_{im}$$

$$+ g_{jp} g^{mq} C^p_{qk} C^i_{im} + g_{mp} g^{iq} C^p_{kq} C^m_{ij}$$

$$+ \frac{1}{2} g^{mr} g^{qi} g_{kp} g_{js} C^p_{ir} C^s_{qm}). \tag{26}$$

The last formula is very useful for the study of the spatially homogeneous cosmological models.

3 Killing Vector Fields and Symmetry

3.1 Lie Derivative of Tensor Fields

Let us suppose that on the manifold $\mathcal{M}$ is defined a vector field $X(P)$. In any point of the manifold this field defines a curve to which this field is tangent. One can represent this field as

$$X = \frac{d}{d\lambda}, \tag{27}$$

where λ parametrizes the curve. Then, making infinitesimal shifts along the curves, one can define the Lie derivative L_X.

For a function f:

$$L_X f = Xf = \frac{df}{d\lambda}. \tag{28}$$

For a vector field Y:

$$L_X Y = [X, Y]. \tag{29}$$

For other tensor fields the Lie derivative is defined by means of the Leibniz rule:

$$L_X(U \otimes V) = L_X U \otimes V + U \otimes L_X V \tag{30}$$

and by the rule

$$L_X CT = CL_X T, \tag{31}$$

where C means a contraction with respect to some indices.

The Lie derivative has also another remarkable property:

$$[L_X, L_Y] = L_{[X,Y]}. \tag{32}$$

3.2 Killing Vector Fields, Killing Equations and Isometries

Let us consider a manifold $\mathcal{M}$, possessing a metric tensor g. Then, if for some vector field X:

$$L_X g = 0, \tag{33}$$

$$g_{ij,k} X^k + g_{ik} X^k_{,j} + g_{kj} X^k_{,i} = 0, \tag{34}$$

$$X_{i;j} + X_{j;i} = 0, \tag{35}$$

then X is called Killing vector field and the equation is called Killing equation. In Eq. (35) the symbol ";" signifies the covariant derivative.

The transformations of the manifold generated by the Killing fields do not change the form of the metric and are called isometries.

We can ask ourselves how many Killing vectors can have a n-dimensional manifold? To begin with let us consider the manifold $\mathbb{R}^n$. This manifold is isometric with respect to n translations, generated by the Killing vectors

$$X_i = \frac{\partial}{\partial x^i}, \ i = 1, \ldots, n \tag{36}$$

and with respect to $n(n-1)/2$ rotations, generated by the Killing vector fields

$$X_{ij} = x_i \frac{\partial}{\partial x^j} - x_j \frac{\partial}{\partial x_i}, \ i < j. \tag{37}$$

Thus, the manifold $\mathbb{R}^n$ has $n(n+1)/2$ Killing vector fields. One can show that the maximal number of the Killing vectors for a manifold with the dimensionality n is $\frac{n(n+1)}{2}$ (see, e.g., [35]).

3.3 Four-Dimensional Pseudo-Riemannian Spacetimes with the Signature $+ - --$

Let us consider now the four-dimensional pseudo-Riemannian spacetimes with the signature $+ - --$. In this case the maximal number of the Killing vector fields is 10. It is known that only three types of spacetimes of maximal symmetry with 10 Killing fields exist. First of the them is the Minkowski spacetime of the special relativity with the metric

$$ds^2 = (dx^0)^2 - (dx^1)^2 - (dx^2)^2 - (dx^3)^2. \tag{38}$$

Then, the de Sitter spacetime, which is a hyperboloid

$$(X^1)^2 + (X^2)^2 + (X^3)^2 + (X^4)^2 - (X^0)^2 = R^2, \tag{39}$$

embedded into the five-dimensional spacetime with the metric

$$ds^2 = (dX^0)^2 - (dX^1)^2 - (dX^2)^2 - (dX^3)^2 - (dX^4)^2. \tag{40}$$

Finally, there exists the Anti-de Sitter spacetime, which is a hyperboloid

$$(X^0)^2 + (X^4)^2 - (X^1)^2 - (X^2)^2 - (X^3)^2 = R^2, \tag{41}$$

embedded into the five-dimensional spacetime with the metric

$$ds^2 = (dX^0)^2 + (dX^4)^2 - (dX^1)^2 - (dX^2)^2 - (dX^3)^2. \tag{42}$$

3.4 Synchronous Reference Frame

Now we would like to study the four-dimensional spacetimes, which do not possess the maximal symmetry but still have some Killing vector fields. It will be convenient first to introduce the synchronous reference frame [16].

Let us consider a universe with the metric

$$ds^2 = dt^2 - \gamma_{\alpha\beta} dx^\alpha dx^\beta, \quad \alpha, \beta = 1, 2, 3. \tag{43}$$

In Eq. (43) we have used the coordinate basis of vector fields and one-forms. This metric can be written down in a more general basis:

$$ds^2 = dt^2 - \gamma_{ab} \omega^a \otimes \omega^b, \quad a, b = 1, 2, 3, \tag{44}$$

where ω^a are three basis one-forms on the spatial submanifold.

Now we can introduce some useful definitions and formulae. First of all, let us define the extrinsic curvature or the second fundamental form:

$$\kappa_{ab} \equiv \frac{\partial \gamma_{ab}}{\partial t}, \quad \kappa_a^b = \gamma^{bc} \kappa_{ac}, \quad \kappa = \kappa_a^a. \tag{45}$$

Then we shall need the three-dimensional affine connection λ_{ab}^c. This connection is compatible with the spatial metric γ_{ab}. Using κ_{ab} and λ_{ab}^c we can represent the components of the Ricci tensor as follows:

$$R_{00} = -\frac{1}{2} \frac{\partial \kappa}{\partial t} - \frac{1}{4} \kappa_a^b \kappa_b^a,$$

$$R_{0a} = \frac{1}{2} (\kappa_{a|b}^b - \kappa_{b|a}^b),$$

$$R_{ab} = P_{ab} + \frac{1}{2} \frac{\partial \kappa_{ab}}{\partial t} + \frac{1}{4} (\kappa_{ab}\kappa - 2\kappa_a^c \kappa_{bc}). \tag{46}$$

Here | is the three-dimensional covariant derivative and P_{ab} is the three-dimensional Ricci tensor.

Implementation of the synchronous reference frame and of the formulae given above represent a particular case of the $3 + 1$-decomposition (see, e.g., the book [36] and references therein).

Now we can ask ourselves how many Killing vector fields can have three-dimensional spatial submanifolds of the spacetime? The answer follows immediately from the general formula for the maximal number of the Killing vector fields for an n-dimensional manifold: for $n = 3, n(n + 1)/2 = 6$. The universes with six Killing vector fields defined on the three- dimensional spatial submanifolds are nothing but the Friedmann–Lemaître universes: flat, closed and open. Their metrics are

$$ds^2 = dt^2 - a^2(t)(dx^2 + dy^2 + dz^2), \tag{47}$$

$$ds^2 - a^2(t)(d\chi^2 + \sin^2 \chi (d\theta^2 + \sin^2 \theta d\phi^2)), \tag{48}$$

$$ds^2 = dt^2 - a^2(t)(\sinh^2 \chi (d\theta^2 + \sin^2 \theta d\phi^2)). \tag{49}$$

In what follows we shall consider a more wide class of the cosmological models—the spatially homogeneous universes. Under homogeneous manifold we mean such a manifold, where the isometry group acts transitively, i.e. for any two points exists an element of the isometry group which maps one of these points to another one. To be homogenous a three-dimensional manifold should have at least three Killing vector fields. As was already told in the Introduction the classification of all three-dimensional Lie algebras and of the homogenous three-dimensional manifolds was elaborated by Bianchi. In the next section we shall present this classification in its modern form.

4 Bianchi Classification of the Three-Dimensional Lie Algebras and of the Homogeneous Cosmologies

4.1 Reciprocal Basis of Vector Fields

Let us suppose that we have three spatial Killing fields X_a. In this case, our spacetime is spatially homogenous.

The structure functions of these fields are constants. However, it is not convenient to choose these three vector fields as a basis, because the metric components in this basis depend on coordinates.

We can find other three vector fields e_a, which commute with the Killing fields:

$$[e_a, X_b] = 0. \tag{50}$$

Then we can construct the dual basis of one-forms ω^a:

$$\omega^a(e_b) = \delta_b^a. \tag{51}$$

Now, we can construct the cosmological models with the metric

$$ds^2 = dt^2 - g_{ab}(t)\omega^a \otimes \omega^b, \tag{52}$$

where the metric coefficients depend only on the time parameter. This metric is invariant with respect to the Killing vectors X_a.

Indeed, to show that the Lie derivative of the metric (52) with respect to a Killing vector field X_a is equal to zero it is enough to show that this Lie derivative annihilates a one-form ω^b. Let us do it.

We would like to prove that

$$L_{X_a}\omega^b = 0. \tag{53}$$

This statement means that applying the new one-form $L_{X_a}\omega^b$ to an arbitrary vector field Y, we obtain zero:

$$(L_{X_a}\omega^b)(Y) = 0. \tag{54}$$

To show that Eq. (54) is valid it is enough to show that it is true if as vector Y we take a vector field e_c belonging to the reciprocal basis $\{e_d\}$:

$$(L_{X_a}\omega^b)(e_c) = 0. \tag{55}$$

It is convenient to rewrite Eq. (55), using such symbols as the tensor product and the contraction:

$$(L_{X_a}\omega^b)(e_c) = C(L_{X_a}\omega^b \otimes e_b). \tag{56}$$

Now, using the Leibniz rule and the fact that the contraction commute with the Lie derivative, we obtain

$$\begin{aligned}
(L_{X_a}\omega^b)(e_c) &= C(L_{X_a}\omega^b \otimes e_b) \\
&= C(L_{X_a}(\omega^b \otimes e_b) - \omega^b \otimes L_{X_a}e_b) \\
&= L_{X_a}C(\omega^b \otimes e_b) - C(\omega^b \otimes [X_a, e_b]) = 0,
\end{aligned} \tag{57}$$

where we have used the equalities (50) and (51).

We shall show now that the group of transformations generated by the Lie algebra of the reciprocal vector fields $\{e_a\}$ and called reciprocal group is equivalent to the isometry group. This algebra Lie is defined by the relations

$$[e_a, e_b] = D^c_{ab} e_c. \tag{58}$$

We would like to calculate the structure coefficients D^c_{ab}.

One can represent the vector fields e_a as some linear combinations of the Killing vector fields:

$$e_a = a^b_a X_b, \tag{59}$$

where the coefficients a^a_b depend on the point of the manifold. We can look for such a triple of the reciprocal vector fields which in some arbitrary point P of the manifold coincide with the Killing vector fields:

$$a^b_a(P) = \delta^b_a. \tag{60}$$

Then

$$[e_f, X_b] = [a^a_f X_a, X_b] = a^a_f C^c_{ab} X_c - (X_b a^c_f) X_c = 0, \tag{61}$$

i.e.

$$a^a_f C^c_{ab} = (X_b a^c_f). \tag{62}$$

At the point P this condition becomes

$$(X_b a^c_f)(P) = C^c_{fb}. \tag{63}$$

Then,

$$[e_f, e_h] = [a^a_f X_a, a^b_h X_b] = a^a_f a^b_h C^c_{ab} X_c + a^a_f (X_a a^b_h) X_b - a^b_h (X_b a^a_f) X_a$$
$$= D^p_{fh} e_p = D^p_{fh} a^c_p X_c. \tag{64}$$

At the point P, using the preceding equations, we obtain

$$D^p_{fh} = -C^p_{fh}. \tag{65}$$

This means that the Lie algebras of the three Killing vector fields and of the three vector fields belonging to the reciprocal basis coincide.

Now, we are in a position for the construction of the complete classification of the three-dimensional Lie algebras.

4.2 Bianchi–Schucking–Behr Classification of the Three-Dimensional Lie Algebras and of the Three-Dimensional Homogeneous Spaces

As was said above the Lie algebra is defined by its structure constants C_{ab}^c. These structure constants are antisymmetric in its lower indices and should satisfy the Jacobi identities:

$$C_{ab}^f C_{cf}^d + C_{bc}^f C_{af}^d + C_{ca}^f C_{bf}^d = 0. \tag{66}$$

It is convenient to introduce a new 3×3 matrix C^{dc} such that

$$C_{ab}^c = \varepsilon_{abd} C^{dc}, \tag{67}$$

where ε_{abc} is the totally antisymmetric Levi-Civita symbol:

$$\varepsilon_{123} = 1. \tag{68}$$

The Jacobi identity becomes

$$\varepsilon_{bcd} C^{cd} C^{ba} = 0. \tag{69}$$

To prove the validity of the equality (69), let us notice that if in the Jacobi identity (66) we choose the indices a, b and c in such a way that two of them coincide then the equality (66) is satisfied automatically. Thus, let us choose $a = 1, b = 2, c = 3$. Then the Jacobi identity (66), where we have substituted (67) looks as follows

$$\varepsilon_{123} C^{3f} \varepsilon_{3fg} C^{gd} + \varepsilon_{231} C^{1f} \varepsilon_{1fg} C^{gd}$$
$$+ \varepsilon_{312} C^{2f} \varepsilon_{2fg} C^{gd} = 0. \tag{70}$$

Remembering that $\varepsilon_{123} = \varepsilon_{231} = \varepsilon_{312} = 1$, we see that the equality (70) implies

$$\varepsilon_{hfg} C^{hf} C^{gd} = 0, \tag{71}$$

which is nothing but the identity (69).

Now, we shall represent the matrix C^{ab} as a sum of the symmetric and antisymmetric matrices:

$$C^{ab} = n^{ab} + \varepsilon^{abc} a_c, \quad n^{ab} = n^{ba}. \tag{72}$$

Using the identities

$$\varepsilon_{abc} n^{bc} = 0, \quad \varepsilon_{abc} \varepsilon^{bcd} = 2\delta_a^d, \quad \varepsilon^{abc} a_b a_c = 0, \tag{73}$$

we obtain

$$n^{ab} a_b = 0. \tag{74}$$

Thus, the vector a_b is a null vector of the matrix n^{ab}. Let us note that we can always diagonalize the symmetric tensor n^{ab}:

$$n^{ab} = \delta^{ab} n^a. \tag{75}$$

Without losing generality, we can choose

$$a_b = \delta_{b1} a. \tag{76}$$

$$n_1 a = 0. \tag{77}$$

Now using the formulae (67), (72) and (76), we can represent a general three-dimensional Lie algebra in the following form:

$$[X_1, X_2] = -a X_2 + n_3 X_3,$$
$$[X_2, X_3] = n_1 X_1,$$
$$[X_3, X_1] = n_2 X_2 + a X_3. \tag{78}$$

We consider first the so-called class A cosmologies, where $a = 0$.

If $n_1 = n_2 = n_3 = 0$, then all the Killing fields commutate.

This Lie algebra and the corresponding cosmology are called Bianchi-I Lie algebra and Bianchi-I cosmology.

If

$$n_1 = 1, \ n_2 = n_3 = 0. \tag{79}$$

then the only non-vanishing commutator is

$$[X_2, X_3] = X_1. \tag{80}$$

This is the Bianchi-II cosmology.

Let us note that the case $n_1 \neq 0, n_1 \neq 1, n_2 = n_3 = 0$ is equivalent to the case, described above. Indeed, it is enough to change the basis of the algebra, introducing a new vector field $\tilde{X}_1 = n_1 X_1$.

If

$$n_1 = n_2 = 1, \ n_3 = 0, \tag{81}$$

then we have Bianchi-VII ($a = 0$) cosmology.

Also in this case choosing some non-vanishing values for n_1 and n_2, having the same sign, we obtain the same Lie algebra as in Eq. (89). The corresponding change of the basis is $X_1 = \sqrt{n_2}\tilde{X}_1$, $X_2 = \sqrt{n_1}\tilde{X}_2$, $X_3 = \sqrt{n_1 n_2}\tilde{X}_3$.

If

$$n_1 = 1, \ n_2 = -1, \ n_3 = 0, \tag{82}$$

then we have the Bianchi-VI ($a = 0$) cosmology. Naturally, if we have instead of $n_1 = 1, n_2$ a pair of the other non-vanishing numbers, having opposite signs, the Lie algebra is the same.

The choice

$$n_1 = n_2 = n_3 = 1, \tag{83}$$

corresponds to the Bianchi-IX cosmology.

Choosing three arbitrary non-vanishing constants n_1, n_2, n_3, having the same sign, we reproduce the Lie algebra (83), making the following substitution:

$$X_1 = \sqrt{n_2 n_3}\tilde{X}_1, \ X_2 = \sqrt{n_1 n_3}\tilde{X}_2, \ X_3 = \sqrt{n_1 n_2}\tilde{X}_3.$$

The choice

$$n_1 = n_2 = 1, \ n_3 = -1, \tag{84}$$

gives the Bianchi-VIII cosmology. As before the choice of three non-vanishing constants, two of which have the same sign again reproduce the Lie algebra (84).

We shall consider now the class B cosmologies, i.e. the cosmologies where $a \neq 0$. As follows immediately from Eq. (77) for the class B Lie algebras

$$n_1 = 0.$$

If we choose

$$a = 1, \ n_1 = n_2 = n_3 = 0. \tag{85}$$

We have Bianchi-V cosmology. The Lie algebra is

$$[X_1, X_2] = -X_2,$$
$$[X_2, X_3] = 0,$$
$$[X_3, X_1] = X_3. \tag{86}$$

Having a non-vanishing value of $a \neq 1$, we can reproduce the Lie algebra (86) by choosing $X_1 = a\tilde{X}_1$.

The next choice is the Bianchi-IV universe with

$$a = 1, \; n_1 = n_2 = 0, \; n_3 = 1, \tag{87}$$

whose Lie algebra is

$$[X_1, X_2] = -X_2 + X_3,$$
$$[X_2, X_3] = 0,$$
$$[X_3, X_1] = X_3. \tag{88}$$

For arbitrary non-vanishing numbers a and n_3, we can show that the corresponding Lie algebra coincides with the Lie algebra (88) by choosing the basis with

$$X_1 = a\tilde{X}_1, X_3 = \frac{\tilde{X}_3}{n_3}.$$

The so-called Bianchi-VII ($a \neq 0$) cosmology arises when

$$n_1 = 0, \; n_2 = n_3 = 1, a \neq 0. \tag{89}$$

One can easily check that in this case the value of $a \neq 0$ is important and we cannot choose such a basis, which transforms the value of the parameter a into $a = 1$.

The same is true for the Bianchi-VI ($a \neq 0, a \neq 1$) Lie algebra with

$$n_1 = 0, n_2 = 1, n_3 = -1, a \neq 0, a \neq 1. \tag{90}$$

For historical reasons a particular case

$$n_1 = 0, n_2 = 1, n_3 = -1, a = 1 \tag{91}$$

is known as the Bianchi-III cosmology.

4.3 Einstein Equations in the Synchronous Reference System for the Bianchi Universes

Let us now write down the full set of the Einstein equations in the synchronous reference system for the Bianchi universes.

$$\kappa_{ab} = \dot{\gamma}_{ab}, \kappa_a^b = \dot{\gamma}_{ac}\gamma^{cb}. \tag{92}$$

$$P_a^b = \frac{1}{2\gamma}[2C^{bd}C_{ad} + C^{db}C_{ad} + C^{bd}C_{da} - C_d^d(C_a^b + C_a^{\;b})$$
$$+\delta_a^b((C_d^d)^2 - 2C^{df}C_{df})], \tag{93}$$

$$C_a{}^b = \gamma_{ac}C^{cb}, \; C_{ab} = \gamma_{ac}\gamma_{bd}C^{cd}. \tag{94}$$

$$R_0^0 = -\frac{1}{2}\dot{\kappa}_a^a - \frac{1}{4}\kappa_a^b\kappa_b^a, \tag{95}$$

$$R_a^0 = -\frac{1}{2}\kappa_b^a(C_{ca}^b - \delta_a^b C_{dc}^d), \tag{96}$$

$$R_a^b = -\frac{1}{2\sqrt{\gamma}}(\sqrt{\gamma}\kappa_a^b)\dot{} - P_a^b. \tag{97}$$

Substituting the expressions for the components of the Ricci tensor (95)–(97) into the Einstein equations we obtain the Einstein equations in the synchronous reference system for the Bianchi universes.

5 Bianchi-I Universe

5.1 Geometry of the Bianchi-I Universe

As we know all the Killing vector fields in the Bianchi-I universe commute amongst them and we can choose them as a coordinate holonomic basis:

$$X_a = \frac{\partial}{\partial x^a}, \; a = 1, 2, 3. \tag{98}$$

The reciprocal basis coincides with the basis of the Killing vector fields

$$e_a = X_a. \tag{99}$$

The dual basis of one-forms is

$$\omega^a = dx^a, \tag{100}$$

and the metric is defined as

$$ds^2 = dt^2 - a^2(t) \sum_{a=1}^{3} e^{2\beta_a(t)}(dx^a)^2, \tag{101}$$

$$\sum_{a=1}^{3} \beta_a = 0, \tag{102}$$

where β_a are the anisotropy parameters.

Let us note that the ranges of the coordinates x^a can coincide with the straight line $\mathbb{R}$, but can also be compact. In the last case the spatial geometry of the Bianchi-I universe represents a three-dimensional torus.

5.2 The Dynamics of the Bianchi-I Universe Filled with an Isotropic Perfect Fluid

Studying the dynamics of the Bianchi-I universe filled with an isotropic perfect fluid we shall follow the approach, presented in paper [37]. First, we shall write down the expressions for the components of the external curvature and of the Ricci tensor for the case of the Bianchi-I universe:

$$\kappa_{ab} = 2(\dot{a}a + \dot{\beta}_a a^2)e^{2\beta_a}\delta_{ab}, \tag{103}$$

$$\kappa_a^b = 2\left(\frac{\dot{a}}{a} + \dot{\beta}_a\right)\delta_a^b, \tag{104}$$

$$\kappa = 6\frac{\dot{a}}{a}, \tag{105}$$

$$P_a^b = 0, \tag{106}$$

$$R_0^0 = -3\frac{\ddot{a}}{a} - \sum_{a=1}^{3}\dot{\beta}_a^2, \tag{107}$$

$$R_a^0 = 0, \tag{108}$$

$$R_a^b = -\left(\frac{\ddot{a}}{a} + 2\frac{\dot{a}^2}{a^2} + 3\frac{\dot{a}}{a}\dot{\beta}_a - \ddot{\beta}_a\right)\delta_a^b, \tag{109}$$

$$R = -\left(6\frac{\ddot{a}}{a} + \frac{\dot{a}^2}{a^2} + \sum_{a=1}^{3}\dot{\beta}_a^2\right). \tag{110}$$

Let us suppose that the universe is filled with some fluid with an isotropic pressure. That means that

$$R_1^1 = R_2^2 = R_3^3. \tag{111}$$

Then

$$R_2^2 + R_3^3 - 2R_1^1 = 0. \tag{112}$$

Hence,

$$\ddot{\beta}_1 + 3\dot{\beta}_1\frac{\dot{a}}{a} = 0, \tag{113}$$

and

$$\dot{\beta}_1 = \frac{\beta_{10}}{a^3}. \tag{114}$$

Analogously,

$$\dot{\beta}_2 = \frac{\beta_{20}}{a^3}, \ \dot{\beta}_3 = \frac{\beta_{30}}{a^3}, \tag{115}$$

where

$$\sum_{a=1}^{3} \beta_{a0} = 0. \tag{116}$$

The 00—component of the Einstein equations

$$R_0^0 - \frac{1}{2}R = \rho, \tag{117}$$

where ρ is the energy density of the fluid, looks now as

$$3\frac{\dot{a}^2}{a^2} - \frac{1}{2}\frac{\bar{\beta}^2}{a^6} = \rho, \tag{118}$$

where

$$\bar{\beta}^2 \equiv \sum_{a=1}^{3} \beta_{a0}^2. \tag{119}$$

5.3 Empty Universe: Kasner Solution

The Kasner solution for an empty Bianchi-I universe was discovered as early as in 1921 [38]. It was the first non-stationary, i.e. time-dependent solution of the Einstein equations, found even before the Friedmann solutions [1, 2]. Here, we present the derivation of this solution, emphasizing its general character and using the formulae, obtained in the preceding subsection.

If $\rho = 0$, then from Eq. (118) follows that

$$a(t) = \left(\frac{3}{2}\right)^{\frac{1}{6}} (\bar{\beta}t)^{\frac{1}{3}}. \tag{120}$$

Correspondingly, three scale factors have the form

$$a_a(t) = a(t)e^{\beta_a(t)} = a_{a0}t^{p_a}, \tag{121}$$

where

$$p_a = \frac{1}{3} + \sqrt{\frac{2}{3}} \frac{\beta_{a0}}{\bar{\beta}} \qquad (122)$$

One can easily check that

$$p_1 + p_2 + p_3 = 1, \qquad (123)$$
$$p_1^2 + p_2^2 + p_3^2 = 1. \qquad (124)$$

The numbers p_a are called Kasner indices, while the relations (124), (160) are the Kasner identities.

5.4 Lifshitz–Khalatnikov Parametrization of the Kasner Indices

In the paper [18] a very convenient parametrization of the Kasner indices was suggested:

$$p_1 = -\frac{u}{1 + u + u^2}, \qquad (125)$$

$$p_2 = \frac{1 + u}{1 + u + u^2}, \qquad (126)$$

$$p_3 = \frac{u(1 + u)}{1 + u + u^2}. \qquad (127)$$

This parametrization reveals a remarkable property of the Kasner indices:

$$p_1\left(\frac{1}{u}\right) = p_1(u), \qquad (128)$$

$$p_2\left(\frac{1}{u}\right) = p_3(u), \qquad (129)$$

$$p_3\left(\frac{1}{u}\right) = p_2(u). \qquad (130)$$

5.5 Universe with Dust: Hechmann–Schucking Solution

An exact solution of the Einstein equations for the Bianchi-I universe filled with dust is also known. It was discovered by Heckmann and Schucking in 1959 [39].

Let us suppose that the dust is present in the Bianchi-I universe. Then, the dependence of the energy density of the dust ρ on the scale factor a is

$$\rho = \frac{\rho_0}{a^3}, \tag{131}$$

where ρ_0 is a positive constant. Substituting the expression (131) into Eq. (118), we obtain

$$3\frac{\dot{a}^2}{a^2} - \frac{1}{2}\frac{\bar{\beta}^2}{a^6} = \frac{\rho_0}{a^3}, \tag{132}$$

The solution of this equation is

$$a^3(t) = \frac{3}{4}\rho_0 t^2 + \sqrt{\frac{3}{2}}\bar{\beta}t. \tag{133}$$

Then, integrating Eq. (115) with a^3 taken from Eq. (133), we obtain

$$\beta_a(t) = \sqrt{\frac{2}{3}}\frac{\beta_{a0}}{\bar{\beta}} \ln \frac{t}{t + \sqrt{\frac{8}{3}\frac{\bar{\beta}}{\rho_0}}} + \beta_{a1}. \tag{134}$$

Hence,

$$a_a(t) = a_{a0}t^{p_a}\left(t + \sqrt{\frac{8}{3}\frac{\bar{\beta}}{\rho_0}}\right)^{\frac{2}{3}-p_a}. \tag{135}$$

Thus, at $t \ll \sqrt{\frac{8}{3}\frac{\bar{\beta}}{\rho_0}}$ the universe behaves as a Kasner universe and at $t \gg \sqrt{\frac{8}{3}\frac{\bar{\beta}}{\rho_0}}$ the universe behaves as a flat Friedmann universe, filled with dust. That means that the phenomenon of the isotropization takes place.

6 Bianchi-II Universe

6.1 Geometry of the Bianchi-II Universe

The Bianchi-II cosmological model is the simplest of the models, where the non-vanishing commutators amongst Killing vector fields are present. The Lie algebra of the Killing vector fields is

$$[X_2, X_3] = X_1,$$
$$[X_1, X_2] = 0,$$
$$[X_3, X_1] = 0. \tag{136}$$

We can construct the Killing vector fields as

$$X_1 = \frac{\partial}{\partial z},$$
$$X_2 = \frac{\partial}{\partial y} - x\frac{\partial}{\partial z},$$
$$X_3 = \frac{\partial}{\partial x}. \tag{137}$$

The vector fields which constitute a reciprocal basis can be chosen as

$$e_1 = \frac{\partial}{\partial z},$$
$$e_2 = \frac{\partial}{\partial y},$$
$$e_3 = \frac{\partial}{\partial x} - y\frac{\partial}{\partial z}. \tag{138}$$

The dual basis of the one-forms is

$$\omega^1 = dz + ydx,$$
$$\omega^2 = dy,$$
$$\omega^3 = dx. \tag{139}$$

The metric of the Bianchi-II universe is

$$dt^2 - a^2(t)(\omega^1)^2 - b^2(t)(\omega^2)^2 - c^2(t)(\omega^3)^2. \tag{140}$$

It is convenient to introduce a new time parameter τ:

$$dt = a(t)b(t)c(t)d\tau, \tag{141}$$

and to represent the scale factors as

$$a = e^{\alpha(\tau)},$$
$$b = e^{\beta(\tau)},$$
$$c = e^{\gamma(\tau)}. \tag{142}$$

6.2 Solutions of the Einstein Equations for an Empty Bianchi-II Universe

Remarkably, it is possible to find the general solution of the Einstein equations for an empty Bianchi-II universe. This solution is particularly interesting because, starting from it one can analyze the behaviour of the more complicated Bianchi-IX universe in the vicinity of the cosmological singularity and to come to the description of the oscillatory approach to this singularity. We shall describe the construction of the solution for an empty Bianchi-II universe, following the presentation given in paper [40].

The spatial components of the Einstein equations in the empty spacetime are now

$$\ddot{\alpha} = -\frac{1}{2}e^{4\alpha}, \tag{143}$$

$$\ddot{\beta} = \frac{1}{2}e^{4\alpha}, \tag{144}$$

$$\ddot{\gamma} = \frac{1}{2}e^{4\alpha}, \tag{145}$$

while the temporal component of the Einstein equations is

$$\ddot{\alpha} + \ddot{\beta} + \ddot{\gamma} = 2(\dot{\alpha}\dot{\beta} + \dot{\alpha}\dot{\gamma} + \dot{\beta}\dot{\gamma}). \tag{146}$$

The first integral of Eq. (226) is

$$\dot{\alpha}^2 = H^2 - \frac{1}{4}e^{4\alpha}, \tag{147}$$

Hence,

$$\dot{\alpha} = \pm\sqrt{H^2 - \frac{1}{4}e^{4\alpha}}. \tag{148}$$

This equation is also integrable and one has

$$\alpha(\tau) = \mp H\tau - \frac{1}{2}\ln(e^{\mp 4H\tau} + 1), \tag{149}$$

where the additional constant of integration is absorbed by a shift of the time parameter.

One can see from Eqs. (226) and (144) that

$$\ddot{\alpha} + \ddot{\beta} = 0. \tag{150}$$

Hence,

$$\dot{\alpha} + \dot{\beta} = B, \tag{151}$$

where B is an integration constant. Using Eqs. (151) and (149) we obtain

$$\beta(\tau) = B\tau + B_0 - \alpha(\tau). \tag{152}$$

We shall omit the constant B_0 because its role is reduced to the constant rescaling of the factor $b(t)$.

Analogously,

$$\gamma(\tau) = C\tau + C_0 - \alpha(\tau). \tag{153}$$

Substituting the obtained solutions into the Einstein 00 Eq. (146), we find the following relation between the constants H, B and C:

$$BC = H^2. \tag{154}$$

We can choose the sign "plus" in Eq. (149). Then, when $\tau \to -\infty$ the logarithmic scale factors behave as

$$\alpha(\tau) = H\tau,$$
$$\beta(\tau) = (B - H)\tau,$$
$$\gamma(\tau) = (C - H)\tau. \tag{155}$$

The spatial volume of the universe behaves as

$$abc = \exp(\alpha + \beta + \gamma) = \exp[(B + C - H)\tau]. \tag{156}$$

When $\tau \to \infty$,

$$\alpha(\tau) = -H\tau,$$
$$\beta(\tau) = (B + H)\tau,$$
$$\gamma(\tau) = (C + H)\tau. \tag{157}$$

The spatial volume of the universe is

$$abc = \exp(\alpha + \beta + \gamma) = \exp[(B + C + H)\tau]. \tag{158}$$

Choosing all three constants H, A and B positive we see that at the moment $\tau \to -\infty$ the universe is closer to the singularity, than at the moment $\tau \to \infty$. At the moment, corresponding to $\tau \to \infty$,

$$a(t) \sim t^{-\frac{H}{B+C+H}} = t^{p_1},$$
$$b(t) \sim t^{\frac{B+H}{B+C+H}} = t^{p_2},$$
$$c(t) \sim t^{\frac{C+H}{B+C+H}} = t^{p_3}. \tag{159}$$

Introducing the Lifshitz–Khalatnikov parameter $u = H/B$ and taking into account the relation $H^2 = BC$ we see that the Kasner indices have the standard parametrization

$$p_1 = -\frac{u}{1 + u + u^2},$$
$$p_2 = \frac{1 + u}{1 + u + u^2},$$
$$p_3 = \frac{u(1 + u)}{1 + u + u^2}. \tag{160}$$

When the universe moves closer to the singularity $\tau \to -\infty$ we have

$$a(t) \sim t^{\frac{H}{B+C-H}} = t^{p'_1},$$
$$b(t) \sim t^{\frac{B-H}{B+C-H}} = t^{p'_2},$$
$$c(t) \sim t^{\frac{C-H}{B+C-H}} = t^{p'_3}. \tag{161}$$

Here the Kasner indices are given by

$$p_1' = \frac{u}{1 - u + u^2} = p_2(u - 1),$$

$$p_2' = \frac{1 - u}{1 - u + u^2} = p_1(u - 1),$$

$$p_3' = \frac{u(u - 1)}{1 - u + u^2} = p_3(u - 1). \tag{162}$$

Thus, we have reproduced the known law of the transition from one Kasner epoch to another [19–22].

We have considered the situation when at $\tau \to \infty$ the Kasner index corresponding to the scale factor a, which stays in the numerator of the potential term, has a sign different from that of the two other scale factors b and c, which stay in the denominator of the potential term. We shall consider now also the situation when at $\tau \to \infty$ the Kasner index of the scale factor a is positive while the Kasner index of the scale factor b is negative. It can be realized if the integration constant H is negative: $H = -|H|$, while the constant B is positive and $B < |H|$. It is convenient to introduce the parameter u as follows:

$$u = \frac{|H|}{B} - 1. \tag{163}$$

Then

$$a(t) \sim t^{\frac{|H|}{B + C - |H|}} = t^{p_2},$$

$$b(t) \sim t^{\frac{B - |H|}{B + C - |H|}} = t^{p_1},$$

$$c(t) \sim t^{\frac{C - |H|}{B + C - |H|}} = t^{p_3}. \tag{164}$$

When $\tau \to -\infty$, i.e. when the universe is closer to the singularity, we have

$$a(t) \sim t^{\frac{-|H|}{B + C + |H|}} = t^{p_2'},$$

$$b(t) \sim t^{\frac{B + |H|}{B + C + |H|}} = t^{p_1'},$$

$$c(t) \sim t^{\frac{C + |H|}{B + C + |H|}} = t^{p_3'}. \tag{165}$$

Now the relations between the values of the Kasner indices at the asymptotic regimes are different and are given by

$$p_1' = \frac{(u + 1) + 1}{(u + 1)^2 + (u + 1) + 1} = p_2(u + 1),$$

$$p_2' = \frac{-(u+1)}{(u+1)^2 + (u+1) + 1} = p_1(u+1),$$

$$p_3' = \frac{(u+1)[(u+1)+1]}{(u+1)^2 + (u+1) + 1} = p_3(u+1). \tag{166}$$

In this case, instead of the standard shift of the Lifshitz–Khalatnikov parameter $u \to u - 1$, one has the shift $u \to u + 1$.

7 Bianchi-IX Universe, the Oscillatory Approach to the Singularity (BKL) and the Mixmaster Universe

7.1 Geometry of the Bianchi-IX Universe

The Bianchi-IX Lie algebra has the form

$$[X_1, X_2] = X_3,$$
$$[X_2, X_3] = X_1,$$
$$[X_3, X_1] = X_2. \tag{167}$$

This is nothing but the algebra Lie of the Lie group $SU(2)$.

The elements of the $SU(2)$ group can be represented as

$$g = x^0 \begin{pmatrix} 1 & 0 \\ 0 & 1 \end{pmatrix} + x^1 \begin{pmatrix} 0 & 1 \\ 1 & 0 \end{pmatrix} + x^2 \begin{pmatrix} 0 & -i \\ i & 0 \end{pmatrix} + x^3 \begin{pmatrix} 1 & 0 \\ 0 & -1 \end{pmatrix}, \tag{168}$$

where

$$(x^0)^2 + (x^1)^2 + (x^2)^2 + (x^3)^2 = 1, \tag{169}$$

i.e. the group manifold of $SU(2)$ is the three-dimensional sphere of the unit radius, embedded into the four-dimensional Euclidean space $\mathbb{R}^4$.

We can choose the three Killing vectors in the space $\mathbb{R}^4$ as follows:

$$X_1 = \frac{1}{2} \left(x^0 \frac{\partial}{\partial x^1} - x^1 \frac{\partial}{\partial x^0} - x^2 \frac{\partial}{\partial x^3} + x^3 \frac{\partial}{\partial x^2} \right),$$

$$X_2 = \frac{1}{2} \left(x^0 \frac{\partial}{\partial x^2} - x^2 \frac{\partial}{\partial x^0} + x^1 \frac{\partial}{\partial x^3} - x^3 \frac{\partial}{\partial x^1} \right),$$

$$X_3 = \frac{1}{2} \left(x^0 \frac{\partial}{\partial x^3} - x^3 \frac{\partial}{\partial x^0} - x^1 \frac{\partial}{\partial x^2} + x^2 \frac{\partial}{\partial x^1} \right). \tag{170}$$

It is easy to check that these three vector fields constitute the Bianchi-IX Lie algebra in $\mathbb{R}^4$ and being restricted on the sphere $\mathbb{S}^3$ their commutation relations as those from Eq. (167) as well. The triple of the reciprocal vector fields is

$$
\begin{aligned}
e_1 &= \frac{1}{2}\left(x^0\frac{\partial}{\partial x^1} - x^1\frac{\partial}{\partial x^0} + x^2\frac{\partial}{\partial x^3} - x^3\frac{\partial}{\partial x^2}\right), \\
e_2 &= \frac{1}{2}\left(x^0\frac{\partial}{\partial x^2} - x^2\frac{\partial}{\partial x^0} - x^1\frac{\partial}{\partial x^3} + x^3\frac{\partial}{\partial x^1}\right), \\
e_3 &= \frac{1}{2}\left(x^0\frac{\partial}{\partial x^3} - x^3\frac{\partial}{\partial x^0} + x^1\frac{\partial}{\partial x^2} - x^2\frac{\partial}{\partial x^1}\right).
\end{aligned}
\tag{171}
$$

We shall use the Hopf parametrization for the three-dimensional unit sphere:

$$
\begin{aligned}
x^0 &= \cos\chi\cos\phi, \\
x^1 &= \cos\chi\sin\phi, \\
x^2 &= \sin\chi\cos\psi, \\
x^3 &= \sin\chi\sin\psi, \\
0 &\le \phi < 2\pi,\ 0 \le \psi < 2\pi,\ 0 \le \chi \le \frac{\pi}{2}.
\end{aligned}
\tag{172}
$$

The fields e_i from Eq. (171) being restricted on the three-sphere $\mathbb{S}^3$ have the form:

$$
e_1 = \frac{1}{2}\left(\frac{\partial}{\partial\phi} + \frac{\partial}{\partial\psi}\right),
$$

$$
\begin{aligned}
e_2 = \frac{1}{2}\Bigg(&\tan\chi\sin(\phi + \psi)\frac{\partial}{\partial\phi} - \cot\chi\sin(\phi + \psi)\frac{\partial}{\partial\psi} \\
&+ \cos(\phi + \psi)\frac{\partial}{\partial\chi}\Bigg),
\end{aligned}
\tag{173}
$$

$$
\begin{aligned}
e_3 = \frac{1}{2}\Bigg(&-\tan\chi\cos(\phi + \psi)\frac{\partial}{\partial\phi} + \cot\chi\cos(\phi + \psi)\frac{\partial}{\partial\psi} \tag{174} \\
&+ \sin(\phi + \psi)\frac{\partial}{\partial\chi}\Bigg).
\end{aligned}
\tag{175}
$$

The dual basis of one-forms is

$$
\begin{aligned}
\omega^1 &= 2(\cos^2\chi\, d\phi + \sin^2\chi\, d\psi), \\
\omega^2 &= \sin 2\chi\sin(\phi + \psi)d\phi - \sin 2\chi\sin(\phi + \psi)d\psi \\
&\quad + 2\cos(\phi + \psi)d\chi,
\end{aligned}
\tag{176}
$$

$$\omega^3 = -\sin 2\chi \cos(\phi + \psi)d\phi + \sin 2\chi \cos(\phi + \psi)d\psi$$
$$+2\sin(\phi + \psi)d\chi. \tag{177}$$

Using these forms we can construct the metric of a Bianchi-IX universe.
    It is easy to check that

$$\omega^1 \otimes \omega^1 + \omega^2 \otimes \omega^2 + \omega^3 \otimes \omega^3 \tag{178}$$

is the metric of the three-dimensional sphere with the radius equal to 2.

7.2 Dynamics of an Empty Bianchi-IX Universe: Oscillatory Approach to the Cosmological Singularity—BKL

We have seen that the presence of the term $\frac{a^2}{b^2c^2}$ in the spatial curvature of the Bianchi-II universe, where $a \to \infty$, while $b, c \to 0$ when the universe tends to the singularity, induces the transition from one Kasner regime (epoch) to another. As a result, the functions $a(t)$ and $b(t)$ exchange their roles, while the parameter undergoes a shift:

$$u \to u - 1. \tag{179}$$

The term $\frac{a^2}{b^2c^2}$ becomes small and cannot influence the dynamics of the universe. Thus, all the evolutions of the Bianchi-II universe can be seen as a transition from one asymptotic Kasner regime to another.
    However, in the Bianchi-IX universe three terms are present in the three-dimensional scalar curvature, which plays the role of a potential:

$$\frac{a^2}{b^2c^2}, \quad \frac{b^2}{a^2c^2}, \quad \text{and} \quad \frac{c^2}{a^2b^2}. \tag{180}$$

When the term $\frac{a^2}{b^2c^2}$ becomes small, the term $\frac{b^2}{a^2c^2}$ becomes large.
    It implies the next change of the Kasner epoch (a and b exchange their roles) and again

$$u \to u - 1. \tag{181}$$

If we start from some value

$$u > 1 \tag{182}$$

then we have $[u]$ changes of the Kasner epochs, where $[u]$ is an integer part of the number u.

After these $[u]$ changes of the Kasner epochs, when the scale factors a and b exchange their roles, the universe arrives to the regime when the parameter u becomes smaller than one:

$$u < 1. \tag{183}$$

That means that

$$p_2 > p_3$$

and the functions $b(t)$ and $c(t)$ exchange their roles.

It is called the change of the Kasner era and can be described by the transformation

$$u \rightarrow \frac{1}{u}. \tag{184}$$

The number $\frac{1}{u} > 1$ and the new series of the changes of the Kasner epochs inside the new Kasner era begin. All this is called "Oscillatory approach to the cosmological singularity" or "BKL" (Belinski–Khalatnikov–Lifshitz).

7.3 A Particular Case: u is a Rational Number

If

$$u = \frac{p}{q}, \quad \text{a rational number,} \tag{185}$$

then the process of the changes of Kasner epochs and eras will bring us to the value of the Lifshitz–Khalatnikov parameter

$$u = 1 \tag{186}$$

and, then to

$$u = 0.$$

It means that

$$p_1 = 0, \tag{187}$$

$$p_2 = 1, \tag{188}$$

$$p_3 = 0. \tag{189}$$

Hence, the metric is

$$ds^2 = dt^2 - t^2 dx^2 - dy^2 - dz^2. \tag{190}$$

What kind of spacetime describes this metric? It is simply a disguised Minkowski spacetime.

Indeed, if we take the Minkowski spacetime with metric

$$ds^2 = dT^2 - d\xi^2 - dy^2 - dz^2 \tag{191}$$

and make the change of variables:

$$T = t \cosh x,$$

$$\xi = t \sinh x, \tag{192}$$

we shall obtain it.

However, the rational numbers are the measure zero subset of the set of real numbers. Thus, a typical evolution of the Bianchi-IX universe does not bring it to the Minkowski spacetime.

7.4 Mixmaster Universe

For the description of the behaviour of the Bianchi-IX universe in the vicinity of the cosmological singularity, one can use the Hamiltonian formalism as was done by Misner [24]. In this formalism it is convenient to parametrize the scale factors as

$$a = e^{\Omega + \frac{\beta_+}{2} + \frac{\sqrt{3}}{2}\beta_-},$$

$$b = e^{\Omega + \frac{\beta_+}{2} - \frac{\sqrt{3}}{2}\beta_-},$$

$$c = e^{\Omega - \beta_+}. \tag{193}$$

Introducing the conjugate momenta, one arrives to the gauge-fixed Hamiltonian

$$H = \frac{1}{2}\left(p_+^2 + p_-^2 - \frac{1}{4}p_\Omega^2\right) + V(\beta_+, \beta_-, \Omega). \tag{194}$$

If we consider a Bianchi-II universe, then the potential V is

$$V = V_0 e^{4(\Omega - \beta_+)}. \tag{195}$$

In the case of the Bianchi-IX universe the structure of the potential is more complicated and the dynamics of the system can be represented as a motion of a ball in a billiard with moving walls. The billiard is two-dimensional (β_+, β_-) and the parameter Ω plays the role of a time variable. The dependence of the potential on the Ω means that the walls of the billiard are moving.

A bounce of the ball from one of the walls is the change of the Kasner regime. The relations connecting the Kasner indices with the conjugate momenta:

$$p_1 = \frac{p_\Omega + 4p_+}{3p_\Omega},$$

$$p_2 = \frac{p_\Omega - 2p_+ - 2\sqrt{3}p_-}{3p_\Omega},$$

$$p_3 = \frac{p_\Omega - 2p_+ + 2\sqrt{3}p_-}{3p_\Omega}. \tag{196}$$

Inversely,

$$\frac{p_+}{p_\Omega} = \frac{3p_1 - 1}{4},$$

$$\frac{p_-}{p_\Omega} = \frac{\sqrt{3}(p_3 - p_2)}{4}. \tag{197}$$

The Kasner indices satisfy the condition $p_1^2 + p_2^2 + p_3^2 = 1$ due to the Hamiltonian constraint

$$p_\Omega^2 = 4(p_+^2 + p_-^2),$$

which in this form is valid far away from the wall, where the potential is negligible.

Let us introduce now the Misner parametrization of the Kasner indices:

$$p_1 = -\frac{(s-3)(s+3)}{3(s^2+3)},$$

$$p_2 = \frac{2s(s-3)}{3(s^2+3)},$$

$$p_3 = \frac{2s(s+3)}{3(s^2+3)}. \tag{198}$$

We can make a change of variables, eliminating the dependence of the potential on the spatial volume Ω:

$$\bar{\beta}_+ = \frac{1}{\sqrt{3}}(\beta_+ - \Omega),$$

$$\bar{\Omega} = \frac{1}{2\sqrt{3}}(4\Omega - \beta_+). \tag{199}$$

$$\bar{p}_+ = \frac{1}{\sqrt{3}}(4p_+ + p_\Omega),$$

$$\bar{p}_\Omega = \frac{2}{\sqrt{3}}(p_+ + p_\Omega). \tag{200}$$

Now, the Hamiltonian looks as

$$H = \frac{1}{2}\left(\frac{1}{4}\bar{p}_+^2 + p_-^2 - \frac{1}{4}\bar{p}_\Omega^2\right) + V_0 e^{-4\sqrt{3}\bar{\beta}_+}. \tag{201}$$

The conjugate momenta p_- and $\bar{p}_\Omega$ are constant because the potential does not depend on β_- and $\bar{\Omega}$.

Then,

$$\frac{\bar{p}_\Omega}{p_-} = \frac{2}{\sqrt{3}}\frac{\frac{p_+}{p_\Omega} + 1}{\frac{p_-}{p_\Omega}} = const. \tag{202}$$

This relation connect the values of the momenta at two Kasner regimes. It can be rewritten as

$$\frac{s}{3} + \frac{3}{s} = const. \tag{203}$$

The only change of the variable s, leaving the left-hand side of this relation intact is

$$s \to \frac{9}{s}. \tag{204}$$

The change of the Kasner regime described in terms of the Misner parameter s coincides with that, described in terms of the Lifshitz–Khalatnikov parameter u.

8   Chaos in Cosmology

8.1   Lengths of the Kasner eras and Stochasticity

We have seen that the evolution of the universe towards a singular point consists of successive periods (called eras) in which distances along two axes oscillate and along the third axis decrease monotonically, the volume decreases according to a law which is near to $\sim t$. In the transition from one era to another, the axes along which the distances decrease monotonically are interchanged. The order in which the pairs of axes are interchanged and the order in which eras of different lengths follow each other acquire a stochastic character.

To every sth era corresponds a decreasing sequence of values of the parameter u. This sequence has the form

$$u_{max}^{(s)}, u_{max}^{(s)} - 1, \ldots, u_{min}^{(s)},$$

where

$$u_{min}^{(s)} < 1.$$

Let us introduce the following notation:

$$u_{min}^{(s)} = x^{(s)}, \quad u_{max}^{(s)} = k^{(s)} + x^{(s)} \tag{205}$$

i.e.

$$k^{(s)} = [u_{max}^{(s)}].$$

The number $k^{(s)}$ defines the eras length.

For the next era we obtain

$$u_{max}^{(s+1)} = \frac{1}{x^{(s)}}, \quad k^{(s+1)} = \left[\frac{1}{x^{(s)}}\right]. \tag{206}$$

The ordering with respect to the length of $k^{(s)}$ of the successive eras (measured by the number of Kasner epochs contained in them) acquires asymptotically a stochastic character.

The random nature of this process arises because of the rules which define the transitions from one era to another in the infinite sequence of values of u. If all this infinite sequence begins since some initial value

$$u_{max}^{(0)} = k^{(0)} + x^{(0)},$$

then the lengths of series

$$k^{(0)}, k^{(1)}, \ldots$$

are numbers included into an expansion of a continued fraction (see, e.g., [41]):

$$k^{(0)} + x^{(0)} = k^{(0)} + \cfrac{1}{k^{(1)} + \cfrac{1}{k^{(2)} + \cdots}}. \tag{207}$$

We can describe statistically this sequence of eras if we consider instead of a given initial value

$$u_{max}^{(0)} = k^{(0)} + x^{(0)}$$

a distribution of $x^{(0)}$ over the interval $(0, 1)$ governed by some probability law [25].

Then we also obtain some distributions of the values of $x^{(s)}$ which terminate every sth series of numbers. It can be shown that with increasing s, these distributions tend to a stationary (independent of s) probability distribution $w(x)$ in which the initial value $x^{(s)}$ is completely "forgotten":

$$w(x) = \frac{1}{(1+x)\ln 2}. \tag{208}$$

$$W(k) = \frac{1}{\ln 2} \ln \frac{(k+1)^2}{k(k+2)}. \tag{209}$$

The source of stochasticity arising at the oscillatory approach to the cosmological singularity can be described in such terms: the transition from one Kasner era to another is described by the Gauss transformation of the interval $[0, 1]$ into itself by the formula

$$Tx = \left\{ \frac{1}{x} \right\}, \quad \text{i.e., } x_{s+1} = \left\{ \frac{1}{x_s} \right\}, \tag{210}$$

where $\{x\}$ is a fractional part of the number x. This transformation is expanding and possesses the property of exponential instability.

It is not a one-to-one transformation. Its inverse is not unique. In other words, fixing the value of the parameter u we can predict the evolution towards singularity, but we cannot describe the past.

8.2 Proof of the Formula for the Stationary Probability Distribution of the Parameter $x = \{u\}$

Let us suppose that there is a real number which stays in the interval between x and $x + \Delta x$ such that $[x, x + \Delta x] \subset [0, 1]$. Then the Gauss transformation (210) transforms this interval into the interval, which length is

$$\Delta x \rightarrow \frac{\Delta x}{x(x+\Delta x)} \sim \frac{\Delta x}{x^2}, \tag{211}$$

provided

$$\frac{\Delta x}{x^2} < 1.$$

Let us suppose that the $(s + 1)$th era is characterized by some number x. What can we say about the preceding era? As we have already mentioned the Gauss transformation is not a one-to-one transformation and every number has an infinite set of the inverse images. Namely, to the preceding sth era correspond the following numbers:

$$\frac{1}{x + n}, \quad n = 1, 2, \cdots \tag{212}$$

Taking into account the "dilatation" of the interval between x and $x + \Delta x$ during the Gauss transformation (211), we can write down the formula expressing the probability distribution $w_{s+1}(x)$ by means of the preceding probability expansion $w_s(x)$:

$$w_{s+1}(x) = \sum_{n=1}^{\infty} w_s \left(\frac{1}{x + n} \right) \frac{1}{(x + n)^2}. \tag{213}$$

Let us suppose that this distribution after some number of Gauss transformations becomes stable

$$w_{s+1}(x) = w_s(x) = w(x). \tag{214}$$

The equation for this stable probability distribution is

$$w(x) = \sum_{n=1}^{\infty} w \left(\frac{1}{x + n} \right) \frac{1}{(x + n)^2}. \tag{215}$$

Let us look for the solution of Eq. (215) in the form

$$w(x) \sim \frac{1}{1 + x}. \tag{216}$$

Substituting the expression (216) into Eq. (215), we see that the latter is satisfied. Then taking into account the condition

$$\int_0^1 w(x)dx = 1,$$

we come to the formula (208).

9   Mixmaster Universes and Infinite-Dimensional Lie Algebras

9.1   Presence of Matter

This section is devoted to further developments of the investigations, concerning the behaviour of the universe in the vicinity of the cosmological singularity. We shall begin with a relatively simple question about of the influence of the presence of matter on the BKL dynamics [42].

If one considers the universe filled with a perfect fluid with the equation of state

$$p = w\rho, \; w < 1, \; w = \text{constant},$$

then the presence of this matter cannot change the dynamics in the vicinity of the singularity.

$$\rho = \frac{\rho_0}{(abc)^{w+1}} = \frac{\rho_0}{t^{w+1}}. \tag{217}$$

This term is weaker than the terms of the geometrical origin coming from the time derivatives of the metric, which behave like $1/t^2$, let alone the perturbations due to the presence of spatial curvature, responsible for changes of a Kasner regime, which behave like $1/t^{2+4|p_1|}$. The situation changes, if the parameter $w = 1$. Such kind of matter is called "stiff matter" and can be represented by a massless scalar field. In this case $\rho \sim 1/t^2$ and the contribution of matter is of the same order as main terms of geometrical origin. The Kasner indices satisfy now the relations

$$p_1 + p_2 + p_3 = 1, \;\; p_1^2 + p_2^2 + p_3^2 = 1 - q^2, \tag{218}$$

where the number q^2 reflects the presence of the stiff matter and

$$q^2 \leq \frac{2}{3}. \tag{219}$$

If $q^2 > 0$, then there exist combinations of three positive Kasner indices, satisfying the above relations.

If $q^2 \geq \frac{1}{2}$ then only sets of three positive Kasner indices are acceptable. If a universe finds itself in a Kasner regime with three positive indices, the perturbative terms, existing due to the spatial curvatures, are too weak to change this Kasner regime, and thus, it becomes stable. In the presence of the stiff matter, the universe after a finite number of changes of Kasner regimes finds itself in a stable regime and oscillations stop. The massless scalar field plays "anti-chaotizing" role in the process of the cosmological evolution.

One can show that after one bounce (change of the Kasner regime) the parameter q^2 changes as follows:

$$q^2 \to q'^2 = q^2 \times \frac{1}{(1+2p_1)^2} > q^2. \tag{220}$$

Thus, the value of the parameter q^2 grows and the probability to find all the three Kasner indices to be positive increases.

9.2 Multidimensional Cosmology

In this subsection we shall consider the so-called multidimensional cosmologies, where the number of the spatial dimensions is greater than 3 [43–45].

First of all consider a multidimensional analog of a Bianchi-I universe for which exist a generalized Kasner solution with the metric

$$ds^2 = dt^2 - \sum_{i=1}^{d} t^{2p_i} dx^{i2}, \tag{221}$$

where the generalized Kasner indices satisfy the relations

$$\sum_{i=1}^{d} p_i = \sum_{i=1}^{d} p_i^2 = 1. \tag{222}$$

Just like in the three-dimensional case, in the presence of spatial curvature terms the transition from one Kasner epoch to another occurs:

$$p_1', p_2', \ldots, p_d' = \text{ordering of } (q_1, q_2, \ldots, q_d), \tag{223}$$

Here

$$q_1 = \frac{-p_1 - P}{1 + 2p_1 + P}, \quad q_2 = \frac{p_2}{1 + 2p_1 + P}, \ldots,$$

$$q_{d-2} = \frac{p_{d-2}}{1 + 2p_1 + P}, \quad q_{d-1} = \frac{2p_1 + P + p_{d-1}}{1 + 2p_1 + P},$$

$$q_d = \frac{2p_1 + P + p_d}{1 + 2p_1 + P}, \tag{224}$$

$$P = \sum_{i=2}^{d-2} p_i. \tag{225}$$

However, such a transition from one Kasner epoch to another occurs if at least one of the numbers α_{ijk} is negative. These numbers are defined as

$$\alpha_{ijk} \equiv 2p_i + \sum_{l \neq j,k,i} p_l, \ (i \neq j, i \neq k, j \neq k). \tag{226}$$

For the spacetimes with $d < 10$ one of the factors α is always negative and one change of Kasner regime is followed by another one, implying the oscillatory behaviour of the universe in the neighbourhood of the cosmological singularity.

For the spacetimes with $d \geq 10$ exist such combinations of Kasner indices, for which all the numbers α_{ijk} are positive. If a universe enters into the Kasner regime with such indices, (so-called Kasner stability region) its chaotic behaviour disappears and this Kasner regime conserves itself.

The discovery of the fact that the chaotic character of the approach to the cosmological singularity disappears in the spacetimes with $d \geq 10$ was unexpected and looked as an accidental result of a game between real numbers satisfying the generalized Kasner relations.

Later it became clear that behind this fact there is a deep mathematical structure, namely, the hyperbolic Kac–Moody algebras.

9.3 Infinite-Dimensional Lie Algebras

As is well-known the theory of finite-dimensional Lie algebras was developed at the end of XIX century in works by Killing and Cartan. The same Cartan has begun studying infinite-dimensional Lie algebras and has given the classification of the simple infinite-dimensional Lie algebras of vector fields on the finite-dimensional spaces. The general theory of the infinite-dimensional Lie groups and algebras do not yet exist. However, there are four classes of the infinite-dimensional Lie groups and Lie algebras which are under investigation now. First class includes the Lie algebras of vector fields and the corresponding groups of diffeomorphisms of manifolds. The second class includes Lie groups of smooth maps from the manifold under consideration to the finite-dimensional Lie group. The third class consists of the classical Lie groups and Lie algebras of the operators acting on the Hilbert or Banach spaces. Finally, the fourth class is the class of the Kac–Moody algebras [27], which we are interested in here.

Let us give some basic definitions.

Every Lie algebra is defined by its generators $h_i, e_i, f_i, i = 1, \ldots, r$, where r is the rank of the Lie algebra, i.e. the maximal number of its generators h_i which commute each other (these generators constitute the Cartan subalgebra). The commutation relations between the generators are

$$[e_i, f_j] = \delta_{ij} h_i,$$
$$[h_i, e_j] = A_{ij} e_j,$$
$$[h_i, f_j] = -A_{ij} f_j,$$
$$[h_i, h_j] = 0. \tag{227}$$

The coefficients A_{ij} constitute the generalized Cartan $r \times r$ matrix such that $A_{ii} = 2$, its off-diagonal elements are non-positive integers and $A_{ij} = 0$ for $i \neq j$ implies $A_{ji} = 0$.

One can say that the e_i are rising operators, similar to well-known operator $L_+ = L_x + i L_y$ in the theory of angular momentum, while f_i are lowering operators like $L_- = L_x - i L_y$. The generators h_i of the Cartan subalgebra could be compared with the operator L_z. The generators should also obey the Serre's relations

$$(\text{ad } e_i)^{1 - A_{ij}} e_j = 0,$$
$$(\text{ad } f_i)^{1 - A_{ij}} f_j = 0, \tag{228}$$

where $(\text{ad} A) B \equiv [A, B]$.

The Lie algebras $\mathcal{G}(A)$ constructed on a symmetrizable Cartan matrix A have been classified according to the properties of their eigenvalues:

if A is positive definite, $\mathcal{G}(A)$ is a finite-dimensional Lie algebra;

if A admits one null eigenvalue and the others are all strictly positive, $\mathcal{G}(A)$ is an affine Kac–Moody algebra; if A admits one negative eigenvalue and all the others are strictly positive, $\mathcal{G}(A)$ is a Lorentz KM algebra.

There exists a correspondence between the structure of a Lie algebra and a certain system of vectors in the r-dimensional Euclidean space, which simplifies the classification of the Lie algebras.

These vectors called roots represent the rising and lowering operators of the Lie algebra.

The vectors corresponding to the generators e_i and f_i are called simple roots.

The system of simple positive roots (i.e. the roots, corresponding to the rising generators e_i) can be represented by nodes of their Dynkin diagrams, while the edges connecting (or non-connecting) the nodes give an information about the angles between simple positive root vectors.

An important subclass of Lorentz KM algebras can be defined as follows: A KM algebra such that the deletion of one node from its Dynkin diagram gives a sum of finite or affine algebras is called an hyperbolic KM algebra. These algebras are all known. In particular, there exists no hyperbolic algebra with a rank higher than 10.

Another important notion is the Weyl group.

The reflections with respect to hyperplanes orthogonal to simple roots leave the systems of roots invariant.

The corresponding finite-dimensional group is called Weyl group.

The hyperplanes mentioned above divide the r-dimensional Euclidean space into regions called Weyl chambers. The Weyl group transforms one Weyl chamber into another.

9.4 Connection Between the Hyperbolic Lie Algebras and the Chaos in the Cosmological Models

Now we are in a position to discuss the connection between the cosmological models (cosmological billiards) and infinite-dimensional Lie algebras, discovered in [28].

The links between the billiards describing the evolution of the universe in the neighbourhood of singularity and its corresponding Kac–Moody algebra can be described as follows:

- the Kasner indices describing the "free" motion of the universe between the reflections from the wall correspond to the elements of the Cartan subalgebra of the KM algebra;
- the dominant walls, i.e. the terms in the equations of motion responsible for the transition from one Kasner epoch to another, correspond to the simple roots of the KM algebra;
- the group of reflections in the cosmological billiard is the Weyl group of the KM algebra;
  the billiard table can be identified with the Weyl chamber of the KM algebra.

One can imagine two types of billiard tables:

- infinite such where the linear motion without collisions with walls is possible (non-chaotic regime) and
- those where reflections from walls are inevitable and the regime can be only chaotic.
- Remarkably, the Weyl chambers of the hyperbolic KM algebras are designed in such a way that infinite repeating collisions with walls occur. It was shown that all the theories with the oscillating approach to the singularity such as Einstein theory in dimensions $d < 10$ and superstring cosmological models correspond to hyperbolic KM algebras.

10 Concluding Remarks

The study of the homogenous and anisotropic Bianchi cosmological models appeared to be very fruitful from the mathematical point of view. It opens also some interesting perspectives for direct physical applications. The phenomenon of the oscillatory approach to the cosmological singularity in the Mixmaster universe looks especially attractive. There is a broad activity in this field. Here, in the concluding remarks we shall mention only some works and directions without pretending to be complete.

The existence of links between the BKL approach to the singularities and the structure of some infinite-dimensional Lie algebras has inspired some authors [31] to declare a new program of development of quantum gravity and cosmology. They propose "to take seriously the idea that near the singularity (i.e. when the curvature gets larger than the Planck scale) the description of a spatial continuum and spacetime based (quantum) field theory breaks down, and should be replaced by a much more abstract Lie algebraic description. Thereby the information previously encoded in the spatial variation of the geometry and of the matter fields gets transferred to an infinite tower of Lie-algebraic variables depending only on 'time'. In other words we are led to the conclusion that space–and thus, upon quantization also spacetime—actually disappears (or "de-emerges") as the singularity is approached."

In papers [46, 47] the direct inclusion of fermions (which are unnecessary element in the cosmological models, based on the superstring and supergravity models) was studied in detail.

In paper [48] the cosmological billiards in multidimensional models, mimicking some supergravity models and including a huge amount of branes, i.e. antisymmetric covariant tensor field, were studied.

A broad review of rigorous results obtained in the description of the Mixmaster models was presented in paper [49].

There has been a long debate about the existence of chaos in the Mixmaster cosmological model. It was connected with the fact that some standard chaotic indicators like Lyapunov exponents are coordinate dependent. In papers [50, 51] the coordinate-independent methods, using such notions as fractals and topological entropy, the authors have shown that the Mixmaster universe is indeed chaotic.

In some recent works quantum cosmology of the Mixmaster type models was also considered. In paper [52] it was stated that 1-epoch BKL-eras were the most probable configuration at which wave functions have to be evaluated, while BKL eras containing $n \gg 1$ epochs are shown to be a less probable configuration. In paper [53] the singularity avoidance mechanism in a quantum model of the Mixmaster universe was studied. The authors of [53] used a compound quantization procedure: an affine coherent state quantization for isotropic variables and a Weyl quantization for anisotropic ones. The resolution of the classical singularity occurred due to a repulsive potential generated by the affine quantization. This procedure showed that during contraction towards singularity the quantum energy of anisotropic degrees of freedom grows much slower than the classical one.

Finally, let us mention the study of the Mixmaster type models in the Hořava-Lifshitz gravity [54]. As is well-known the quantum gravity is non-renormalizable theory. In paper [55] it was suggested a generalization of the General Relativity, based on the Lagrangians, which are quadratic in external curvature, but include the three-dimensional curvature higher-order terms. Such models can be renormalizable but loose the Lorentz-invariance because the temporal and spatial coordinates are not treated on equal footing. As a by-product of more complicated structure of the equations of motion some additional scenarios can arise in cosmological models inspired by the Hořava-Lifshitz gravity (see, e.g., [56, 57]). The question of the chaoticity of the cosmological models of the Hořava-Lifshitz type is not resolved definitely, but one can understand that the laws of the oscillatory approach to the singularity in this models differ from those which one encounters in the General Relativity.

Concluding, we would like to express an opinion that the study of the homogeneous, but anisotropic cosmological models reserves a lot of surprises.

References

1. A. Friedman, Über die Krümmung des Raumes. Z. Phys. **10**, 377 (1922)
2. A. Friedman, Über die Möglichkeit einer Welt mit konstanter negativer Krümmung des Raumes. Z. Phys. **21**, 326 (1924)
3. G. Lemaître, Un Univers homogène de masse constante et de rayon croissant rendant compte de la vitesse radiale des nèbuleuses extra-galactiques. Ann. de la Soc. Scien. de Bruxelles **47**, 49 (1927)
4. G. Lemaître, The expanding universe. Mon. Not. Roy. Astron. Soc. **91**, 490 (1931)
5. A.A. Starobinsky, Stochastic De sitter (inflationary) stage in the early universe, in *Field Theory, Quantum Gravity and Strings*, ed. by H.J. DeVega, N. Sanchez (Springer-Verlag, Berlin, 1986)
6. A.D. Linde, *Particle Physics and Inflationary Cosmology* (Harward Academic Publishers, Brighton, 1990)
7. A.G. Riess et al. [Supernova Search Team], Observational evidence from supernovae for an accelerating universe and a cosmological constant. Astron. J. **116**, 1009 (1998)
8. S. Perlmutter et al. [Supernova Cosmology Project Collaboration], Measurements of Omega and Lambda from 42 high redshift supernovae. Astrophys. J. **517** 565 (1999)
9. E.M. Lifshitz, On the gravitational stability of the expanding universe. J. Phys. (USSR) **10**, 116 (1946)
10. V.F. Mukhanov, G.V. Chibisov, Quantum fluctuations and a nonsingular universe. JETP Lett. **33**, 532 (1981)
11. L. Bianchi, Sugli spazi a tre dimensioni che ammettono un gruppo continuo di moviment. Memorie di Matematica e di Fisica della Societa Italiana delle Scienze, Serie Terza **11**, 267 (1898)
12. A. Einstein, the foundation of the general theory of relativity. Annalen Phys. **49**(7), 769 (1916)
13. A.H. Taub, Empty space-times admitting a three parameter group of motions. Annals Math. **53**, 472 (1951)
14. A. Krasinski, C.G. Behr, E. Schucking, F.B. Estabrook, H.D. Wahlquist, G.F.R. Ellis, R. Jantzen, W. Kundt, The Bianchi classification in the Schucking-Behr approach. Gen. Rel. Grav. textbf35, 475 (2003)
15. G.F.R. Ellis, M.A.H. MacCallum, A class of homogeneous cosmological models. Commun. Math. Phys. **12**, 108 (1969)
16. L.D. Landau, E.M. Lifshitz, *The Classical Theory of Fields* (Pergamon Press, Oxford, 1979)
17. M.P. Ryan, L.C. Shepley, *Homogeneous Relativistic Cosmologies* (Princeton University Press, Princeton, 1975)
18. E.M. Lifshitz, I.M. Khalatnikov, Investigations in relativistic cosmology. Adv. Phys. **12**, 185 (1963)
19. V.A. Belinskii, I.M. Khalatnikov, On the nature of the singularities in the general solutions of the gravitational equations. Sov. Phys. JETP **29**(5), 911 (1969)

20. I.M. Khalatnikov, E.M. Lifshitz, General cosmological solution of the gravitational equations with a singularity in time. Phys. Rev. Lett. **24**, 76 (1970)
21. V.A. Belinsky, I.M. Khalatnikov, E.M. Lifshitz, Oscillatory approach to a singular point in the relativistic cosmology. Adv. Phys. **19**, 525 (1970)
22. V.A. Belinsky, I.M. Khalatnikov, E.M. Lifshitz, A general solution of the Einstein equations with a time singularity. Adv. Phys. **31**, 639 (1982)
23. V. Belinski, M. Henneaux, *The Cosmological Singularity* (Cambridge University Press, 2018)
24. C.W. Misner, Mixmaster universe. Phys. Rev. Lett. **22**, 1071 (1969)
25. E.M. Lifshitz, I.M. Lifshitz, I.M. Khalatnikov, Asymptotic analysis of oscillatory mode of approach to a singularity in homogeneous cosmological models. Sov. Phys. JETP **32**(1), 173 (1971)
26. I.M. Khalatnikov, E.M. Lifshitz, K.M. KhaninL, N. Shchur, Y.G. Sinai, On the stochasticity in relativistic cosmology. J. Stat. Phys. **38**, 97 (1985)
27. V.G. Kac, *Infinite Dimensional Lie Algebras* (Cambridge University Press, Cambridge, 1990)
28. T. Damour, M. Henneaux, Oscillatory behavior in homogeneous string cosmology models. Phys. Lett. B **488**108 (2000); Erratum: [Phys. Lett. B **491**, 377 (2000)]
29. T. Damour, M. Henneaux, E(10), BE(10) and arithmetical chaos in superstring cosmology. Phys. Rev. Lett. **86**, 4749 (2001)
30. T. Damour, M. Henneaux, B. Julia, H. Nicolai, Hyperbolic Kac–Moody algebras and chaos in Kaluza–Klein models. Phys. Lett. B **509**, 323 (2001)
31. T. Damour, H. Nicolai, Symmetries, singularities and the de-emergence of space. Int. J. Mod. Phys. D **17**, 525 (2008)
32. S. Kobayashi, K. Nomizu, *Foundations Of Differential Geometry*, vol. 1 (Wiley, Hoboken, 1996)
33. B. Schutz, *Geometrical Methods of Mathematical Physics* (Cambridge University Press, Cambridge, 1999)
34. M. Fecko, *Differential Geometry and Lie Groups for Physicists* (Cambridge University Press, Cambridge, 2011)
35. S. Weinberg, *Gravitation and Cosmology: Principles and Applications of the General Theory of Relativity* (Wiley, Hoboken, 2008)
36. E. Gourgoulhon, *3+1 Formalism in General Relativity: Bases of Numerical Relativity* (Springer, Berlin, 2012)
37. I.M. Khalatnikov, A.Y. Kamenshchik, A generalization of the Heckmann-Schucking cosmological solution. Phys. Lett. B **553**, 119 (2003)
38. E. Kasner, Geometrical theorems on Einstein's cosmological equations. Am. J. Math. **43**, 217 (1921)
39. O. Heckmann, E. Schucking, Newtonsche und Einsteinsche Kosmologie. Handbuch der Physik, **53**, 489 (1959)
40. L. Giani, A.Y. Kamenshchik, Hořava-Lifshitz gravity inspired Bianchi-II cosmology and the mixmaster universe. Class. Quant. Grav. **34**(8), 085007 (2017)
41. A.Ya Khinchin, *Continued Fractions* (Dover, Downers Grove, 1997)
42. V.A. Belinski, I.M. Khalatnikov, Effect of scalar and vector fields on the nature of the cosmological singularity. Sov. Phys. JETP **36**, 591 (1973)
43. J. Demaret, M. Henneaux, P. Spindel, Nonoscillatory behavior in vacuum Kaluza-Klein cosmologies. Phys. Lett. **164B**, 27 (1985)
44. J. Demaret, J.L. Hanquin, M. Henneaux, P. Spindel, A. Taormina, The fate of the mixmaster behavior in vacuum inhomogeneous Kaluza-Klein cosmological models. Phys. Lett. B **175**, 129 (1986)
45. J. Demaret, Y. De Rop, M. Henneaux, Are Kaluza-Klein models of the universe chaotic? Int. J. Theor. Phys. **28**, 1067 (1989)
46. T. Damour, P. Spindel, Quantum supersymmetric Bianchi IX cosmology. Phys. Rev. D **90**(10), 103509 (2014)
47. T. Damour, P. Spindel, Quantum supersymmetric cosmological billiards and their hidden Kac–Moody structure. Phys. Rev. D **95**(12), 126011 (2017)

48. V.D. Ivashchuk, V.N. Melnikov, Quantum billiards with branes on product of Einstein spaces. Eur. Phys. J. C **76**(5), 287 (2016)
49. J.M. Heinzle, C. Uggla, Mixmaster: fact and belief. Class. Quant. Grav. **26**, 075016 (2009)
50. N.J. Cornish, J.J. Levin, The mixmaster universe is chaotic. Phys. Rev. Lett. **78**, 998 (1997)
51. N.J. Cornish, J.J. Levin, Mixmaster universe: a chaotic Farey tale. Phys. Rev. D **55**, 7489 (1997)
52. O.M. Lecian, BKL maps, Poincaré sections, and quantum scars. Phys. Rev. D **88**, 104014 (2013)
53. H. Bergeton, E. Czuchry, J.P. Gazeau, P. Małkiewicz, W. Piechocki, Singularity avoidance in a quantum model of the Mixmaster universe. Phys. Rev. D **92**, 124018 (2015)
54. I. Bakas, F. Bourliot, D. Lust, M. Petropoulos, Mixmaster universe in Horava–Lifshitz gravity. Class. Quant. Grav. **27**, 045013 (2010)
55. P. Horava, Quantum gravity at a Lifshitz point. Phys. Rev. D **79**, 084008 (2009)
56. S. Mukohyama, Horava-Lifshitz cosmology: a review. Class. Quant. Grav. **27**, 223101 (2010)
57. I.M. Khalatnikov, A.Y. Kamenshchik, Stochastic cosmology, perturbation theories, and Lifshitz gravity. Phys. Usp. **58**(9), 878 (2015)

The Physics of LIGO–Virgo

Giancarlo Cella

To Elena

Abstract I give a synthetic review of the physics involved in the direct detection of gravitational waves, both from the point of view of what can be learned and what is needed to obtain and interpret the measurements.

Keywords Gravitational waves · Detectors

1 Introduction

The Einstein general theory of relativity [1] allows for the existence of gravitational waves, namely perturbation of the geometry of spacetime which propagates at the speed of light. This was immediately realized by Einstein himself [2–4].

The last few years, just a century after these first papers, have been characterized by a series of experimental milestones in gravitational physics. The LIGO/Virgo collaboration has obtained the first direct detection of gravitational waves [5] together with the first evidence of a coalescence of a pair of black holes [6], the first three-detectors coincident detection of a black holes merger [7] and the first direct detection of a coalescence between a pair of neutron stars [8] with an associated short gamma ray burst.

These results have been made possible by a progressive improvement of detectors' sensitivity, which was enabled by more and more sophisticated techniques for the reduction of several fundamental and technical noises.

G. Cella (✉)
INFN Sez. Pisa, Pisa, Italy
e-mail: giancarlo.cella@pi.infn.it

© Springer Nature Switzerland AG 2019
S. Cacciatori et al. (eds.), *Einstein Equations: Physical and Mathematical Aspects of General Relativity*, Tutorials, Schools, and Workshops in the Mathematical Sciences, https://doi.org/10.1007/978-3-030-18061-4_4

1.1 Some History

The direct search of gravitational waves has a long story. First attempts were done by J. Weber in the 1960s, at the University of Maryland in College Park. He used an aluminum bar of 1400 kg as a detector. The idea is that the lower normal mode of oscillation of the bar can be excited by the interaction with a gravitational wave. The response of the bar is significant only in a small interval of frequencies around the resonance.

Weber claimed several detections [9–11] that he interpreted as gravitational waves coming from the center of the galaxy. These were never confirmed in an independent way and were not accepted by the scientific community.

Almost at the same time a different idea was proposed by Pustovoit and Gertenshtein [12], who noted that a Michelson interferometer could be used, obtaining a large detection bandwidth and potentially a higher sensitivity than compared with the bar's one. This possibility was considered by Weber himself, together with his student R. Forward. Forward constructed a first prototype of this alternative detector after moving to Hughes Research Laboratories [13].

A careful and extensive study about interferometric detectors of gravitational wave was started in this period by R. Weiss at MIT. Weiss characterized quantitatively and systematically the noise sources coupled to the apparatus [14]. Weiss proposed and constructed a prototype of interferometric detector in the early 1980s, changing the basic optical scheme by introducing along the arms of a Michelson interferometer two "delay lines." In this setup the light moves several times back and forth with the result of increasing the sensitivity.

This can be seen as the starting point for the design and realization of large scale earth-bound interferometric detectors which are active today, and will be described in somewhat more detail in Sect. 2. At the present time there are three interferometers of this kind, which form the LIGO–Virgo network. The Virgo interferometer is located in Italy near Pisa: we will give some detail about it later. The two LIGO interferometers are located in USA, one in Louisiana and one in the State of Washington. Their size (4 km arms for LIGO and 3 km arms for Virgo) and their basic scheme are similar.

An initial version of LIGO and Virgo operated from 2002 through 2010, without detecting any kind of gravitational wave signals. A new generation of detectors, Advanced LIGO [15] and Advanced Virgo [16], started to get data in 2015 with an improved sensitivity which allowed for the first detections. Soon KAGRA, another detector with similar sensitivity currently in construction in Japan, will start to be operative. On a time scale of few years another 4 km LIGO interferometer, LIGO-India [17], located in India will also join the network.

Fig. 1 The data points indicate the observed change in the epoch of periastron of PSR B1913+16 with date. Continuous line shows the prediction of general relativity (data from [18]). Hulse and Taylor were awarded with the Nobel prize for the discovery of this system in 1993

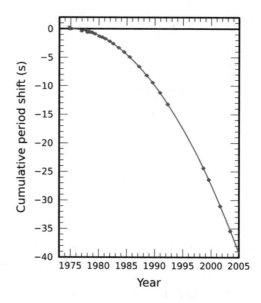

1.2 Indirect Evidences

Here we will describe mainly results connected with direct detection of gravitational waves. This means that the evidence is given by a measure obtained with a man-made detector, which couples directly to the metric perturbation.

However we must remember that a convincing evidence for the existence of gravitational waves has been provided by the study of binary pulsar systems. In particular by carefully analyzing the orbital period of the binary pulsar 1913+16 discovered by Hulse and Taylor in 1974 [19], Taylor and Weisberg [20] found a law for the variation of the orbital period of the system in agreement with the prediction of general relativity about the amount of energy and angular momentum carried away by the emitted gravitational waves (see Fig. 1).

2 Gravitational Wave Detectors

The basic principle of an interferometric detector of gravitational waves is the accurate comparison between two round trip times of light, measured by the same observer, between freely falling masses. The simplest optical scheme is the one of a Michelson interferometer depicted in Fig. 2.

The free falling test masses are realized with mirrors. Two of them (labeled as TM1 and TM2) are perfectly reflecting, while the third one (called *beam splitter*) separates the incoming light beam in two parts of equal intensity. Ideally the arm lengths ℓ_1 and ℓ_2 should be chosen in such a way that in absence of any disturbance

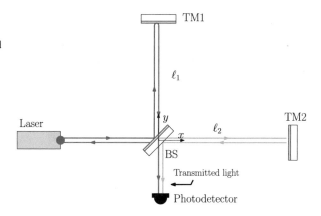

Fig. 2 Optical scheme of a Michelson gravitational waves' detector. The two end mirrors TM1 and TM2 and the beam splitter BS are the test masses, and the round trip time BS-TM1-BS and BS-TM2-BS are compared

there is a complete destructive interference for the transmitted light that reach the photodetector, which is given by the superposition of the beams that moved back and forth the two arms.

More explicitly, the amplitude of the transmitted light can be written as

$$A_{tr} = \sqrt{I_0} r_{BS} t_{BS} \left(e^{2ik_\ell \ell_1} - e^{2ik_\ell \ell_2} \right) e^{i\omega_\ell t} + c.c. \qquad (2.1)$$

where r_{BS} and t_{BS} are the reflectivity and transmissivity of the beam splitter $r_{BS}^2 + t_{BS}^2 = 1$ in absence of mirror losses, I_0 is the intensity of the incoming laser beam, $k_\ell = \omega_\ell/c$ and ω_ℓ is the laser's angular frequency. When $\ell_1 = \ell_2 = \bar{\ell}$ $A_{tr} = 0$, and for small variations we can write $\ell_i = \bar{\ell} + \delta\ell_i$ and

$$A_{tr} \simeq \sqrt{I_0} r_{BS} t_{BS} e^{2ik_\ell \bar{\ell}} 2ik_\ell (\delta\ell_1 - \delta\ell_2) + c.c \qquad (2.2)$$

where we neglected $O(k_\ell^2 \delta\ell_i^2)$ corrections. Apparently the amplitude of transmitted light is proportional to the difference between the lengths of the two arms of the interferometer. If the effect of a gravitational wave can be understood as a variation of these length, we have a detector. Now we will see that this is the case when a particular approximation is valid, and when an appropriate set of coordinates are chosen. On the other hand, it is clear that the light intensity measured by a photodetector is coordinate independent.

2.1 The Coupling Between Detector and Signal

We give now a more detailed analysis of the Michelson interferometer, introducing the gravitational wave field. We can write the metric in the form

$$g_{\mu\nu} = \eta_{\mu\nu} + h_{\mu\nu} \qquad (2.3)$$

where $\eta_{\mu\nu}$ is the Minkowski metric and $h_{\mu\nu}$ a small correction, $|h_{\mu\nu}| \ll 1$. The smallness of the correction term in Eq. (2.3) is expected to be true when gravitational effects are small and an appropriate set of nearly Cartesian coordinates have been chosen: this should be a good approximation near the detector.

By inserting this expression in the Riemann tensor we can expand in powers of $h_{\mu\nu}$. At the linear order we obtain

$$R_{\mu\nu\rho\sigma} = \frac{1}{2}\left(h_{\sigma\mu,\nu,\rho} - h_{\nu\sigma,\mu,\rho} - h_{\rho\mu,\nu,\sigma} + h_{\rho\nu,\mu,\sigma}\right).$$

(2.4)

When the $h_{\mu\nu}$ field changes slowly on the length scale of the detector if we assume that the test mass is freely falling we can use the geodesic equation to evaluate the separation between them, namely

$$\frac{d^2\xi^\mu}{d\tau^2} = R^\mu{}_{\nu\rho\sigma}u^\nu u^\rho \xi^\sigma.$$

(2.5)

In the proper detector frame this equation can be simplified considerably. Note that this is a very convenient reference frame from the point of view of the description of the experimental apparatus, because it is possible to use Newtonian physics inside it. We do not repeat the derivation, that can be found on standard textbooks (see for example [21]) and we give the final result which is

$$\ddot{\xi}^i = -R^i{}_{0j0}\xi^j$$

(2.6)

where ξ^i is the separation between two test masses.

The Riemann tensor contains all the gravitational effects, which are not only the gravitational ones but also, for example, tidal forces generated by masses nearby. So we need to separate the gravitational waves' contributions. We step back in a general reference frame, and starting from the Einstein equation

$$G_{\mu\nu} = \frac{8\pi G}{c^4}T_{\mu\nu}$$

(2.7)

we rewrite it at linear order in $h_{\mu\nu}$ as

$$-\bar{h}_{\mu\nu,\alpha}{}^{,\alpha} - \eta_{\mu\nu}\bar{h}_{\rho\sigma}{}^{,\rho,\sigma} + \bar{h}_{\mu\rho,\nu}{}^{,\rho} + \bar{h}_{\nu\rho,\mu}{}^{,\rho} = \frac{16\pi G}{c^4}T_{\mu\nu}.$$

(2.8)

Here for convenience we introduced the trace reversed strain tensor field $\bar{h}_{\mu\nu} = h_{\mu\nu} - 1/2\eta_{\mu\nu}h$ and the trace $h = h_\alpha{}^\alpha$. We can use the freedom of choosing the coordinate system to simplify this expression. This freedom is somewhat limited in this context, because we supposed $h_{\mu\nu}$ to be a small correction and we must preserve this property. For this reason we will consider only coordinate transformations of the form

$$x^\mu \rightarrow x^\mu + \xi^\mu(x) \tag{2.9}$$

which gives the transformation rule $h_{\mu\nu} \rightarrow h_{\mu\nu} - \xi_{\mu,\nu} - \xi_{\nu,\mu}$. Note that $\xi_{\mu,\nu} + \xi_{\nu,\mu}$ must be small. It can be proved easily that using this freedom we can impose the Lorentz condition

$$\overline{h}_{\mu\nu}{}^{,\nu} = 0 \tag{2.10}$$

in such a way that Eq. (2.8) becomes

$$\Box \overline{h}_{\mu\nu} = -\frac{16\pi G}{c^4} T_{\mu\nu} \tag{2.11}$$

which is a wave equation with a source proportional to the energy-momentum tensor.

We can still perform a transformation like (2.9) provided that $\Box \xi^\mu = 0$, in such a way to preserve the Lorentz condition (2.10). If $T_{\mu\nu} = 0$ we can set in particular $h_{0\mu} = 0$ and $h = 0$: this peculiar and useful coordinate system is named transverse traceless, and we can write

$$\Box \mathsf{h}_{ij} = 0 \tag{2.12}$$

$$\mathsf{h}_{\mu 0} = 0 \tag{2.13}$$

$$\mathsf{h}_i{}^i = 0. \tag{2.14}$$

We indicate with $\mathsf{h}_{\mu\nu}$ the strain tensor in this peculiar coordinate system. Note that $\mathsf{h}_{\mu\nu} = \overline{h}_{\mu\nu}$, and that only the purely spatial components, labeled with Latin indices, are nonzero.

The solution of the wave equation can be written as a superposition of plane waves

$$\mathsf{h}_{ij} = \sum_a \int \frac{d^3k}{(2\pi)^3} a^a(\vec{k}) \varepsilon^a(\hat{k})_{ij} e^{i(\vec{k}\cdot\vec{x} - kct)} + c.c. \tag{2.15}$$

where the integral is over all the wave vectors $\vec{k}$ and the sum over all the elements of a basis for the polarization tensors. These should be symmetric, transverse, and traceless, namely $k^i \varepsilon^a(\hat{k})_{ij} = 0$ and $\varepsilon^a(\hat{k})_i^i = 0$. By defining a orthonormal basis $(\hat{m}, \hat{n})$ in the space normal to $\vec{k}$ we see that there are two independent polarization tensors that can be written as

$$\varepsilon^+ = \hat{m} \otimes \hat{m} - \hat{n} \otimes \hat{n} \tag{2.16}$$

$$\varepsilon^\times = \hat{m} \otimes \hat{n} + \hat{n} \otimes \hat{m}. \tag{2.17}$$

In this way we characterized the physical degrees of freedom associated to gravitational waves. Coming back to the geodesic deviation equation, we see that in the transverse and traceless reference frame we can write

$$R_{i0j0} = -\frac{1}{2}\ddot{h}_{ij}$$

but up to the linear order $R_{\mu\nu\rho\sigma}$ is invariant, so we can rewrite Eq. (2.6) as

$$\ddot{\xi}^i = \frac{1}{2}\ddot{h}_{ij}\xi^j.$$

We see that in the proper detector frame we can describe the effect of a gravitational wave by a Newtonian force. For the Michelson detector in Fig. 2 we can write the motion equations

$$m_{\text{TM1}}\ddot{r}^i_{\text{TM1}} = \frac{1}{2}m_{\text{TM1}}\ell_1\ddot{h}_{iy} + f^i_{\text{TM1}}$$

$$m_{\text{TM2}}\ddot{r}^i_{\text{TM2}} = \frac{1}{2}m_{\text{TM2}}\ell_2\ddot{h}_{ix} + f^i_{\text{TM2}}$$

$$m_{\text{BS}}\ddot{r}^i_{\text{BS}} = f^i_{\text{BS}}$$

where f^i_X are additional forces which act on the test mass X. Note that the reference frame has been centered on the beam splitter, so there are no forces generated by the gravitational wave which act on it. Another observation is that the force is proportional to the distance between the beam splitter and the test mass. This is the reason why a large interferometer is more sensitive. The field of the forces associated to a plane wave is represented in Fig. 3.

If we compare the field on the right side with the position of the detector in Fig. 2, we see that it describes a gravitational wave which is optimally coupled to the instrument. In fact the mirrors respond to the force by oscillating with an opposite phase: when the arm 1 is stretched, the arm 2 is dilated. As the detector output is proportional to $\ell_1 - \ell_2$ the two effects add their contributions. On the other hand the gravitational wave represented on the left side is uncoupled to the detector, as both arms do not change their length.

A general plane wave couples to the detector in a way which depends by its direction and polarization. The coupling can be described by introducing a *detector tensor* D_{ij}, in such a way that the measured signal is proportional to

$$h(t) \equiv D^{ij}h_{ij}. \tag{2.18}$$

For a Michelson interferometer, but also for the more general interferometric detectors which we will describe later, we have

$$D^{ij} = u^i u^j - v^i v^j \tag{2.19}$$

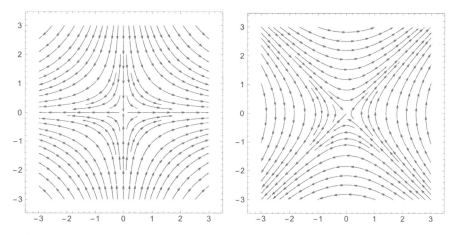

Fig. 3 On the left site, the force field which describes the action of a plane gravitational wave on test masses, in a plane $z = 0$ perpendicular to the direction of propagation. A linear polarization has been chosen, with $\varepsilon^+ = \hat{x} \otimes \hat{x} - \hat{y} \otimes \hat{y}$. On the right side, the force field for the polarization $\varepsilon^\times = \hat{x} \otimes \hat{y} + \hat{y} \otimes \hat{x}$. Left plot can be obtained from the right one by a $\pi/4$ rotation. The real force fields are time dependent, and the represented one is modulated by an oscillating term

where $\hat{u}$ and $\hat{v}$ are versors with the same direction of the arms. The coupling of two different linear polarization is represented as a function of the wave's direction in Fig. 4.

It can be seen that an interferometer is not a very directional detector: there are for sure completely blind directions (in the plane of the interferometer) but there is a large solid angle where the coupling is good.

2.2 More Refined Optical Schemes

The Michelson interferometer is a quite simple detector, which is good enough to understand the basic principle of detection. However the real detectors, such as Advanced Virgo and Advanced LIGO, adopt a more complicate scheme. A simplified representation of Advanced Virgo is presented in Fig. 5.

The basic configuration of an interferometer based on the beam splitter is still present, together with the mirrors EMY and EMX at the end of the two arms which correspond to TM1 and TM2 in Fig. 2. The most relevant difference is the insertion of a pair of additional mirrors IMY and IMX at the end of the arms. In this way the two arms become resonant optical cavities. IMX and IMY are not completely reflective, and they allow for some of the light coming from the beam splitter to enter the cavity and also to exit from it.

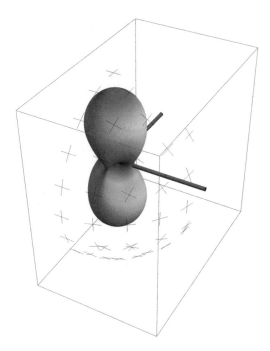

Fig. 4 The coupling of a plane wave with linear polarization to an interferometric detector described by the detector tensor $\hat{x} \otimes \hat{x} - \hat{y} \otimes \hat{y}$, as a function of the wave's direction. The interferometer's arms are shown, and the polar plot shows $\left| D_{ij}h^{ij} \right|$ as a function, the wave direction. For each direction the optimal linear polarization $\varepsilon = \hat{e}_1 \otimes \hat{e}_1 - \hat{e}_2 \otimes \hat{e}_2$ is chosen, represented by the segments oriented as the $\hat{e}_{1,2}$

Given the amplitude A_i of the field impinging on the input mirror, the reflected field will be

$$A_r = -rA_i + t^2 \sum_{n=1}^{\infty} r^{n-1} e^{in\phi} A_i = e^{i\phi} \frac{1 - re^{-i\phi}}{1 - re^{i\phi}} A_i \equiv e^{i\Xi} A_i$$

where r and t are the reflectivity and the transmissivity of the mirror and ϕ the phase acquired in a cavity round trip. Here we are summing over all the possible paths of the beam: the amplitude of the light which bounces back and forth n times inside the cavity takes into account a factor t when it enters and another one when it exits, $n - 1$ reflections on the input mirror and the total phase shift. We see that the phase difference between the input and the reflected field is given by

$$\Xi = \phi + 2 \arctan\left(\frac{r \sin \phi}{1 - r \cos \phi} \right). \tag{2.20}$$

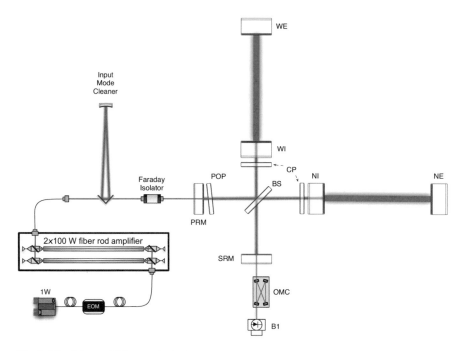

Fig. 5 The Advanced Virgo optical scheme

When $\phi = 0$ after a round trip there is no phase shift, and the cavity is resonating. Note that

$$\frac{d\Xi}{d\phi}\bigg|_{\phi=0} = \frac{1}{1-r}$$

and if we take r very near to one we see that the phase difference between input and output is very big. But $\phi = 2\omega_\ell \ell/c$, so Ξ is very sensitive to the length variation of the cavity around the resonance. Note that when $r = 0$ we recover the Michelson case, in other words a resonant cavity of length L couples to gravitational radiation as a Michelson with an arm length of $L(1-r)^{-1} \gg L$.

Another difference between Figs. 2 and 5 is the presence of the mirror PRM1. Its role can be explained by observing that PRM1 can be considered the input mirror of another optical cavity, where the role of the end mirror is played by all the interferometer beyond the beam splitter. If we evaluate the amplitude of the field inside a cavity we find

$$\mathbf{A}_c = t\sum_{n=1}^{\infty} r^{n-1} e^{in\phi} \mathbf{A}_i = \frac{te^{i\phi}}{1-re^{i\phi}} \mathbf{A}_i$$

the field intensity is obtained taking the square modulus of the amplitude, obtaining

$$I_c = \frac{t^2}{(1 - r\cos\phi)^2 + r\sin^2\phi} I_i \stackrel{\phi=0}{\rightarrow} \frac{1 + r}{1 - r} I_i.$$

When the cavity is at the resonance for $r \simeq 1$ the intensity of the light inside is much larger that the incoming one. In other words, the PRM1 mirror *recycles* the light reflected by the interferometer, and this is equivalent to increase by a factor $\simeq 2(1 - r)^{-1}$ the power of the laser. As we will see in Sect. 3.3, this is a way to improve the detector sensitivity.

Finally another addition in Fig. 2 is the mirror SRM1. Once again effect is that of a sensitivity enhancement, as will be explained in Sect. 3.3 too.

2.3   *Advanced Detectors*

As we said in the introduction the currently operative interferometers are called *Advanced*. These are second generation instruments that are expected to gain a final design sensitivity in $h_{\mu\nu}$ an order of magnitude larger that first generation one. Comparing first and second generation detectors we see that the basic optical scheme is not very different, with the exception of the insertion of a *signal recycling mirror* (SRM1 in Fig. 5).

What makes the difference is a set of improvement that have the effect of reducing the noise level. In particular in Advanced Virgo

- laser beams with a larger transverse size (a factor 2.5 compared with first generation ones) have been adopted;
- larger and heavier mirror (42 kg masses, increased by a factor 2) are used;
- improved optical elements, with very low levels of absorption ($<0.5 \times 10^{-6}$) and scattering ($<10^{-5}$) have been obtained;
- the factor $(1-r)^{-1}$ discussed previously has been increased by a factor 3 reaching the value of 420, and thus increasing the sensitivity of the cavity;
- a strategy for thermal control of optical aberrations has been implemented;
- the mitigation of diffused light was improved;
- the vacuum level inside the apparatus has been increased reducing the pressure by a factor 10^{-2}. Currently the residual pressure is of about 10^{-9} mbar.

Finally at the final design sensitivity a large laser power (200 W) is foreseen.

The amplitude of a signal is inversely proportional to its distance. This means that when we gain an order of magnitude in $h_{\mu\nu}$ sensitivity we increase by a factor 10^3 the volume of the universe where a source can be detected with some level of confidence. For this reason a seemingly modest improvement in sensitivity can give a large scientific gain. In Sect. 3 we will discuss the fundamental sources of noise that limit the sensitivity of a detector and the strategies used to mitigate them.

3 Fundamental Noises

A large coupling between the detector and the gravitational wave is for sure a good thing, but by itself it is not a particularly significant figure of merit. What is really important is a comparison between the signal and the noise amplitudes.

The simplest characterization of the sensitivity of a gravitational wave detector can be given by its strain equivalent amplitude noise spectrum. The idea is to normalize the detector's output $s(t)$ in such a way that

$$s(t) = h(t) + n(t) \tag{3.1}$$

where $h(t)$ is the measured strain defined by Eq. (2.18) and $n(t)$ the noise. If the noise is stationary the power spectrum of $n(t)$ can be written as

$$\langle \tilde{n}(\omega)^* \tilde{n}(\omega') \rangle = 2\pi \delta \left(\omega - \omega' \right) S_n(\omega) \tag{3.2}$$

where $\tilde{X}$ indicate a Fourier transform. The strain equivalent noise spectrum is just given by $\sqrt{S_n}$. The design noise budget planned for Advanced Virgo is plotted in Fig. 6. The best sensitivity window is between $20 \, \text{Hz}$ and $4 \times 10^3 \, \text{kHz}$. We see

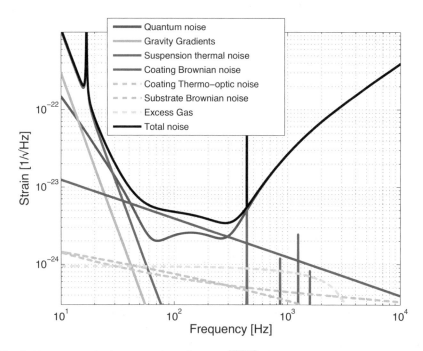

Fig. 6 The strain equivalent spectral amplitudes $\sqrt{S_n(\omega)}$ of several noises of Virgo Advanced, as a function of the frequency. Most relevant contributions come from quantum noise and from thermal noise of mirror coating and suspensions. Seismic noise dominates below $10 \, \text{Hz}$ and it is not shown

also that the dominant source of noises are the thermal one (below 300 Hz) and the quantum one, in the higher frequency region.

3.1 Seismic Noise

Seismic noise does not appear in the picture, but becomes the main limit below 10 Hz. Below 1 Hz it is called *microseism noise*, and it is originated mainly by natural sources, depending on oceanic and large scale meteorological conditions. Around 1 Hz the main contribution comes from wind effects and local meteorological conditions, while at higher frequencies it is mainly anthropogenic seismic noise (which means originated by human activities).

The mean seismic fluctuation of the ground is site dependent, but its order of magnitude is about 10^{-6} m, the main contributions coming from low frequencies.

Special care is taken to attenuate it, the reason being that the typical spectral amplitude of the motion of the surface of the earth is at least nine orders of magnitude larger than the expected signal in the detection bandwidth. In Virgo an hybrid active and passive system called *superattenuator* [22] (see Fig. 7) is used for seismic noise attenuation. The passive part is based on a chain of oscillators which are interposed between the ground and the suspended mirror. They are designed in such a way that the proper frequencies ω_p of the chain are below 1 Hz. Above these the oscillator chain spectral transfer function is depressed by a factor $\sim \left(\frac{\omega_p}{\omega}\right)^{2n}$, where n is the number of stages. With a large enough n the required attenuation factor can be reached. Currently $O(10^{-15})$ attenuation factors can be obtained in this way.

This passive attenuation system is not effective below few Hz, which is the scale set by ω_p, as can be seen from Fig. 7 (right). This is not directly relevant for the sensitivity, because the detection bandwidth starts at higher frequencies. But the residual low frequency motion is much larger than the range where Eq. (2.20) can be linearized around the resonance $\phi = 0$. In order to appreciate the problem we can expand Eq. (2.20) up to the second order in ϕ, obtaining

$$\Xi = \frac{1+r}{1-r}\phi - \frac{1}{3}\frac{r(1+r)}{(1-r)^3}\phi^3 + O\left(\phi^5\right) \tag{3.3}$$

and we see that the linear approximation is good when

$$|\phi| = 4\pi\frac{\delta\ell}{\lambda_\ell} \ll \sqrt{\frac{3}{r}}(1-r)$$

which is a small value when r is near to one as is the case. In Virgo and LIGO the laser's wavelength is $\lambda_\ell = 10^{-6}$ m, of the same order of the mean residual motion,

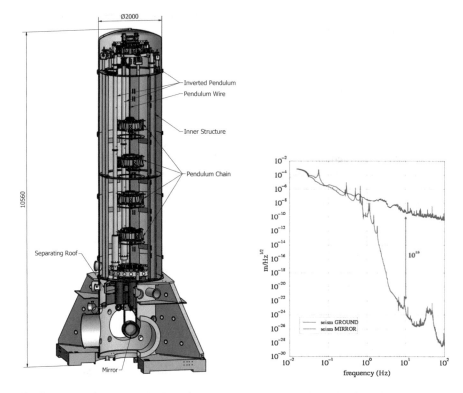

Fig. 7 On the left, a schematic view of a Virgo *superattenuator*. The chain of oscillators is built by an inverted pendulum fixed on the ground, followed by several pendula, which give the attenuation of horizontal motion. Oscillators integrated in the body of the filters attenuate in a similar way the vertical motion. On the right, the spectral amplitude of the seismic motion measured on the ground, and the spectral amplitude of the residual motion at the mirror

so $\delta\ell$ must be small also at very low frequencies. This is obtained with a quite elaborated active control strategy:

- a feedback strategy called *inertial damping* is applied on the inverted pendulum, to reduce the drifts and the effect of tidal strains;
- another active control acting on the *marionette* reduces the angular mirror displacements down to a fraction of μrad;
- a local active damping procedure is finally applied to the test mass.

3.2 Thermal Noise

The main target for second generation (advanced) detectors was the improvement of the best sensitivity of one order of magnitude in h. This will be obtained (when the design sensitivity will be reached) mainly by reducing thermal and quantum noise. I will give some details about quantum noise in the next subsection, and I concentrate here on thermal noise.

Thermal noise is the unavoidable result of the coupling between the apparatus and the environment. When the system is in thermodynamical equilibrium its thermal fluctuations can be evaluated by using the fluctuation dissipation theorem. This theorem tell us that the spectral amplitude of thermal noise is proportional to dissipative effects.

In an interferometric detector we·are interested in the phase fluctuations induced by the thermal agitation of the mirror surface on the laser beam. There are essentially three different possibilities for reducing them:

1. The thermodynamical way. Thermal noise amplitude is proportional to the square root of the temperature, so it is possible to reduce it with a cryogenic setup. This is not easy, especially because the cryogenic infrastructure can reintroduce other kind of noises, such as mechanical vibration. Beside that, a non-trivial problem is the elimination of the heat which is transferred by the laser to the mirrors. This is a small fraction of the total power, as a large effort is done to use mirrors with a very low absorption, but can become significant if the laser power is very high. Neither Virgo Advanced nor LIGO Advanced are cryogenic experiments, but this approach will be attempted in the Japanese detector KAGRA which will be operative soon [23] joining the LIGO–Virgo network.
2. The statistical way. The total phase shift of the laser beam is an average of the mirror displacement weighted by the laser intensity profile. A direct way to reduce thermal fluctuations is to use larger beams, which average more. This is the reason why the beam sizes have been increased in advanced detectors. There are however some limitations, because the dimension of a beam is limited by the mirror size and it is not possible to increase this by a large factor without a deep redesign of infrastructures.
3. The material improvement way. By optimizing the material and the mechanical construction we can reduce the sources of dissipation. Dissipation is a measurement of the coupling between the system and the environment. The largest contribution to thermal noise comes currently from the optical coating of the mirrors, which is needed to obtain the requested reflectivities (very large or very small, depending on the applications). But the mechanical properties of the materials used for the coating are not good enough, and there is an intense activity in searching for better materials and solutions.

One of the main difficulty in the thermal noise reduction program is about the strong interdependence between possible solutions and other requirements, such as mechanical and optical properties of the apparatus.

3.3 Quantum Noise

At higher frequencies in the detection bandwidth the most relevant limit is the laser shot noise, which is connected with the quantum nature of laser's light.

The most direct way to decrease the shot noise is to increase the laser power. However this solution is not a free lunch: a more powerful laser can introduce problems connected with thermal lensing and power absorption which can be quite difficult to solve, such as the parametric resonance one that I will describe at the end. An alternative is based on the use of squeezed light, and I will discuss extensively about this.

As the program of thermal noise reduction will become more and more effective the impact of optical noise will be more and more important also at intermediate frequencies, and more refined techniques will be needed to reduce it.

A large reduction of thermal noise is foreseen for the so-called *third generation detectors*, such as the *Einstein Telescope* [24]. In this case it is possible that optical noise will become the most fundamental limit of sensitivity for the detector both at higher and lower frequencies [25], and a new aspect of the problem will appear. As light carries momentum, it generates radiation pressure. Radiation pressure increases with the power of the laser, and when we try to reduce shot noise we generate a radiation pressure noise via *ponderomotive effects*. It comes out that at a given frequency there is an optimal sensitivity which is determined by the best compromise between shot noise and radiation pressure noise, and is called *Standard Quantum Limit*. This can be seen as a direct consequence of the Heisenberg uncertainty principle. The possibilities of evading this limit have been extensively studied in these years and will be discussed.

An Example: The Optical Cavity A very crude model for a laser beam, which however is good enough to understand several peculiarities of ponderomotive effect in interferometers, is a plane wave with some given frequency ω_ℓ. When this plane wave interact with optical elements such as mirrors it is reflected and modulated. We will write the classical electric field at a given point as

$$E(t) = E_1(t) \cos \omega_\ell t + E_2(t) \sin \omega_\ell t \tag{3.4}$$

where $E_i(t)$ are functions with a slow variation on the timescale ω_0^{-1}, which contain the information about modulations. When modulation effects are small it is useful to write Eq. (3.4) in a slightly different way

$$E(t) = \left[\overline{E}_1 + \delta E_1(t) \right] \cos \omega_\ell t + \left[\overline{E}_2 + \delta E_2(t) \right] \sin \omega_\ell t \tag{3.5}$$

with $\delta E_i \ll \overline{E}_i$. In this way we emphasize the separation between the ideal, monochromatic plane wave contribution $\overline{E} \cos \omega_\ell t$, and the small corrections δE_i which contain the relevant information coming from the detector (and the noise).

This is the basic idea beyond the graphical representation in Fig. 8. The "average" part $\overline{E}_i$ is represented by the black sticks, rotating with angular frequency ω_ℓ, the "fluctuating" one $\delta E_i(t)$ by the gray "balls", and the real field is obtained after a projection over the q axis. From this graphical representation it is easy to understand that the component of the fluctuating vector δE_i parallel to $\overline{E}_i$ represents the amplitude fluctuation of the field, and the perpendicular component is proportional to the phase fluctuation, namely $\delta\phi(t) \sim \delta E_\perp / \overline{E}$.

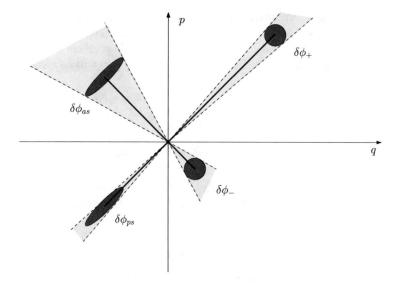

Fig. 8 A graphic representation of the field described by Eq. (3.5). The black stick represents the "average" part of the field (which is monochromatic with angular frequency ω_ℓ). The dark gray areas represent the fluctuating parts of the field: they can be seen as the typical region where the field value can be found after measurement. It follows that the light gray regions are estimates of phase fluctuation

It is easy to understand how the operators which describe the field at different points are related. For example, after a free propagation over a length ℓ the field is simply delayed by ℓ/c, which means

$$\overline{E}' + \delta E'(t) = \mathbb{R}\left(\frac{\omega_\ell}{c}\ell\right)\left[\overline{E} + \delta E'\left(t - \frac{\ell}{c}\right)\right] \tag{3.6}$$

where $\mathbb{R}(\theta)$ is a rotation matrix of an angle θ, $V = \begin{pmatrix} V_1 & V_2 \end{pmatrix}^T$. Note that the fluctuating part is not simply rotated, but it gets an additional delay.

The most important rule from our point of view is the one that gives the field reflected on a moving mirror. If the mirror's displacement from his reference position $\delta z(t)$ is small and of the same order of field's fluctuation in the linear approximation we can obtain from (3.6)

$$\overline{E}' + \delta E'(t) = \mathbb{R}(\omega_\ell \delta \tau)\left[\overline{E} + \delta E(t - \delta \tau)\right]$$

$$\simeq \overline{E} + \delta E(t) + \frac{2\omega_\ell \delta z(t)}{c}\mathbb{R}\left(\frac{\pi}{2}\right)\overline{E} \tag{3.7}$$

where $\delta\tau = 2\delta z(t)/c$ and we neglected $O(\delta^2)$ effects. Note that the average part is unchanged, while the fluctuating one gets an additional contribution proportional to the average field and perpendicular to it: as expected, the mirror fluctuations generate phase fluctuations.

The mirror moves for several reasons: it can be affected by noises such as seismic or thermal, and hopefully it can respond to the tidal effect of a gravitational wave. The field itself acts on the mirror with a radiation pressure force, which is proportional to the square of the field, whose fluctuating part will be given, after an average over ω_ℓ^{-1} timescales, by

$$\overline{F}_{RP} + \delta F_{RP} = k\overline{E^2(t)} = \frac{k}{2}\left|\overline{E}\right|^2 + \frac{k}{2}\left|\overline{E}'\right|^2$$

$$+ k\overline{E}^T \delta E\,(t) + k\overline{E}'^T \delta E'\,(t)\,.$$

Note that δF_{RP} is proportional to the field's fluctuations parallel to $\overline{E}$, namely to amplitude fluctuations. The mirror will respond to external forces with something like

$$\delta z(t) = (G \star \delta F_{RP} + G \star \delta F)\,(t)$$

$$= k\left[2G \star \overline{E}^T \delta E + G \star \delta F\right](t) \qquad (3.8)$$

where G is the appropriate mechanical Green function. Together with Eq. (3.7) this gives

$$\delta E'(t) \simeq \delta E(t) + \frac{4k\omega_\ell \left|\overline{E}\right|^2}{c}\mathbb{R}\left(\frac{\pi}{2}\right)\left[G \star \mathbb{P}_\| \delta E\right](t)$$

$$+ \frac{2k\omega_\ell}{c}\mathbb{R}\left(\frac{\pi}{2}\right)\overline{E}\,(G \star \delta F)\,(t) \qquad (3.9)$$

where $\mathbb{P}_\|$ is the projector in the direction of the average field. When external forces δF are not present inside an optical cavity the fields near the two mirrors are related by

$$\delta E_A = \hat{T}\left[\mathbb{R}\,(\phi)\,(\mathbb{I} + \mathbb{M}_B\star)\right]\delta E_B$$

$$\delta E_B = t\hat{T}\mathbb{R}\,(\phi)\,\delta E_{IN} + r\hat{T}\left[\mathbb{R}\,(\phi)\,(\mathbb{I} - \mathbb{M}_A\star)\right]\delta E_A\,.$$

Here for brevity we defined a matrix which describe the modulation introduced by the mirror X

$$\mathbb{M}_X \equiv \frac{4k\omega_\ell \left|\overline{E}_X\right|^2}{c}\mathbb{R}\left(\frac{\pi}{2}\right)G_X\mathbb{P}_{\|X}$$

and a time delay operator

$$\hat{T}f(t) = f\left(t - \frac{L}{c}\right)\,.$$

The front mirror A, which is used to inject an external field δE_{IN}, is described by the reflection and transmission coefficients r, t while the other is supposed to be completely reflective. By solving for the fields we get

$$\delta E_A = \mathbb{C}^{-1} t \hat{T} \left[\mathbb{R}(\phi)(\mathbb{I} + \mathbb{M}_B \star) \right] \hat{T} \mathbb{R}(\phi) \delta E_{IN}$$

where

$$\mathbb{C} = \mathbb{I} - r \hat{T} \left[\mathbb{R}(\phi)(\mathbb{I} + \mathbb{M}_B \star) \right] \hat{T} \left[\mathbb{R}(\phi)(\mathbb{I} - \mathbb{M}_A \star) \right]$$

and finally we can write the field exiting from the cavity as a function of the incoming one

$$\delta E_{OUT} = t \mathbb{C}^{-1} t \hat{T} \left[\mathbb{R}(\phi)(\mathbb{I} + \mathbb{M}_B \star) \right] \hat{T} \mathbb{R}(\phi) \delta E_{IN}$$
$$- r(\mathbb{I} + \mathbb{M}_A \star) \delta E_{IN}. \tag{3.10}$$

This relatively compact expression contains a very rich phenomenology, as we will sketch in the following.

Squeezed States A quantum description of the system can be obtained by promoting the fluctuating fields $\delta E(t)$ to operators $\delta \hat{E}(t)$ with appropriate commutation relations. As a consequence of these the two components of $\delta \hat{E}(t)$ cannot be measured with arbitrary precision at the same time: quantum fluctuations will always be present and will be characterized by the symmetrized expectation values

$$\mathbb{K}(t - t') = \left\langle \delta \hat{E}(t) \delta \hat{E}(t') \right\rangle_{symm}. \tag{3.11}$$

At least up to a first approximation, the laser beam used in gravitational detectors can be modeled as a single mode of the electromagnetic field in a coherent state. Coherent states are minimum uncertainty states, because they saturate the Heisenberg uncertainty relations. In particular they are a "fair" compromise between phase and amplitude indeterminacy of the field, and in this sense they can be seen as the "most classical" states of light.

A particular class of squeezed states share with the coherent ones the minimum uncertainty property, however they can have a reduced phase uncertainty at the expense of a larger amplitude one (see Fig. 8). This means that for a given value of the laser power the shot noise (connected with phase indeterminacy) can be reduced.

A qualitative representation of coherent and squeezed state in the quadrature space, which can be seen as the analogous of the classical phase space for an harmonic oscillator, is given in Fig. 8. It comes out that fluctuation of coherent and squeezed states are Gaussian. For this reason the fluctuation of a coherent state is represented by a circle which indicate the uncertainty region of the field (the indeterminations of both quadratures δp and δq are the same and are uncorrelated). A squeezed state is represented by an ellipse, with a fluctuation reduced along one quadrature and increased along the other. The phase fluctuation $\delta\phi_-$ can be reduced both by increasing the laser power ($\delta\phi_+ < \delta\phi_-$) or by appropriately squeezing the state ($\delta\phi_{ps} < \delta\phi_-$).

It is important to realize that the squeezing generated by optomechanical effects can be frequency dependent. In other words, if we evaluate the Fourier transform of the correlation matrix (3.11) we obtain a frequency dependent spectral covariance matrix. Typically when the mechanical response of the apparatus is large (for example, near a resonance or in the low frequency region for a free mass) the ponderomotive effect becomes large and we obtain a strongly stretched ellipse. The appropriate tool to give a well defined meaning to the error ellipse is the Wigner quasi probability functions, which for coherent and minimum uncertainty squeezed states is Gaussian.

The idea of using squeezed light to improve the sensitivity of an interferometric detector of gravitational is around since a long time [26]. Squeezed states of light can be produced routinely today, with increasing levels of squeezing. The standard generation technique uses parametric amplification in nonlinear crystals, where the polarization P is related to the field in a nonlinear way

$$P = \varepsilon_0 \left(\chi_1 E + \chi_2 E^2 + \chi_3 E^3 + \cdots \right). \tag{3.12}$$

Here the χ_2 term induces a cubic coupling between photons, the χ_3 one a quartic coupling and so on.

The use of squeezed light is foreseen in the next generation of gravitational wave detectors. The main difficulty is to produce squeezing in the frequency range of interest, which is the audio region for earth-bound detectors. But the basic idea is conceptually quite simple. As we said a reduced phase fluctuation corresponds to a higher value of radiation pressure. However the response of the mirrors to the radiation pressure force $\delta\tilde{F}_{RP}$ is given in frequency domain, supposing a free mirror of mass M, by

$$\delta\tilde{x}_{\text{mirror}} = -\frac{1}{M(2\pi f)^2}\delta\tilde{F}_{RP}$$

so it is depressed in the high frequency region and can be less important in the low frequency one, where other noises can dominate.

Optomechanical Effects When the optical noise is relevant in all the frequency interval of interest the problem of standard quantum limit evasion must be solved. This possibility has been demonstrated theoretically in several papers. It can be shown that in the low photon number regime the optimal states are the Fock ones [27, 28]. When the number of photons involved is high, the optimal approach is based on squeezed states [29].

A general observation is that the final sensitivity depends on both the optical degrees of freedom of the apparatus and the mechanical one. These are coupled by radiation pressure, which acts in a dual way:

- the quantum state of the optical modes can be changed by the back action of the mechanical degrees of freedom. For example, a mirror recoiling under the action of the radiation pressure fluctuation can induce a phase fluctuation on the reflected beam correlated with the first, generating a squeezed state (*ponderomotive squeezing*). This mechanism can be described by an Hamiltonian which is completely analogous to the one generated by the χ_2 term in Eq. (3.12), though the physical mechanism is completely different. In the linearized approximation for a simple cavity this effect is completely described by Eq. (3.10), once the appropriate parameters of the cavity are inserted.
- The dynamic of the mirrors can be changed. For example, the radiation present inside a slightly out of resonance (detuned) cavity can produce a considerable effective mechanical stiffness, the so-called *optical spring effect*. This effect is also implicitly contained inside Eq. (3.10), but can be understood better if we rewrite Eq. (3.8) as

$$\delta z_X(t) = k G_X \star \delta F_X + 2k G_X \star \overline{\boldsymbol{E}}_X^T \delta \boldsymbol{E}_X.$$

When an external force is present, in the second member will depend from it in the general case, namely

$$\overline{\boldsymbol{E}}_A \delta \boldsymbol{E}_A = \chi_{AA} \star \delta F_A + \chi_{AB} \star \delta F_B + \cdots$$

$$\overline{\boldsymbol{E}}_B \delta \boldsymbol{E}_B = \chi_{BA} \star \delta F_A + \chi_{BB} \star \delta F_B + \cdots$$

where the dots indicate terms depending only on the incoming fields. We will not write explicitly this dependence, but it is clear that at the end we will obtain equation of motion for the mirrors of the form

$$\delta z_A = k G_A \star \delta F_A$$
$$+ 2k G_A \star (\chi_{AA} \star \delta F_A + \chi_{AB} \star \delta F_B + \cdots)$$
$$\delta z_B = k G_B \star \delta F_B$$
$$+ 2k G_B \star (\chi_{BA} \star \delta F_A + \chi_{BB} \star \delta F_B + \cdots)$$

which represent the modified dynamics of the mirrors. In particular a force on the mirror A can have an effect, mediated by the radiation pressure, on the mirror B. Delay effects can also induce damping (or anti-damping) effects, and instabilities can appear.

By injecting squeezed light it is possible to evade the standard quantum limit [30–32]. It is also possible to completely eliminate radiation pressure effects by a combination of squeezed state injection and output filtering [32].

In order to use squeezed states of light a particular care should be used in order to avoid optical losses due to absorption or scattering. These deteriorate the squeezing level and the problem is particularly relevant because the implementation of standard quantum limit evasion requires often additional optical elements.

The study of optomechanical effects is an active field of research, which is connected with fundamental questions about the foundations of quantum mechanics and quantum information. It is also expected to produce an increasing quantity of technological applications. Usually the most accessible examples of optomechanical devices are small scale ones (e.g., microcavities). A relevant point to be stressed is that in the case of gravitational waves detectors we deal with macroscopic masses, and in principle we have the possibility of studying quantum mechanical effects, such as entanglement [33], in this regime.

The direct observation of the ponderomotive squeezing effect is of great interest. Several attempts have been made, and a number of experiments are currently in various phases of their development. The effect has been observed on membranes [34] and using an optomechanical cavity coupled to a wave guide [35]. Direct evidence of ponderomotive squeezing is still lacking in macroscopic optical cavities.

There are other possible schemes for standard quantum limit evasion. Of particular interest are optical schemes based on the Sagnac effect [36] which are intrinsically unaffected by back action effects, and can be seen as examples of *Quantum Nondemolition* measurements.

Other examples are schemes which exploit the *optical spring effect* previously mentioned [37–39]. Both approaches are starting to become technically feasible.

Squeezing and Thermal Noise The interaction of the system with the environment can be modeled, via the fluctuation dissipation theorem, by introducing appropriate external stochastic forces. It is intuitive (and can be explicitly verified by looking at Eq. (3.9)) that these external forces will introduce an additional phase noise. The interplay between optical noise and thermal noise is not completely trivial when squeezed light is used, and the optimal strategies to be used in order to improve the sensitivity must be carefully designed.

A point worth to be mentioned is that in presence of an optical spring the evaluation of thermal noise must be done with care. In particular, the mechanical response of the system that must be used to apply the fluctuation dissipation theorem must not take into account the optical stiffness. The reason for that can be understood intuitively because the source of optical stiffness is the laser electromagnetic field inside the cavity, which cannot be seen as a system in thermal equilibrium with the environment.

This can be exploited to reduce thermal noise in the frequency domain of interest. For example it is possible to push the resonance of a system to higher frequencies adding an optical spring constant to the mechanical one. Modeling the mechanical damping with a loss angle ϕ we can write

$$K_{eff} = K_{mech} \left(1 + i\phi\right) + K_{opt}$$

and when $K_{opt} \gg K_{mech}$ we obtain a large Q factor by a kind of "dilution effect." As a side note, the application of the fluctuation dissipation effect in a situation which cannot be strictly defined of thermodynamic equilibrium is by itself an interesting subject in my opinion worth of further theoretical and experimental investigation.

Squeezing and Losses When there are optical losses inside the apparatus a quantum description becomes somewhat subtle. The main issue is that quantum mechanics must be unitary, and this is not compatible with a naive model.

It is easy to understand the point using a simple model. Suppose that a field described by an operator $\hat{a}$ is transmitted and partially absorbed through a lens. One could be tempted to describe the transmitted field with an operator

$$\hat{b} = \sqrt{1 - \eta^2}\hat{a}$$

where η parameterize the loss. However the commutation rules of this operator cannot be canonical:

$$\left[\hat{b}, \hat{b}^\dagger\right] = \left(1 - \eta^2\right)\left[\hat{a}, \hat{a}^\dagger\right] \neq \left[\hat{a}, \hat{a}^\dagger\right].$$

In the correct modelization we must look at the absorbed field of the mode $\hat{a}$ as to a scattered one in another mode $\hat{a}_s$. And the same mechanism which produces this scattering will be able to scatter a mode $\hat{b}_s$ into the mode $\hat{b}$. The correct description will be

$$\hat{b} = \sqrt{1 - \eta^2}\hat{a} - \eta\hat{b}_s$$

$$\hat{a}_s = \eta\hat{a} + \sqrt{1 - \eta^2}\hat{b}_s$$

and it is easy to verify that the commutation rule is preserved.

A practical consequence of this is that if $\hat{a}$ describes a carefully prepared squeezed state, after the loss the squeezing of the field $\hat{b}$ will be reduced. The reason is that $\hat{b}_s$ will describe a non-squeezed vacuum state, and its fluctuations will add to the ones of $\hat{a}$.

There are many techniques which give very interesting perspectives for the evasion of the standard quantum limit in next generations gravitational wave detectors. The direct experimental study of systems where quantum ponderomotive squeezing effects are relevant opens up the possibility of seeing quantum mechanics at works on really macroscopic physical systems, and this is by itself something whose importance cannot be underestimated.

There are several interesting issues which are in my opinion worth of further investigation, that I mentioned only very briefly in this talk. I would like to mention in particular

- ideas for looking at optomechanical interactions between several modes (both mechanical and optical). States with interesting correlations could be produced in this way and maybe used to allow for very sensitive measurements;
- the use of feedback in macroscopic systems, at the quantum level;
- investigations about the role of entanglement.

3.4 Parametric Instabilities

An important issue connected to the use of higher power lasers in interferometric gravitational wave detectors is the one of parametric instability [40]. Radiation pressure couples the optical modes not only to the mirror motion, but also to several of its internal mechanical modes.

If the laser power is large and the quality factor of the mechanical modes is high enough an instability can show up. Several ideas have been proposed to solve this problem. For example, the geometry of the mirror, in particular its curvature radius, can be controlled up to some level using thermal expansion. As mirror's internal modes depend on the geometry, one could try to move them out of the resonance condition. Another possibility is the reduction of the Q factor of the modes using a lossy coating, but this should be done in such a way not to increase the thermal noise. It is also possible to try to actively control the mechanical modes by injecting in an appropriate way higher order modes of an auxiliary laser.

3.5 Gravity Fluctuations

As discussed in Sect. 3.1, great care is taken in order to isolate test masses from the ground motion. But seismic motion is one of the sources of another kind of noise that cannot be attenuated, called *gravity gradient noise* or *Newtonian noise*.

Gravity gradient noise is the effect of the direct coupling between environmental fluctuations of mass density and test masses. The fluctuations can be induced by seismic motion, but also by atmospheric motion, infrasound in the infrastructures and moving objects.

The fluctuating gravitational potential can be written as

$$\delta \Phi \left(\vec{r}, t \right) = G \int \frac{\vec{\nabla} \cdot \left[\rho_0 \left(\vec{r}', r \right) \vec{u} \left(\vec{r}', r \right) \right]}{|\vec{r} - \vec{r}'|} d^3 \vec{r}' \tag{3.13}$$

where ρ_0 is the unperturbed mass density and $\vec{u}$ the velocity field of the medium. In the seismic case the gradient of the mass current density at the numerator in Eq. (3.13) is different from zero when the velocity field has a nonzero divergence (compressional effects, for example, in a seismic wave with a longitudinal component) or when the unperturbed mass density is inhomogeneous (transport effects, for example, at the surface where ρ_0 has a discontinuity).

Looking at Fig. 6 we see that gravity gradient noise is not expected to be relevant for the sensitivity of advanced detectors, with the exception of periods of strong seismic or meteorological activity. However it will become so for third generation detectors, in the low frequency part of detection bandwidth. For this reason several modelizations have been studied [41, 42], and some mitigation strategies have been proposed.

Analytical models are useful for estimates of the effect, but they must be quite simplified compared with the reality. On the other hand numerical studies, in particular finite element models, can be useful to deal with complex situations and/or complex excitations, that are generated by

- non-trivial geometry or morphology (for example, anisotropy of the medium);
- infrastructure effects;
- effects of localized sources, which are quite common for anthropogenic seismic noise;
- study of short scale non-stationarity.

This approach requires a large computational power, in particular if the dynamics of the medium must be simulated. It also needs validation: analytical models are quite important as test beds for comparison.

About the mitigation of gravity gradient noise, currently two different possibilities have been suggested.

The first option is to put the detector underground [43]. The idea is that typically the seismic noise decreases with the depth, especially if the largest contribution to gravity gradient noise comes from surface waves.

In Fig. 9 an estimate of the effectiveness of this approach is given, for a simple analytical model (homogeneous underground with isotropic excitations of surface waves). Apparently the strategy is effective at not too low frequencies. This can be easily understood: the mass fluctuations are localized near the surface, and the induced fluctuating field is exponentially damped on a length scale of the order of seismic wavelength. But $\lambda_{\text{seism}} = c_s f^{-1}$ where c_s is the sound speed of the wave, so the reduction of a lower frequency fluctuation requires a larger depth.

Another option is to monitor some set of auxiliary quantities correlated with the noise, and design a *subtraction* procedure which can reduce it. The basic idea can be illustrated in a simple and not completely realistic case, where both the gravity gradient noise and the auxiliary measurements are modeled as stationary and Gaussian stochastic processes.

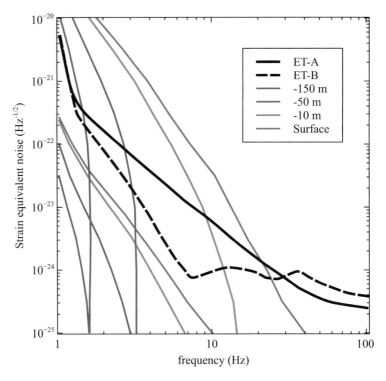

Fig. 9 Estimate of the reduction of gravity gradient noise by going underground. The spectral amplitude of the noise is shown. The area labeled *surface* gives the variability of gravity gradient noise on the surface, depending on the seismic activity of the location. Other areas correspond to different depth of the detector. A comparison is done with two different sensitivity curves for a third generation detector (Einstein Telescope) [25, 44]

We suppose a *subtracted signal* of the form

$$s_s(t) = s(t) - \sum_i \int \Lambda_i \left(t - t'\right) a_i(t')dt'$$

where $s(t)$ is the detector's output, $a_i(t)$ the i-th auxiliary measurements and Λ_i an appropriate set of filters. If we define the subtraction efficiency at a given frequency as

$$\eta(\omega) = 1 - \frac{S_s(\omega)}{S(\omega)}$$

where $S_s(\omega)$ is the power spectrum of the subtracted signal and $S(\omega)$ the power spectrum of the detector's output we can show that the best efficiency is obtained when an appropriate set of filters Λ_i are chosen [43, 45], and is given by

$$\eta_{\mathrm{opt}} = \frac{\sum_{ij} N_i(\omega) C^{-1}(\omega)_{ij} N_j(\omega)}{S(\omega)} \tag{3.14}$$

where $N_i(\omega)$ is the spectral correlation between the detector's output and the i-th auxiliary measurement and $C(\omega)_{ij}$ the spectral correlation between the i-th and the j-th auxiliary measurement.

The efficiency η depends on the sensors which give the auxiliary measurements, i.e., on their positions and orientations. It is apparent from Eq. (3.14) that a large N_i is good. This is not surprising because it is expected that a strong correlation between auxiliary sensors and noise will improve the subtraction procedure. On the other hand a large C_{ij} is bad. This can be understood by considering that a large correlation between the auxiliary measurements reduces the amount of information that can be used for the subtraction.

Simulations and tests show that this procedure can be effective [46], and it is reasonable to suppose that by putting together the two options it will be possible to reduce gravity gradient noise by one or two orders of magnitude.

4 Detections

The main tool for the detection of a gravitational wave signal is the Wiener filter. Suppose that we want to discriminate between two different hypothesis for the observed signal $s(t)$, namely

$$\begin{cases} H_0 & s(t) = n(t) \\ H_1 & s(t) = n(t) + h(t) \end{cases}$$

where $n(t)$ is a stochastic process which represent the interferometer noise, $h(t)$ a signal of know shape. The Neyman Pearson theorem tell us that the *optimal* algorithm is the following: choose H_1 when

$$\frac{P\left(s|H_1\right)}{P\left(s|H_0\right)} > \lambda$$

where λ is a given threshold. In the context of the theorem, *optimal* means the test that maximizes the detection probability P_D (i.e., the probability of choosing the H_1 when the $h(t)$ is present) for a given value of the false alarm probability P_{FA}, which is the probability of choosing H_1 when there is only noise.

For a Gaussian and stationary noise we can write

$$P\left(s|H_0\right) = N \exp\left[-\frac{1}{2}\langle s, s \rangle\right]$$

where N is a normalization constant. We defined the scalar product

$$\langle a, b \rangle = \int \int a\left(t\right) K\left(t - t'\right) b\left(t'\right) dt dt'$$

where K encodes the autocorrelation of the noise. In a similar way we have

$$P\left(s|H_1\right) = N \exp\left\{-\frac{1}{2}\langle s - h, s - h\rangle\right\}$$

and the Neyman–Pearson test is simply[1]

$$\langle s, h\rangle > \lambda'$$

which has a very clear interpretation: we project the measured data on the known signal and we compare the result with a threshold. This is the Wiener filtering procedure, that can be written explicitly in the frequency domain as

$$\int \frac{\tilde{h}(\omega)^* \tilde{s}(\omega)}{S_n(\omega)} \frac{d\omega}{2\pi} > \lambda'.$$

When the signal depends from some set of unknown parameters $\underline{\alpha}$, the scalar product $\langle s, h_{\underline{\alpha}}\rangle$ is sampled over the space of allowed values and the largest one is chosen. This is the detection candidate.

4.1 Coalescences

A *chirp-like* signal is expected in compact binary coalescence events. They can be described as the last part of the evolution for a bound system of two compact objects of masses m_1 and m_2 (black holes or neutron stars). As in the case of the PSR B1913+16 binary system, the two orbiting bodies generate a $\ddot{Q}_{ij} \neq 0$ and radiate gravitational waves losing energy. During this process the relative distance between the bodies decreases, and the orbital frequency increases. If we exclude the final part of the evolution we can neglect relativistic and strong field corrections: the relative energy loss for each orbit is small and we can describe the process in the adiabatic approximation. The result is a quasi-periodic signal, which can be written as

$$h_{ij} \sim \frac{4}{r}\left(\frac{GM_c}{c^2}\right)^{5/3}\left(\frac{\omega_{gw}(t)}{2c}\right)^{2/3}\cos\left[\Phi_0 + \int^t \omega_{gw}(t')dt'\right] \qquad (4.1)$$

neglecting an angular factor which depends on the polarization and on the source position in the sky. Here $M_c = (m_1 m_2)^{3/5}(m_1 + m_2)^{-1/5}$ is the chirp mass of the system and $\omega_{gw}(t)$ is the instantaneous angular frequency of the wave, which is two times the orbital angular frequency, owing to its quadrupolar nature. We see that

---

[1] We absorbed terms independent from the observed $s(t)$ in a redefinition of the threshold.

both the frequency and the amplitude of the signal increase accordingly with the law

$$\omega_{gw}(t) = \frac{1}{4}\left(\frac{GM_c}{c^3}\right)^{-5/8}\left(\frac{5}{t_c - t}\right)^{3/8}. \tag{4.2}$$

When $t = t_c$ both the amplitude and the frequency diverge. Before that the relative distance between the two bodies becomes so small that they merge together. This process cannot be described using a simple analytical approximation but requires numerical simulations which fully take into account strong field effects in general relativity: anyway t_c is called coalescence time and can be used as an estimate for the time of merger. Equation (4.2) will be an acceptable approximation only until[2]

$$f_{gw} \lesssim f_{ISCO} = 2.2 \times 10^3 \text{ Hz} \left(\frac{M_\odot}{m_1 + m_2}\right) \tag{4.3}$$

which we can take as an upper limit for the frequencies emitted by the source.

In the case of the Hulse–Taylor pulsar $\omega_{gw} \simeq 2 \times 10^{-4}$ s: this means that the frequency of the gravitational wave emitted is too small to be detectable by Virgo and LIGO, which as we saw (see Fig. 6) have a good sensitivity in the frequency band between $O(10)$ Hz and $O(10^4)$ Hz. It means also that PSR B1913+16 will merge in $O(10^8)$ years: in the final phase for the system evolution the signal will enter in the detection bandwidth of the detectors, and will stay there until f_{ISCO} will be reached.

Note the inverse proportionality between f_{ISCO} and the mass of the system. This means that if the total mass of the system is large the merger will happen when the signal is inside the detection band. When the total mass increases, the time spent inside the detector's sensitivity window will decrease until it disappear completely.

4.2 Binary Black Holes Coalescences

An extensive campaign of data taking periods for the LIGO–Virgo network, intermixed by commissioning ones, has been planned (see [47] for up-to-date information). The two scientific run O1 (from September 2015 to January 2016) and O2 (from November 2016 to August 2017) have been highly successful. In O1 the first direct detection of gravitational waves was obtained, from the observation of a coalescence of two black holes [5] with mass of about $30M_\odot$ each. The signal, which lasts less than 0.5 s in the sensitivity band, was detected by the LIGO interferometers, as Virgo was not yet commissioned at the time. The event, named

---

[2]This estimate of f_{ISCO} is strictly valid only when one mass is much smaller than the other, and neglects spin effects.

GW150914, was quite evident with a false alarm probability of about $5 \times 10^{-6} \, \text{yr}^{-1}$ and a significativity greater than 5.3σ. Another event, GW151226 [48], was detected in the following. It was interpreted once again as a BBH coalescence.

Several other BBH coalescences were detected by the LIGO/Virgo interferometers during the O2 run (see Figs. 10 and 11). In all cases a detailed study of the signal shape allowed for an estimate of the parameters of the systems: the masses and the spins of the two progenitor black holes and of the final one, the distance of the source and its position in the sky. The accuracy of these estimates depends on the parameter considered. For example, the determination of the "chirp mass"

$$\mathcal{M} = \frac{(m_1 m_2)^{3/5}}{(m_1 + m_2)^{1/5}} \tag{4.4}$$

which is the combination of the progenitor masses m_i which dominates the signal's shape was quite accurate, $28.1^{+1.8}_{-1.5} M_\odot$ for GW150914 and $8.3^{+0.3}_{-0.3} M_\odot$ for GW151226. On the other hand the accuracy on the mass ratio m_1/m_2, which

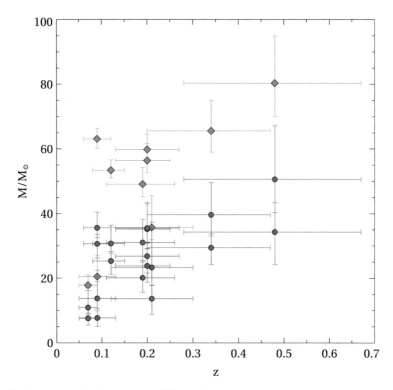

Fig. 10 The masses of all the observed black holes at the present time [49], as a function of z. Circles correspond to the masses in the coalescing binary system, while diamond indicates the masses of the final black hole. Error bars correspond to a 90% interval around the median value

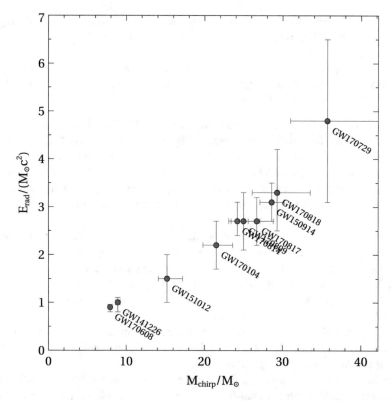

Fig. 11 The total radiated energy for the observed BBH mergers at the present time [49] as a function of the chirp mass of the binary. Error bars correspond to a 90% interval around the median value

contributes only at higher order in the post Newtonian expansion, was lower (around 20% for GW150914).

Using a Bayesian approach it is possible to give an accurate statistical characterization of the source parameters. The basic idea is that, as a consequence of the Bayes theorem, we can write

$$P\left(\underline{\alpha}|s(t)\right) = \frac{P\left(s(t)|\underline{\alpha}\right) P\left(\underline{\alpha}\right)}{P\left(s(t)\right)}. \tag{4.5}$$

Here the quantity of interest is the probability distribution of the source parameters $\underline{\alpha}$, conditioned to the observed data. Accordingly with Eq. (4.5) it can be evaluated from the known probability distribution of the measurement conditioned to given value of the parameters, and from the prior probability distribution of $\underline{\alpha}$.[3] Note that

---

[3] $P(s)$ here play the role of a normalization constant.

in this way it is possible to evaluate joint distributions of parameter and/or integrate away the uninteresting ones. Some examples of the results of this procedure can be found in [50].

Another application is about testing general relativity in the strong coupling regime started. Several investigation of this kind have been done [51], and I will describe shortly only two of them.

Compatibility of Inspiral and Post-Inspiral Signal The signal coming from a coalescence can be partitioned in three different phases. In the first one, called *inspiral*, the two progenitors can be described as orbiting point masses, and the evolution of the system can be evaluated with an accurate perturbative expansion including general relativity effects of high enough order. The perturbative expansion is expected to be accurate because the system is not yet really in the strong gravity regime.

In the second phase, called *merger*, the perturbative expansion breaks down, together with the point mass approximation. The nonlinear general relativity effects become dominant and the evolution of the system can be obtained by solving numerically the Einstein equations. During this phase the two progenitor black holes merge together, radiating a significant fraction of their mass (about $3.0 M_\odot$ for GW150914) as gravitational waves.

In the third phase, called *ring down*, the final compact object, which can be described as a perturbed black hole, radiates its excitation energy evolving toward its final stationary Kerr configuration.

An estimate of the mass and final spin of the source can be done using only the inspiral signal (with a signal-to-noise ratio of about 19.5 for GW150914), or only the post-inspiral signal (with $SNR \simeq 16$ for GW150914). The two probability distributions can be tested for compatibility. In this way strong gravity effects not explained by the Einstein's theory can be put in evidence. Results show agreement with general relativity predictions [51].

Quasi Normal Modes In the *ring down* phase the excitation of the final black hole can be described as a superposition of *quasi normal modes*, namely harmonic damped oscillations, each parameterized by a frequency ω_m and a decay time τ_m. A relevant fact is that the values of ω_m and Q_m are expected to depend only on the mass and spin of the black hole.

This prediction can be tested by comparing the measured values of ω_m, τ_m with the ones expected from the estimated final mass and spin. This is a non-trivial procedure. If we analyze the signal starting from a time t_0 too near to the merger we cannot be sure that we entered the linear regime. Also if this is true, we expect a complicate expression

$$h(t) \simeq \sum_m A_m \theta(t - t_0) e^{-(t-t_0)/\tau_m} \cos[\omega_m(t - t_0) + \phi_m] \qquad (4.6)$$

with relevant contributions coming from several modes.

Moving the start time t_0 far from the merger we expect only the contribution of the mode with the largest τ_m to survive. But the amplitude of the signal will decay giving a low signal-to- noise ratio. Anyway tests using only the dominant mode have been attempted [51], obtaining results consistent once again with general relativity.

Localization Toward the end of O2 run Virgo was commissioned and it joined the network. Immediately after, the first three interferometers detection of a BBH coalescence, GW170814, was obtained [7].

The inclusion of Virgo produced a qualitative improvement for the accuracy of the localization of the source in the sky. This can be understood easily: the largest amount of information about the direction of propagation of the gravitational wave signal is encoded in the relative delays between the signal detected by the different interferometers in the network. With two detectors, we have just a single delay. For example GW150914 was seen by the two LIGO interferometers with a delay of about 7 ms, while the light time of flight between them was 10 ms. This gave a constraint for the angle between the propagation direction and the line which connects the two detectors, which allowed for a determination of a ring shaped region in the sky where the source must be. The width of the ring was determined by the accuracy on the delay estimate and additional, weaker constraints about the compatibility of the two signal shapes allowed for some reduction of the region.

When three detectors are available three different delays can be used: in principle this allows for a complete determination of the source position, modulo a reflection degeneracy. Once again the final accuracy depends on the error on the delays and on additional constraint, but a large improvement was expected and found [7]: the typical determination for the initial LIGO only network was about $1160\,\mathrm{deg}^2$, to be compared with about $60\,\mathrm{deg}^2$ for the LIGO–Virgo case.

This improvement was not really important for GW170814, as the only signal expected from a BBH coalescence is a gravitational one. But it was the demonstration that the era of multi-messenger observations could start.

4.3 GW170817

In the very final part of O2 run a different kind of coalescence was jointly detected by the LIGO–Virgo collaboration, with a false alarm probability of about $10^{-5}\,\mathrm{yr}^{-1}$. While previous observations remained inside the sensitivity window for less than two seconds, in this case the signal last around a minute, and reached much higher frequencies.

The largest instantaneous frequency of the signal, and the time interval of the observation, depends mainly on the dimension of progenitors. The frequency of emitted gravitational waves is two times the orbital one, so the progenitors need to reach a small orbital period. This is possible only (think about the third Kepler law for a rough estimate) if they can orbit one very near to the other before merger. Black holes with reasonable mass cannot be compact enough to allow for this, so the simplest interpretation is that the observed signal is the result of a coalescence between two neutron stars [8]. This interpretation was confirmed by the detailed determination of the parameters, which gives for the progenitor masses $1.36M_\odot < m_1 < 1.60M_\odot$ and $1.17M_\odot < m_2 < 1.36M_\odot$.

Another support to the hypothesis came from the observation of several electromagnetic counterparts. Fermi/GBM detected a gamma radiation peak 1.7 s after the merger. The localization done with the gravitational detectors was consistent with the one provided by Fermi and INTEGRAL, and the signal was interpreted as a short GRB (GRB 170817A) generated by the merger [52].

Nuclear Matter Equation of State Looking at the gravitational signal only, it was possible for the first time to measure the tidal deformability of the progenitors. Information about that are contained in the waveform: a high deformability can enhance the production of gravitational radiation in the final inspiral phase and introduce an early cutoff during the merger. The tidal deformability gives information about the nuclear matter equation of state, and this open the possibility of ruling out some of the existing models [53]. With the current sensitivity in the high frequency region which is sensitive to tidal effects few models (with very low stiffness) have been excluded.

Hubble Parameter Determination Another interesting result was the independent estimation of the Hubble parameter, using only the gravitational signal. The constraint is currently not competitive with the ones from Planck and SHoES [54], but is conceptually important because it does not need the use of a distance ladder. And in the future, when n similar events will be available, it will be possible to combine the results of all the analysis reducing the error essentially by a factor $\sqrt{n}$.

Electromagnetic Counterparts Beyond the mentioned gamma signal from Fermi, GW170817 was followed by an extensive set of electromagnetic emissions in the X, UV, optical, infrared, and radio band. These were extensively observed by a large number of collaborations [55], owing to the good localization provided by LIGO–Virgo and Fermi+INTEGRAL.

An important result obtained in this way was the detailed observation of the evolution of the spectra, which are consistent with the kilonova scenario: matter ejected in the post-merger phase undergoes r-process in a neutron rich environment, providing a mechanism for the production of elements heavier than iron.

5 Other Sources

A coalescing binary is not the only possible source of gravitational waves. In this section we discuss other promising ones.

5.1 Continuous Sources

Continuous signals can be produced by a rotating neutron star. This very compact object (a mass typically of about $1.4 M_\odot$ is confined in a region with a radius of few kilometers) can generate gravitational waves if a small deviation from sphericity is present. This can be allowed by elastic or magnetic effects in the crust, and the expected deviation is quite small. However a signal of this kind could help, if detected, in clarifying the nature of the matter inside the star, and in particular its equation of state [56].

5.2 Burst Signals

A *burst* signal is characterized by its small duration, typically of the $O(10^{-2})$ s. They can be produced in violent explosive events, such as a *supernovae* explosions but also in some more exotic cases, for example, in cosmic string dynamics.

In the case of *supernovae*, a requirement for the generation of gravitational wave is that the core collapse must be asymmetric, otherwise no quadrupole or higher time dependent moment can be generated. There is no clear consensus about the asymmetry one can expect, as the physics involved is quite intricate and large scale numerical simulations are needed. An *supernova* event in our galaxy is expected once in 10–100 y, so the chances of observing it are not very high. However if this will happen the study of the emitted signal could add a lot to the knowledge of this still poorly understood physical system.

5.3 Stochastic Signals

A stochastic background of gravitational waves can be seen as a gravitational wave field which evolves from an initially random configuration, or as the result of a superposition of many uncorrelated and unresolved sources. There are potentially several kind of gravitational wave stochastic backgrounds, and accordingly with their origin they can be broadly classified as astrophysical and cosmological.

In the first case, the background's sources are astrophysical events (such as a compact binaries coalescence, a supernova, a rotating NS, and so on) since the beginning of stellar activity. The study of an astrophysical SB can give important information about the evolution of stellar populations. Though a large number of potential sources have been studied, it is true that the predicted amplitudes are typically too small in order to be seen with current generation detectors. There are exceptions however, as we will see.

A cosmological SB (see [57] for a short review) carries in principle very important information about the early universe, the reason being that gravitational interaction is so weak (compared, for example, with the electromagnetic one) that gravitational waves could be considered decoupled since the very beginning of the universe evolution. There are several mechanism that can generate a cosmological SB, and we will sketch briefly some of them. It must be admitted that also in this case the detection could be considered not guaranteed with current sensitivities. Once again there are interesting exceptions where the task does not seem hopeless. And the perspective for third generation earth-bound detectors and space ones like LISA seems quite interesting.

The Detection of a Stochastic Background The description of a stochastic background can be summarized by a mode expansion of the gravitational strain field in the frequency domain like

$$\tilde{h}_{ab}(\vec{x}, \omega) = \sum_P \int d\Omega_{\hat{n}} \, A^P_{\hat{n},\omega} \, \varepsilon^P_{ab}(\hat{n}) e^{i\frac{\omega}{c}\hat{n}\cdot\vec{x}}. \tag{5.1}$$

Here each mode is parameterized by a polarization P and a propagation direction $\hat{n}$, and its amplitude is represented by a stochastic variable $A^P_{\hat{n},\omega}$. The statistical properties of the stochastic background depend on the generation mechanism and can be described by the cumulants of the A's amplitudes.

A stochastic background can be safely considered stationary, because the typical physical time scales involved are much larger that the observation time. It is also expected to be typically Gaussian,[4] and in this case it is completely described by its second order cumulants. As a consequence form a statistical point of view the only relevant quantity is the second order correlation $\langle \tilde{h}_{I_1} \tilde{h}_{I_2} \rangle$ between the signal of a pair of detectors.

The optimal detector can be obtained starting from the specific form of this correlation. As an example, if the stochastic background is also isotropic and non-polarized it comes out that the optimal statistic is given by

$$Y = \int d\omega \frac{\tilde{x}_1^*(\omega)\tilde{x}_2(\omega) \, \gamma_{12}(\omega) S_h(\omega)}{S_1(\omega)S_2(\omega)} \tag{5.2}$$

where S_i is the power spectrum of the i-th detector noise, S_h the theoretical power spectrum of the stochastic background, and $\tilde{x}_i$ the output of the i-th detector, which is expected to be a sum of the signal $\tilde{h}_i$ and of the noise. The function γ_{12} is called *overlap reduction function* and is equal to the spectral coherence between the signals coupled to the two detectors. A loss of coherence is expected, since two detectors are coupled in a different way to the modes of the expansion (5.1), unless they are in the same place and oriented in the same way. In particular, a loss of coherence

---

[4]We will see later that exceptions are possible.

for the signal at a frequency f is expected when the separation d between the two detectors becomes of the same order or larger than the corresponding wavelength, $f > cd^{-1} = \tau^{-1}$. For example the separation between the two LIGO detectors if about 10^{-2} light-seconds, and the coherence is reduced by 50% around 50 Hz. Equation (5.2) tells us that in order to detect a stochastic background the output of two different detectors must be correlated. The signal-to-noise ratio can be defined as

$$\text{SNR}^2 \propto T \int_0^\infty S_h^2(\omega) \frac{\gamma_{12}^2(\omega)}{S_1(\omega) S_2(\omega)} d\omega \tag{5.3}$$

and as expected it is reduced by a loss of coherence. This means that it is good for a pair of detectors to be aligned, and in principle it should be good for them to be one near the other. However it must be taken into account that the analysis which leads to the optimal statistic (5.2) is based on the hypothesis that the noise of the two detectors is uncorrelated, and it is difficult to be sure about that if they are nearby. Note also the proportionality $\text{SNR} \propto \sqrt{T}$ which tells us that the probability of detection increases with the observation time, though not particularly fast.

Sources, Sensitivities, and Upper Limits A prediction for the amplitude of a stochastic background can be represented by its differential energy spectrum, which is defined as an energy density for logarithmic interval of frequency normalized to the critical energy density, and is connected to the observed power spectrum of the gravitational strain signal by

$$h_0^2 \Omega_{GW}(f) = \frac{1}{\rho_c} \frac{d\rho_{gw}}{d \log f} = \frac{4\pi^2 h_0^2}{3H_0^2} f^3 S_h(f). \tag{5.4}$$

Probably the most fundamental production mechanism for a cosmological gravitational waves stochastic background is the parametric amplification. Basically it can be understood by looking at the effect of the time dependence of the scale parameter a in a Friedmann–Robertson–Walker metric which describes an expanding universe. The evolution of the amplitudes of gravitational waves modes in (5.1) is given by an equation like

$$\ddot{A}_\omega + \left(\frac{\omega^2}{c^2} - \frac{\ddot{a}}{a} \right) A_\omega = 0 \tag{5.5}$$

which describes an harmonic oscillator with a time dependent parameter. The transition between two different evolution regime, for example, between the inflationary epoch and the radiation dominated one, can produce an amplification of the amplitudes. However the predicted final background predicted in the standard inflationary scenario is too small to be detected with the current and foreseen sensitivities [63] (see blue line in Fig. 12). Non-standard models for universe

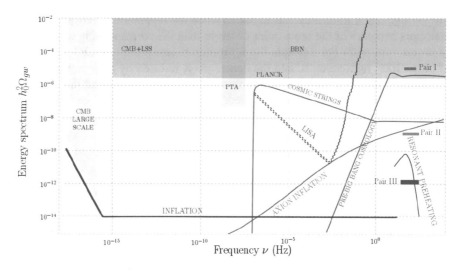

Fig. 12 Some expected spectra of cosmological gravitational wave stochastic backgrounds, compared with upper limits and current sensitivities. Continuous lines correspond to energy spectrum predicted for a selection of production mechanism and are described in the text. Dashed lines correspond to the sensitivity level which can be obtained using few months of data with a pair of first (LIGO, Virgo) second (LIGO Advanced, Virgo Advanced [58]), and third (ET [59]) generation detectors, and with LISA. Filled area represents current upper limits. BBN is an integral bound obtained from the observed abundances of light elements [60]. CMB bounds are related to cosmic microwave background observations [60, 61]. PTA is obtained from pulsar timing array observations [62]

evolution can produce more interesting results. Alternative cosmologies like pre-big bang scenarios (red line in Fig. 12) can evade the CMB large anisotropy upper limit at lower frequency and the BBN-CMB integral bound at higher ones and can produce a significant background at the frequencies of interest for earth-bound detectors [64]. In inflation models which include couplings between axion and gauge fields [65] (green line in Fig. 12) the gauge field back-reaction on the inflaton can extend the inflationary epoch, in such a way that gravitational waves production at higher frequencies is increased. During a resonant preheating phase at the end of the inflation, inflaton energy can be transferred efficiently to other particles. This mechanism can also be described by Eq. (5.5), and the prediction for this kind of background can be interesting for third generation detectors (purple line in Fig. 12). Spectrum peak depends on energy scale (which is chosen at 10 GeV in the plot, higher frequencies correspond to a larger scale).

A completely different production mechanism is connected with the presence of a dynamical network of cosmic (super)strings. String entering in the horizon can evolve dynamically, and a population of string loops can be generated via interconnections. These loops can emit gravitational waves, especially when cusps and kinks are present, and generate a stochastic background similar to the one represented in Fig. 12 (magenta line) [66]. There is a large parameter space to explore for these models, and observations started to give constraint about it.

The LIGO–Virgo collaboration produced upper limits on gravitational wave stochastic background. The current one [67] is the best bound on the interferometric observation frequency window. The direct observation of a stochastic background will allow to test for fundamental physics models and is starting to give constraint for them in synergy with CMB and pulsar timing array observations [68].

Implication of Detections for Stochastic Background The coalescence events discussed in Sect. 4 were not only the first direct detections of gravitational waves, but also the first direct evidences of BBH and BNS coalescences. Their observation allows to put constraint on their rates, and indirectly on black hole and neutron star populations. The gravitational wave luminosity of a BNS event is much lower (about $0.025 M_\odot$ for GW170817) compared with the BBH one, so the range of detection is also smaller. Both for BBH and BNS coalescences distant events can be too weak to be detected with a good enough statistical confidence. But if we look far, we expect to see an increasing number of events.

An interesting possibility is that the statistical superposition of a large number of events, too overlapped and weak to be resolvable, can generate a *stochastic background* of gravitational waves. This could be detected in principle by looking at the correlations between outputs of interferometers in the network.

Estimate based on the detected events and population models shows that in principle this kind of background could be seen with the current generation of detectors, at their design sensitivity [69]. Stochastic backgrounds of BBH and BNS coalescences are just a promising example of a large set of *astrophysical stochastic background*. I will not discuss here other possible astrophysical backgrounds, and the interested reader can refer to reviews [70].

Future Perspectives Earth-bound gravitational waves detector measurements started to set interesting upper limits [71] and constraints on models, especially when combined with other observations [68]. These limits are expected to improve when the data coming from the LIGO and Virgo advanced detectors will be analyzed, and accordingly with theoretical predictions in some particular cases a detection could be possible.

When the sensitivity will further improve, for example, with third generation detectors such as ET [59], which is expected to gain an order of magnitude in strain compared with advanced detectors, several more refined investigations will become of interest. In the case of a detection it will be possible to estimate parameters and disentangle contributions coming from different polarizations. This will be particularly interesting to test for models beyond general relativity. These typically predict additional polarizations with respect to the standard plus and cross one.

Data analysis techniques that make possible these studies are known [72] and currently applied to the available data [73]. It will be also possible to test for anisotropies, using techniques which are also well defined and applied in the LIGO–Virgo collaboration [74]. Finally, it will be possible to test non-Gaussian models [75]. Many astrophysical stochastic backgrounds can potentially be non-Gaussian, and in some cases (for example, cosmic strings) this can be true for cosmological backgrounds. The reason for this is that the background can be generated by a number of events (for example, compact binaries coalescences) whose signals overlap in such a way that it is not possible to separate the single contribution. But the number of overlapped events can be not very large, and in this case the central limit theorem cannot be applied. The optimal detection of these non-Gaussian backgrounds is a non-trivial problem, and interesting information about the model can be in principle extracted. This is an open issue.

In all the three cases mentioned (polarization disentanglement, study of anisotropies, study of non-Gaussianities) the availability of an extended network of detectors is crucial in order to apply the relevant data analysis techniques.

6 Conclusions

In the last years the LIGO–Virgo collaboration obtained several important scientific discoveries, such as the first direct detection of gravitational waves, the first evidences of BBH and BNS coalescences, the first evidence of an association between a BNS coalescence, short GRB, and kilonova. The detection of a BNS coalescence opened the era of multi-messenger astronomy.

Several other sources are expected in the future, for example, rotating neutron stars and supernovae.

The next O3 scientific run is expected in the early 2019, with three detectors in operation with improved performances. As these will approach the design sensitivity the rate of events will grow quite fast.

Event with a very high signal-to-noise ratio will become available and will allow for better parameter estimation. For example with an improved high frequency sensitivity it will be possible to obtain severe constraint on the nuclear matter equation of state.

The detection of a cosmological stochastic background is probably something that we can expect in third generation detectors (see Fig. 13 for a comparison between the sensitivities of initial, advanced and third generation detectors), but in the meantime for sure several upper limits and constraint on models will be obtained. A new way of looking at the universe is born.

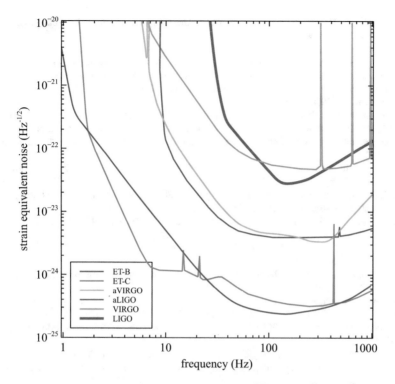

Fig. 13 The strain equivalent noise spectral amplitude $\sqrt{S_n(\omega)}$ of initial, second generation (Advanced) and third generation detectors. See Eq. (3.2) for a definition

Acknowledgement I would like to thank the organizers of the Domoschool 2018 for inviting me, and the editors of this book for their patience in waiting for my contribution.

References

1. A. Einstein, Die Grundlagen der Allgemeinene Relativitätstheorie (The foundations of the theory of general relativity). Ann. Phys. (1900) (series 4) **354**(7), 769–822 (1916). ISSN 0003-3804. https://doi.org/10.1002/andp.19163540702
2. A. Einstein, Näherungsweise Integration der Feldgleichungen der Gravitation (Approximate integration of the field equations of gravitation), in *Ständiger Beobachter der Preussischen Akademie der Wissenschaften: Part 1* (1916), pp. 688–696
3. A. Einstein, *Über Gravitationswellen* (On gravitational waves), in *Ständiger Beobachter der Preussischen Akademie der Wissenschaften, Part 1* (1918), pp. 154–167
4. A. Einstein, N. Rosen, On gravitational waves. J. Frankl. Inst. **223**(1), 43–54 (1937). ISSN 0016-0032 (print), 1879–2693 (electronic). https://doi.org/10.1016/S0016-0032(37)90583-0
5. B.P. Abbott, R. Abbott, T.D. Abbott, M.R. Abernathy, F. Acernese, K. Ackley, C. Adams, T. Adams, P. Addesso, et al., Observation of gravitational waves from a binary black hole merger. Phys. Rev. Lett. **116**, 061102 (2016). https://doi.org/10.1103/PhysRevLett.116.061102

6. B.P. Abbott, R. Abbott, T.D. Abbott, M.R. Abernathy, F. Acernese, K. Ackley, C. Adams, T. Adams, P. Addesso, et al. Properties of the binary black hole merger GW150914. Phys. Rev. Lett. **116**, 241102 (2016). https://doi.org/10.1103/PhysRevLett.116.241102

7. B.P. Abbott, R. Abbott, T.D. Abbott, F. Acernese, K. Ackley, C. Adams, T. Adams, P. Addesso, R.X. Adhikari, et al., GW170814: a three-detector observation of gravitational waves from a binary black hole coalescence. Phys. Rev. Lett. **119**, 141101 (2017). https://doi.org/10.1103/PhysRevLett.119.141101

8. B.P. Abbott, R. Abbott, T.D. Abbott, F. Acernese, K. Ackley, C. Adams, T. Adams, P. Addesso, R.X. Adhikari, et al., GW170817: observation of gravitational waves from a binary neutron star inspiral. Phys. Rev. Lett. **119**, 161101 (2017). https://doi.org/10.1103/PhysRevLett.119.161101

9. J. Weber, Gravitational-wave-detector events. Phys. Rev. Lett. **20**(23), 1307 (1968). https://doi.org/10.1103/PhysRevLett.20.1307

10. J. Weber, Gravitational radiation from the pulsars. Phys. Rev. Lett. **21**, 395–396 (1968). https://doi.org/10.1103/PhysRevLett.21.395

11. J. Weber, Evidence for discovery of gravitational radiation. Phys. Rev. Lett. **22**, 1320–1324 (1969). https://doi.org/10.1103/PhysRevLett.22.1320

12. M.E. Gertsenshtein, V.I. Pustovoit, On the detection of low frequency gravitational waves. Sov. Phys. JETP **16**, 433 (1962).

13. R.L. Forward, Wide band laser interferometer gravitational radiation experiment. Phys. Rev. **D17**, 379–390 (1978). https://doi.org/10.1103/PhysRevD.17.379

14. R. Weiss, *Electronically Coupled Broadband Gravitational Antenna*. Quarterly Progress Report (Research Laboratory of Electronics (MIT), Cambridge, 1972), p. 105

15. J. Aasi et al., Advanced LIGO. Class. Quant. Grav. **32**, 074001 (2015). https://doi.org/10.1088/0264-9381/32/7/074001

16. F Acernese, M Agathos, K. Agatsuma, D. Aisa, N. Allemandou, A. Allocca, J. Amarni, P. Astone, G. Balestri, et al., Advanced Virgo: a second-generation interferometric gravitational wave detector. Classical Quantum Gravity **32**(2), 024001 (2015)

17. C.S. Unnikrishnan, Indigo and LIGO-India: scope and plans for gravitational wave research and precision metrology in India. Int. J. Mod. Phys. D **22**(1), 1341010 (2013). https://doi.org/10.1142/S0218271813410101

18. J.M. Weisberg, J.H. Taylor, Relativistic binary pulsar B1913+16: thirty years of observations and analysis. ASP Conf. Ser. **328**, 25 (2005)

19. R.A. Hulse, J.H. Taylor, A High-sensitivity pulsar survey. Astrophys. J. **191**, L59 (1974). https://doi.org/10.1086/181548

20. J.H. Taylor, J.M. Weisberg, Further experimental tests of relativistic gravity using the binary pulsar PSR 1913+16. Astrophys. J. **345**, 434–450 (1989). https://doi.org/10.1086/167917

21. M. Maggiore *Gravitational Waves* (Oxford University Press, Oxford, 2007). ISBN 9780198570745

22. T. Accadia et al., The seismic Superattenuators of the Virgo gravitational waves interferometer. Low Freq. Noise Vibrat. Act. Control **30**(1), 63–79 (2011) https://doi.org/10.1260/0263-0923.30.1.63

23. T. Akutsu et al., KAGRA: 2.5 generation interferometric gravitational wave detector. Nat. Astron. **3**(1), 35–40 (2019). https://doi.org/10.1038/s41550-018-0658-y

24. M. Punturo et al., The third generation of gravitational wave observatories and their science reach. Classical Quantum Gravity **27**, 084007 (2010). https://doi.org/10.1088/0264-9381/27/8/084007

25. S. Hild, S. Chelkowski, A. Freise, J. Franc, N. Morgado, R. Flaminio, R. DeSalvo, A xylophone configuration for a third generation gravitational wave detector. Classical Quantum Gravity **27**, 015003 (2010). https://doi.org/10.1088/0264-9381/27/1/015003

26. C.M. Caves, quantum mechanical noise in an interferometer. Phys. Rev. **D23**, 1693–1708 (1981). https://doi.org/10.1103/PhysRevD.23.1693

27. A.M. Steinberg, M.W. Mitchell, J.S. Lundeen, Super-resolving phase measurements with a multiphoton entangled state. Nature **429**, 161–164 (2004)

28. J.G. Rarity, P.R. Tapster, E. Jakeman, T. Larchuk, R.A. Campos, M.C. Teich, B.E.A. Saleh, Two-photon interference in a Mach-Zehnder interferometer. Phys. Rev. Lett. **65**, 1348–1351 (1990). https://doi.org/10.1103/PhysRevLett.65.1348

29. R. Demkowicz-Dobrzanski, K. Banaszek, R. Schnabel, Fundamental quantum interferometry bound for the squeezed-light-enhanced gravitational wave detector GEO 600. Phys. Rev. **A88**(4), 041802 (2013). https://doi.org/10.1103/PhysRevA.88.041802

30. H.P. Yuen, Contractive states and the standard quantum limit for monitoring free-mass positions. Phys. Rev. Lett. **51**, 719–722 (1983). https://doi.org/10.1103/PhysRevLett.51.719

31. M.T. Jaekel, S. Reynaud, Quantum limits in interferometric measurements. Europhys. Lett. **13**(4), 301–306 (1990). https://doi.org/10.1209/0295-5075/13/4/003

32. H.J. Kimble, Y. Levin, A.B. Matsko, K.S. Thorne, S.P. Vyatchanin, Conversion of conventional gravitational-wave interferometers into quantum nondemolition interferometers by modifying their input and/or output optics. Phys. Rev. D **65**, 022002 (2001). https://doi.org/10.1103/PhysRevD.65.022002

33. H. Müller-Ebhardt, H. Rehbein, R. Schnabel, K. Danzmann, Y. Chen, Entanglement of macroscopic test masses and the standard quantum limit in laser interferometry. Phys. Rev. Lett. **100**, 013601 (2008). https://doi.org/10.1103/PhysRevLett.100.013601

34. T.P. Purdy, R.W. Peterson, C.A. Regal, Observation of radiation pressure shot noise on a macroscopic object. Science **339**(6121), 801–804 (2013). ISSN 0036-8075. https://doi.org/10.1126/science.1231282

35. A.H. Safavi Naeini et al., Squeezed light from a silicon micromechanical resonator. Nature **500**, 185–189 (2013)

36. T. Eberle, S. Steinlechner, J. Bauchrowitz, V. Händchen, H. Vahlbruch, M. Mehmet, H. Müller-Ebhardt, R. Schnabel, Quantum enhancement of the zero-area Sagnac interferometer topology for gravitational wave detection. Phys. Rev. Lett. **104**, 251102 (2010). https://doi.org/10.1103/PhysRevLett.104.251102

37. A. Buonanno, Y. Chen, Signal recycled laser-interferometer gravitational-wave detectors as optical springs. Phys. Rev. D **65**, 042001 (2002). https://doi.org/10.1103/PhysRevD.65.042001

38. B.S. Sheard, M.B. Gray, C.M. Mow-Lowry, D.E. McClelland, S.E. Whitcomb, Observation and characterization of an optical spring. Phys. Rev. A **69**, 051801 (2004). https://doi.org/10.1103/PhysRevA.69.051801

39. T. Corbitt, C. Wipf, T. Bodiya, D. Ottaway, D. Sigg, N. Smith, S. Whitcomb, N. Mavalvala, Optical dilution and feedback cooling of a gram-scale oscillator to 6.9 mK. Phys. Rev. Lett. **99**, 160801 (2007). https://doi.org/10.1103/PhysRevLett.99.160801

40. V.B. Braginsky, S.E. Strigin, S.P. Vyatchanin, Parametric oscillatory instability in Fabry–Perot interferometer. Phys. Lett. A **287**(5), 331–338 (2001). ISSN 0375-9601. https://doi.org/10.1016/S0375-9601(01)00510-2

41. S.A. Hughes, K.S. Thorne, Seismic gravity gradient noise in interferometric gravitational wave detectors. Phys. Rev. **D58**, 122002 (1998). https://doi.org/10.1103/PhysRevD.58.122002

42. M. Beccaria et al., Relevance of Newtonian seismic noise for the VIRGO interferometer sensitivity. Classical Quantum Gravity **15**, 3339–3362 (1998). https://doi.org/10.1088/0264-9381/15/11/004

43. M.G. Beker et al., Improving the sensitivity of future GW observatories in the 1-Hz to 10-Hz band: Newtonian and seismic noise. Gen. Relativ. Gravit. **43**, 623–656 (2011). https://doi.org/10.1007/s10714-010-1011-7

44. M. Punturo et al., The Einstein Telescope: A third-generation gravitational wave observatory. Classical Quantum Gravity **27**, 194002 (2010). https://doi.org/10.1088/0264-9381/27/19/194002

45. G. Cella, Off-line subtraction of seismic Newtonian noise, in *Recent Developments in General Relativity. Proceedings of the 13th Italian Conference on General Relativity and Gravitational Physics* (Springer, Milano, 1998), pp. 495–503

46. J.C. Driggers, J. Harms, R.X. Adhikari, Subtraction of Newtonian noise using optimized sensor arrays. Phys. Rev. **D86**, 102001 (2012). https://doi.org/10.1103/PhysRevD.86.102001

47. B.P. Abbott, R. Abbott, T.D. Abbott, M.R. Abernathy, F. Acernese, K. Ackley, C. Adams, T. Adams, P. Addesso, et al., Prospects for observing and localizing gravitational-wave transients with advanced LIGO, advanced Virgo and KAGRA. Living Rev. Relativ. **21**(1), 3 (2018). ISSN 1433-8351. https://doi.org/10.1007/s41114-018-0012-9

48. B.P. Abbott et al., GW151226: Observation of gravitational waves from a 22-solar-mass binary black hole coalescence. Phys. Rev. Lett. **116**(24), 241103 (2016). https://doi.org/10.1103/PhysRevLett.116.241103

49. B.P. Abbott et al., GWTC-1: a gravitational-wave transient catalog of compact binary mergers observed by LIGO and Virgo during the first and second observing runs (2018). arXiv:1811.12907

50. B.P. Abbott, R. Abbott, T.D. Abbott, M.R. Abernathy, F. Acernese, K. Ackley, C. Adams, T. Adams, P. Addesso, et al., Improved analysis of GW150914 using a fully spin-precessing waveform model. Phys. Rev. X **6**, 041014 (2016). https://doi.org/10.1103/PhysRevX.6.041014

51. B.P. Abbott, R. Abbott, T.D. Abbott, M.R. Abernathy, F. Acernese, K. Ackley, C. Adams, T. Adams, P. Addesso, et al., Tests of general relativity with GW150914. Phys. Rev. Lett. **116**, 221101 (2016). https://doi.org/10.1103/PhysRevLett.116.221101

52. B.P. Abbott, R. Abbott, T.D. Abbott, F. Acernese, K. Ackley, C. Adams, T. Adams, P. Addesso, R.X. Adhikari, et al., Gravitational waves and gamma-rays from a binary neutron star merger: GW170817 and GRB 170817a. Astrophys. J. **848**(2), L13 (2017). https://doi.org/10.3847/2041-8213/aa920c

53. B.P. Abbott, R. Abbott, T.D. Abbott, F. Acernese, K. Ackley, C. Adams, T. Adams, P. Addesso, R.X. Adhikari, et al., GW170817: measurements of neutron star radii and equation of state. Phys. Rev. Lett. **121**, 161101 (2018). https://doi.org/10.1103/PhysRevLett.121.161101

54. M. Fishbach, R. Gray, I. Magaña Hernandez, H. Qi, A. Sur, A standard siren measurement of the Hubble constant from GW170817 without the electromagnetic counterpart. Astrophys. J. Lett. **871**(1), L13 (2019)

55. B.P. Abbott, R. Abbott, T.D. Abbott, F. Acernese, K. Ackley, C. Adams, T. Adams, P. Addesso, R.X. Adhikari, et al., Multi-messenger observations of a binary neutron star merger. Astrophys. J. **848**(2), L12 (2017). https://doi.org/10.3847/2041-8213/aa91c9

56. D.I. Jones, Gravitational waves from rotating strained neutron stars. Classical Quantum Gravity **19**(7), 1255–1265 (2002). https://doi.org/10.1088/0264-9381/19/7/304

57. C. Caprini, Stochastic background of gravitational waves from cosmological sources. J. Phys. Conf. Ser. **610**(1), 012004 (2015). https://doi.org/10.1088/1742-6596/610/1/012004

58. B.P. Abbott, R. Abbott, T.D. Abbott, M.R. Abernathy, F. Acernese, K. Ackley, C. Adams, T. Adams, P. Addesso, et al., Prospects for observing and localizing gravitational-wave transients with advanced LIGO and advanced Virgo. Living Rev. Relativ. **19**(1), 1 (2016). ISSN 1433-8351. https://doi.org/10.1007/lrr-2016-1

59. M. Punturo, M. Abernathy, F. Acernese, B. Allen, N. Andersson, K. Arun, F. Barone, B. Barr, M. Barsuglia, et al., The Einstein telescope: a third-generation gravitational wave observatory. Classical Quantum Gravity **27**(19), 194002 (2010)

60. M. Maggiore, Gravitational wave experiments and early universe cosmology. Phys. Rep. **331**(6), 283–367 (2000). ISSN 0370-1573. https://doi.org/10.1016/S0370-1573(99)00102-7

61. I. Sendra, T.L. Smith, Improved limits on short-wavelength gravitational waves from the cosmic microwave background. Phys. Rev. D **85**, 123002 (2012). https://doi.org/10.1103/PhysRevD.85.123002

62. Z. Arzoumanian, A. Brazier, S. Burke-Spolaor, S.J. Chamberlin, S. Chatterjee, B. Christy, J.M. Cordes, N.J. Cornish, K. Crowter, et al., The NANOGrav nine-year data set: Limits on the isotropic stochastic gravitational wave background. Astrophys. J. **821**(1), 13 (2016).

63. M.S. Turner, Detectability of inflation-produced gravitational waves. Phys. Rev. D **55**, R435–R439 (1997). https://doi.org/10.1103/PhysRevD.55.R435

64. V. Mandic, A. Buonanno, Accessibility of the pre-big-bang models to LIGO. Phys. Rev. D **73**, 063008 (2006). https://doi.org/10.1103/PhysRevD.73.063008

65. N. Barnaby, E. Pajer, M. Peloso, Gauge field production in axion inflation: consequences for monodromy, non-Gaussianity in the CMB, and gravitational waves at interferometers. Phys. Rev. D **85**, 023525 (2012). https://doi.org/10.1103/PhysRevD.85.023525

66. X. Siemens, V. Mandic, J. Creighton, Gravitational-wave stochastic background from cosmic strings. Phys. Rev. Lett. **98**, 111101 (2007). https://doi.org/10.1103/PhysRevLett.98.111101

67. B.P. Abbott, R. Abbott, T.D. Abbott, M.R. Abernathy, F. Acernese, K. Ackley, C. Adams, T. Adams, P. Addesso, et al., Upper limits on the stochastic gravitational-wave background from advanced LIGO's first observing run. Phys. Rev. Lett. **118**, 121101 (2017). https://doi.org/10.1103/PhysRevLett.118.121101

68. P.D. Lasky et al., Gravitational-wave cosmology across 29 decades in frequency. Phys. Rev. **X6**(1), 011035 (2016). https://doi.org/10.1103/PhysRevX.6.011035

69. B.P. Abbott, R. Abbott, T.D. Abbott, M.R. Abernathy, F. Acernese, K. Ackley, C. Adams, T. Adams, P. Addesso, et al., GW150914: implications for the stochastic gravitational-wave background from binary black holes. Phys. Rev. Lett. **116**, 131102 (2016). https://doi.org/10.1103/PhysRevLett.116.131102

70. T. Regimbau, The astrophysical gravitational wave stochastic background. Res. Astron. Astrophys. **11**(4), 369 (2011)

71. J. Aasi, B.P. Abbott, R. Abbott, T. Abbott, M.R. Abernathy, T. Accadia, F. Acernese, K. Ackley, C. Adams, et al., Improved upper limits on the stochastic gravitational-wave background from 2009–2010 LIGO and Virgo data. Phys. Rev. Lett. **113**, 231101 (2014). https://doi.org/10.1103/PhysRevLett.113.231101

72. A. Nishizawa, A. Taruya, K. Hayama, S. Kawamura, M.-A. Sakagami, Probing nontensorial polarizations of stochastic gravitational-wave backgrounds with ground-based laser interferometers. Phys. Rev. D **79**, 082002 (2009). https://doi.org/10.1103/PhysRevD.79.082002

73. B.P. Abbott, R. Abbott, T.D. Abbott, F. Acernese, K. Ackley, C. Adams, T. Adams, P. Addesso, R.X. Adhikari, et al., Search for tensor, vector, and scalar polarizations in the stochastic gravitational-wave background. Phys. Rev. Lett. **120**, 201102 (2018). https://doi.org/10.1103/PhysRevLett.120.201102

74. J. Abadie, B.P. Abbott, R. Abbott, M. Abernathy, T. Accadia, F. Acernese, C. Adams, R. Adhikari, P. Ajith, et al., Directional limits on persistent gravitational waves using LIGO S5 science data. Phys. Rev. Lett. **107**, 271102 (2011). https://doi.org/10.1103/PhysRevLett.107.271102

75. S. Drasco, É.É. Flanagan, Detection methods for non-Gaussian gravitational wave stochastic backgrounds. Phys. Rev. D **67**:082003 (2003). https://doi.org/10.1103/PhysRevD.67.082003

Part II
Proceedings

Generation of Initial Data for General-Relativistic Simulations of Charged Black Holes

Gabriele Bozzola and Vasileios Paschalidis

Abstract Einstein–Maxwell theory is a description of electromagnetism and gravity from first principles that has yielded several interesting results on black holes with electromagnetic fields. However, since astrophysically-relevant black holes are believed to have negligible electric charge, the theory has been mostly confined within the realm of theoretical investigation. In addition to this, this theory has been studied mostly with analytical tools, which is why the vast majority of available results are restricted to spacetimes endowed with some degree of symmetry (e.g., stationarity and axisymmetry). Consequently, dynamical solutions (where no symmetry is assumed), for which the numerical approach is the only feasible one, represent a largely unexplored territory.

In this paper we present our efforts towards dynamical solutions of the coupled Einstein–Maxwell equations. As a first step, we solve numerically the constraint equations to generate valid initial data for dynamical general-relativistic simulations of generic configurations of black holes that possess electric charge, linear and angular momenta. The initial data are constructed with the conformal transverse-traceless approach, and the black holes are described as punctures within a modified Bowen–York framework. The attribution of physical parameters (mass, charge, and momenta) to the holes is performed by adopting the theory of isolated horizons. We implement our new formalism numerically and find it both to be in agreement with previous results and to show good convergence properties.

Keywords General relativity · Numerical relativity · Einstein–Maxwell · Black holes · 3 + 1 Decomposition

G. Bozzola (✉)
Department of Astronomy and Steward Observatory, University of Arizona, Tucson, AZ, USA
e-mail: gabrielebozzola@email.arizona.edu

V. Paschalidis
Departments of Astronomy and Physics, University of Arizona, Tucson, AZ, USA
e-mail: vpaschal@email.arizona.edu

S. Cacciatori et al. (eds.), *Einstein Equations: Physical and Mathematical Aspects of General Relativity*, Tutorials, Schools, and Workshops in the Mathematical Sciences, https://doi.org/10.1007/978-3-030-18061-4_5

1   Introduction

When the theory of general relativity was first proposed by Albert Einstein in 1915, electromagnetism and gravity were the only two fundamental forces known. It should not come to surprise that several theoretical physicists immediately started working on the interplay between gravity and electromagnetism, charmed by the idea of studying fundamental physics in the new framework of general relativity. For instance, the first efforts toward finding a unified theory of interactions were within this context, e.g., the Kaluza–Klein mechanism. Solutions of the source-free Einstein–Maxwell equations are often referred to as *electrovacuums* (or *electrovacs*) and usually describe black holes with electric[1] charge. These spacetime are the main subject of this paper.

Since the beginning, the tools employed to investigate the Einstein–Maxwell theory have mostly been analytical or perturbative. Only idealized problems can be dealt with using these techniques, mainly when a large degree of symmetry is assumed. The most notable solution is the Kerr–Newman spacetime, which describes a charged and rotating black hole in a stationary state. The Kerr–Newman solution is also famous for the associated *no-hair* or *uniqueness* theorem, which is an important milestone in general relativity that states that in a four-dimensional asymptotically flat spacetime (such as the ones that we consider here), axisymmetric, stationary, and topologically spherical black-hole solutions belong to the Kerr–Newman family. This encompasses the specific cases with no charge, no rotation, or both, respectively, known as Kerr, Reissner–Nordström, and Schwarzschild solutions. Other examples of charged black-hole solutions of the Einstein–Maxwell equations are Majumdar–Papapetrou's static spacetime [8, 10] and its stationary generalization by Israel and Perjes [7]. These systems have N black holes kept in equilibrium by a fine balance between gravitational attraction and electric repulsion.

Little effort has been made to explore electrovacs with lesser degree of symmetry. In this paper we take our first steps in the direction of studying *dynamical electrovacuum spacetimes*, which are solutions of the Einstein–Maxwell equations where no symmetry is assumed. We call them "dynamical" because non-stationary or static solutions do not represent equilibrium configurations, but describe the evolution of the system toward such state.

In particular, in the following we present a method for generating initial data for arbitrary configurations of charged black holes with linear and angular momenta in the $3 + 1$ formalism. We assume that the reader has no familiarity with numerical relativity, and present a "high-level" description of the subject, focusing more on the main ideas than the actual details, which are almost completely omitted. The interested reader can find a much more in-depth and technical discussion in our main paper [6].

---

[1]There is also the possibility of *magnetic* charge, which will be neglected here.

This document is structured as it follows. Section 2 provides some physically interesting examples of applications of dynamical electrovacuum spacetimes. Section 3 briefly describes the 3 + 1 decomposition, and Sect. 4 focuses on the problem of the generation of initial data by solving the constraint equations and our approach to solve them. Conclusions and future directions are detailed in Sect. 5.

2 Dynamical Electrovacuums

Dynamical electrovacuums have much untapped physical interest. This is true not only from a theoretical and fundamental standpoint, but also possibly for the astrophysics of black holes. In this section we report examples of where these spacetimes can be used to improve our understanding of the theory of general relativity and its applications to the physics of the cosmos.

A relevant astrophysical application concerns mergers of black holes. Several dynamical solutions of Einstein's vacuum equations are available in the literature to describe the coalescence and merger of uncharged black holes. These spacetimes were instrumental in the successful discovery of gravitational waves by the LIGO-VIRGO collaboration [11] since they provided invaluable information for the detection and interpretation efforts. Furthermore, dynamical spacetimes in presence of matter are also of great interest for modeling compact relativistic stars or self-gravitating accretion disks. Clearly, the interest in these solutions was mostly driven by their direct astrophysical applications. On the other hand, it is usually assumed that astrophysically-relevant black holes are without electric charge, as if they had charge they would collect oppositely charged particles from the surrounding plasma and quickly discharge. A first important scenario in which dynamical electrovacuums could be applied to is in placing a constraint on the charge of black holes. This can be done by simulating the coalescence and merger of charged black holes and extracting the gravitational emission, which is affected by the presence of the charge (that modifies the structure of the spacetime). This would furnish the first constraint that does not depend on theoretical arguments, but on a true measurement through gravitational waves. Furthermore, some gravitational-wave detections of merger of black holes may have been accompanied by electromagnetic counterparts which could be explained by the presence of accelerating electric charges. More constraints should be put on this scenario to further rule out this possibility and to shed more light in the issue of sources of multi-messenger signals.

An example of a fundamental application of dynamical electrovacuums is related to cosmic censorship and the formation of naked singularities in asymptotically flat spacetimes. The Einstein–Maxwell equations in a dynamical regime offer a laboratory for studying black holes in extreme conditions never explored before. In spite of the fact that much work has already been done to test the robustness of general relativity with respect to "overspun" and "overcharged" black holes, no result is available for the conditions we are considering here with multiple interacting holes in an highly-dynamical spacetime. For instance, little is known regarding the merger of binary black holes both of which are near extremality due to fact that they are highly charged. Another related interesting theoretical problem is the ultra-relativistic collision of charged black holes, and the role that the charges play when the dynamics are dominated by the immense kinetic energy of the two bodies. These regimes have not been explored, yet.

3 3 + 1 Decomposition

Solving analytically the Einstein–Maxwell equations is a non-trivial task even if a high degree of symmetry is imposed. Consequently, the most straightforward approach for accessing the dynamical regime is to rely on numerical techniques. In this respect, it is useful to perform a $3 + 1$ decomposition of the theory in order to restore the concept of *time evolution* of *3D spatial slices* from the full four-dimensional description (the "1" and the "3" in the decomposition refer to these dimensions). The solution of the Einstein–Maxwell equations within the $3 + 1$ approach is obtained by generating valid initial data, and evolving them in a suitable way, effectively casting the system to a Cauchy problem with dynamical degrees of freedom, the electromagnetic fields, and the 3-metric and extrinsic curvature of each spatial slice. This scheme involves several complications such as ill-posedness of evolution equations, issues related to the coordinate freedom intrinsic to general relativity, and physical singularities (when evolving black holes). These problems required about half a century of theoretical and technological advances to be overcome for the case without charges, and a brand new branch of general relativity developed as result of the efforts—the field now known as *numerical relativity* [12]. Nowadays, several groups are able to run numerical-relativity simulations of black holes, and the community has understood which are the key elements to have a stable and reliable evolution. Most importantly to our goals here, Alcubierre proved that the Einstein–Maxwell equations are well-posed in this formulation [1], so it is possible to use directly or extend many of the techniques developed for the uncharged case. We will now use the simpler theory of classical electromagnetism on a flat background as example to briefly illustrate an important consequence of the $3 + 1$ split which remains conceptually the same in the much more involved case of Einstein–Maxwell: valid initial data for a time evolution have to satisfy specific constraints.

The field equations are split in two sets when decomposing the spacetime in spatial and temporal parts. The first is formed by the evolution equations. An example for source-free Maxwell theory in flat background is Faraday's law

$$\partial_t \mathbf{B} = \nabla \times \mathbf{E}, \tag{1}$$

where $\mathbf{B}$ and $\mathbf{E}$ are the electric and magnetic field. Equation (1) can be used in a numerical algorithm to move the configuration of the magnetic field $\mathbf{B}$ from a known state at time t to the one at $t + \Delta t$. The most trivial way to do so is to discretize the time derivative as

$$\partial_t \mathbf{B}(t) = \frac{\mathbf{B}(t + \Delta t) - \mathbf{B}(t)}{\Delta t}, \tag{2}$$

accurate at order Δt. Plugging Equation (2) in Eq. (1) and rearranging the terms, we obtain

$$\mathbf{B}(t + \Delta t) = \mathbf{B}(t) + \Delta t \left(\nabla \times \mathbf{E}(t) \right). \tag{3}$$

It is possible to find the configuration of the magnetic field at time $t + \Delta t$ using information at time t with this equation, for this reason this is an example of evolution law.

The constraints are the other equations that result from the $3 + 1$ decomposition of the spacetime. These are both gravitational and electromagnetic and do not involve any time derivative, which is why they have always to be satisfied. The absence of magnetic monopoles in classical electromagnetism is a familiar example of constraint equation

$$\nabla \cdot \mathbf{B} = 0. \tag{4}$$

It is always possible to find a magnetic field configuration that does not satisfy Eq. (4) and to use it as starting point of an evolution in time. Nonetheless, this would be non-physical as it would not be a valid solution of Maxwell's equations. The same happens in the Einstein–Maxwell case, and this is why it is important to generate initial data that satisfy the constraints. The remaining part of this paper explores a possible way to do this based on the conformal transverse-traceless technique.

4   Constraint Equations

In Einstein–Maxwell theory there are two electromagnetic and four gravitational constraints. These are all coupled elliptic partial differential equations involving a total of eighteen degrees of freedom (the electromagnetic fields, the 3-metric, and the extrinsic curvature). The system is underdetermined, which reflects the fact that the constraints themselves cannot fix all the degrees of freedom and specifying the physical picture lies also in the variables that can be freely set. We apply a technique pioneered by Bowen and York in the 1970s [5] and now commonly used in numerical relativity known as conformal transverse-traceless method to partially decouple the equations and expose some relevant quantities with more direct physical interpretation.

The first step in this approach consists in introducing new variables: the conformal factor ψ and the vector field V^i. The first is defined assuming that each spatial slice is conformally flat, which means that the 3-metric γ_{ij} is proportional to the usual Euclidean metric δ_{ij}:

$$\gamma_{ij} = \psi^4 \delta_{ij} \, . \tag{5}$$

This is a good approximation for several spacetimes, and in any case the time evolution would make the system relax to its natural conformal structure. Then, we apply a similar conformal transformation to the electric and magnetic fields,

$$\mathbf{E} = \psi^6 \bar{\mathbf{E}} \quad \text{and} \quad \mathbf{B} = \psi^6 \bar{\mathbf{B}} \, , \tag{6}$$

with $\bar{\mathbf{E}}, \bar{\mathbf{B}}$ known fields. We will fix these background fields and use the constraints to determine ψ.

The vector V^i is defined from the extrinsic curvature K_{ij} of the slices, as

$$K_{ij} = \psi^{-2} \left(2V_{(i,j)} - \frac{2}{3} \delta_{ij} \partial_k V^k \right) , \tag{7}$$

in other words K_{ij} is obtained applying the vector Laplacian to V^i and scaling by a factor of ψ^2. V^i is in direct relation to the geometry of the spatial slices, for which some assumptions are taken such as the condition of *maximally slicing* that consists in maximizing the spatial volume element and that has proven to be helpful in increasing stability in the subsequent evolution.

The electromagnetic constraints are cast to the usual Maxwell constraints in flat spacetime with these transformations and assumptions, whereas the gravitational ones become four second-order elliptic partial differential equation with source terms that depend only on $\bar{\mathbf{E}}$ and $\bar{\mathbf{B}}$. Schematically, the equations that have to be solved to have valid initial data assume the following form

$$\partial_i \bar{E}^i = 0 \, , \tag{8a}$$

$$\partial_i \bar{B}^i = 0 \, , \tag{8b}$$

$$\mathcal{L}V^i = \bar{S}^i(\bar{\mathbf{E}}, \bar{\mathbf{B}}) \, , \tag{8c}$$

$$\mathcal{D}_V \psi = \bar{\mathcal{E}}(\bar{\mathbf{E}}, \bar{\mathbf{B}}) \, , \tag{8d}$$

where $\mathcal{L}$ is a linear, second-order elliptic differential operator and $\mathcal{D}_V$ a non-linear, second-order elliptic differential operator that depends on V. $\bar{S}^i(\bar{\mathbf{E}}, \bar{\mathbf{B}})$ and $\bar{\mathcal{E}}(\bar{\mathbf{E}}, \bar{\mathbf{B}})$ are source terms and have a known expression.

The conformal method does most of the hard work as now the problem can be easily solved as follows. First, we start with $\bar{\mathbf{E}}$ and $\bar{\mathbf{B}}$ satisfying the flat Maxwell constraints (8a) and (8b) and compute the source terms in Eqs. (8c) and (8d). Next, due to the linearity we split Eq. (8c) in an homogeneous part and an inhomogeneous one:

$$\mathcal{L}V_{GR}^i = 0 \quad \text{and} \quad \mathcal{L}V_{EM}^i = \bar{S}^i(\bar{E}, \bar{B}) \quad \text{with} \quad V^i = V_{GR}^i + V_{EM}^i. \tag{9}$$

Assuming mild conditions on the decay rate of the electromagnetic fields at spatial infinity, it can be shown that it is possible to determine the linear and angular momenta of the black holes with a suitable analytical choice for V_{GR}^i, so we need only solve Eq. (8c) for V_{EM}^i and adopt a standard form for V_{GR}^i that corresponds to uncharged black holes. Then, we solve constraint (8d) for ψ by making an ansatz for ψ as superposition of charged punctures plus a finite correction to be determined with Eq. (8d). Considering a puncture in the approach means removing the physical singularities from the grid by making an ansatz for ψ that removes the $(1/r)$ singular terms from the equations. Finally, as last step in the Bowen–York scheme we revert back to the physical fields needed for the time evolution. This method can be used to describe multiple black holes in different locations as it is possible to exploit the linearity of the first three constraints to superpose solutions. Hence, given the electromagnetic fields and a choice for V_{GR}^i, the machinery outputs all the quantities needed to start an evolution. The most natural choice for the electromagnetic fields is taking $\bar{E}$ and $\bar{B}$ as the superposition of Kerr–Newman fields, since this choice guarantees the electromagnetic constraints are automatically satisfied, and it is the most reasonable due to the aforementioned uniqueness theorem.

The numerical solution of the constraint equations obfuscates the physical content of the initial data so it is important to provide a reliable physical interpretation once the initial data are generated. We use formalism of *isolated horizons* put forth by Ashtekar [3] to locate the horizons and to assign them their physical properties (mass, angular momentum, or electric charge) in a quasi-local way. The main advantage of the approach is that it requires only knowledge of variables on a slice at fixed time, in contrast to the intrinsically global nature of event horizons. Importantly, the framework perfectly reduces to concept of event horizons in the stationary cases and the quasi-local properties take the same values as the global ones defined at infinity.

We implemented our new formalism by modifying existing codes that were designed to solve numerically the case without charge. In particular, our approach extends the `TwoPunctures` code. We called our new software `TwoChargedPunctures` and we validated it by generating the solution for a single black hole and checking against the expected properties. We found that `TwoChargedPunctures` is reliable and accurate and has good convergence properties. An example of a solution is in Fig. 1, which depicts the electric field for a binary system of charged and rotating black holes.

Fig. 1 Electric field lines on the x–z plane for two charged and rotating black holes. The first is rotating about the x axis, the second about the z axis. In both cases, $J = 0.3\,M$, with J angular momentum and M mass of the hole. This is a self-consistent valid initial data for general-relativistic simulations

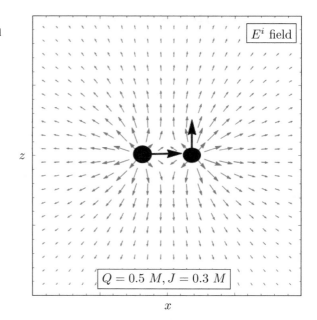

5 Conclusions

This paper describes a method for solving the source-free Einstein–Maxwell equations in cases where no symmetry is assumed. These solutions are called *electrovacuums*. We argued that these spacetimes can be used to improve our understanding of general relativity, for example, opening new possibilities in studying the formation of naked singularities. These spacetimes are extremely challenging to tackle with analytical tools, and numerical techniques provide the most straightforward avenue to investigating them. Thus, we outlined a path to study dynamical electrovacuum spacetimes in the context of numerical relativity.

A solution can be obtained with the 3 + 1 formulation of the Einstein–Maxwell equations by (1) generating valid initial data and (2) performing an evolution in time. In this paper we focused on the first step which corresponds to the solution of the Einstein–Maxwell constraint equations, and the interpretation of the physical content of the resulting initial data. This was done with an extension of Bowen–York's approach and the theory of isolated horizons. The numerical implementation was done by extending the public `TwoPunctures` and the `QuasiLocalMeasures` codes. With our additions, the first solves the constraint equations for two black holes with arbitrary charge, spin, and linear momentum, whereas the second evaluates the physical properties of the black holes and ensures consistency between electromagnetic fields and the gravitational quantities. Our tests indicated that the code is robust and reproduces the expected results with high accuracy.

This is only the beginning of the exploration one of the most fundamental theories available at the moment in a regime previously unexplored. The next natural step is the development of the evolution routines and the first simulations of dynamical Einstein–Maxwell systems.

References

1. M. Alcubierre, J.C. Degollado, M. Salgado, Einstein-Maxwell system in $3 + 1$ form and initial data for multiple charged black holes. Phys. Rev. D **80**(10), 104022 (2009). https://doi.org/10.1103/PhysRevD.80.104022. arXiv: `0907.1151 [gr-qc]`
2. M. Ansorg, B. Brügmann, W. Tichy, A single-domain spectral method for black hole puncture data. Phys. Rev. D **70**, 064011 (2004). https://doi.org/10.1103/PhysRevD.70.064011. Eprint: `arXiv:gr-qc/0404056`
3. A. Ashtekar et al., Generic isolated horizons and their applications. Phys. Rev. Lett. **85**, 3564–3567 (2000). https://doi.org/10.1103/PhysRevLett.85.3564. Eprint: `gr-qc/0006006`
4. T.W. Baumgarte, S.L. Shapiro, *Numerical Relativity: Solving Einstein's Equations on the Computer* (Cambridge University Press, Cambridge, 2010)
5. J.M. Bowen, J.W. York Jr, Time asymmetric initial data for black holes and black hole collisions. Phys. Rev. D **21**, 2047–2056 (1980). https://doi.org/10.1103/PhysRevD.21.2047
6. G. Bozzola, V. Paschalidis, Initial data for general relativistic simulations of multiple electrically charged black holes with linear and angular momenta. Phys. Rev. D **99**(10), 104044. https://doi.org/10.1103/PhysRevD.99.104044
7. W. Israel, G.A. Wilson, A class of stationary electromagnetic vacuum fields. J. Math. Phys. **13**, 865–867 (1972). https://doi.org/10.1063/1.1666066
8. S.D. Majumdar, A class of exact solutions of Einstein's field equations. Phys. Rev. **72**, 390–398 (1947). https://doi.org/10.1103/PhysRev.72.390
9. E.T. Newman et al., Metric of a rotating, charged mass. J. Math. Phys. **6**, 918–919 (1965). https://doi.org/10.1063/1.1704351
10. A. Papapetrou, A static solution of the equations of the gravitational field for an arbitary charge-distribution, in *Proceedings of the Royal Irish Academy. Section A: Mathematical and Physical Sciences*, vol. 51 (1945), pp. 191–204. ISSN: 00358975, http://www.jstor.org/stable/20488481
11. The LIGO Scientific Collaboration and the Virgo Collaboration, GWTC-1: A gravitational-wave transient catalog of compact binary mergers observed by LIGO and Virgo during the first and second observing runs. arXiv e-prints (2018). arXiv: `1811.12907 [astro-ph.HE]`
12. J.W. York Jr, Kinematics and dynamics of general relativity, in *Sources of Gravitational Radiation*, ed. by L.L. Smarr (Cambridge University Press, Cambridge and New York, 1979), pp. 83–126

BMS Symmetries and Holography: An Introductory Overview

Federico Capone

Abstract A short overview of the role of BMS symmetries in various approaches to the flat spacetime holography problem is given. The relevance of BMS symmetries to the infrared structure of gravity is motivated and described at an introductory level with some details, but also the relation between the BMS group and the Carroll group is pointed out. The BMS group—and its proposed extensions—is reviewed in four-dimensional spacetimes in parallel with a discussion of the difficulties related to its definition in higher dimensional spacetimes. The Bondi–Sachs integration scheme is discussed with some details. A short discussion of canonical charges associated to the BMS transformations and the possible relation between superrotations and transition from asymptotically flat to locally asymptotically flat spacetimes is also summarised.

Keywords Asymptotically flat spacetimes · BMS group · Carroll symmetry · Higher dimension · Flat holography

1 Introduction

Asymptotic flatness is a pivotal concept in general relativity because it leads to the definition of conserved charges in the gravitational theory and to the understanding of gravitational radiation [1, 2].

There are also several proposals on how to build a theory of quantum gravity in asymptotically flat (AF) spacetimes. A widespread belief is that quantum gravity in an n-dimensional spacetime is holographic, in the sense that it can be equivalently described in terms of a quantum non-gravitational theory living on an $(n - 1)$-dimensional spacetime. This is the content of the holographic principle, whose

F. Capone (✉)
Mathematical Sciences and STAG Research Centre, University of Southampton, Southampton, UK
e-mail: f.capone@soton.ac.uk

© Springer Nature Switzerland AG 2019
S. Cacciatori et al. (eds.), *Einstein Equations: Physical and Mathematical Aspects of General Relativity*, Tutorials, Schools, and Workshops in the Mathematical Sciences, https://doi.org/10.1007/978-3-030-18061-4_6

first realisation has been AdS/CFT, a duality between a conformal field theory on the conformal boundary of AdS and a particular limit of string theory on AdS. If the holographic principle is to be assumed as a fundamental property of quantum gravity, then it must be valid apart from string theory and asymptotically AdS (AAdS) spacetimes. From this point of view the question on how general is holography naturally arises. The way to answer this question is to analyse the structure and the symmetries of other spacetimes. In the past some results concerning de Sitter (dS) spacetimes have been proposed [3], but the so-called (asymptotically) *flat (spacetime) holography problem* is still open.

In recent years the flat holography problem is again under investigation because several interesting proposals have been discussed, some of them more conservative and others somewhat more radical, in a sense that will be clear later. In all these discussions the asymptotic symmetry group of asymptotically flat spacetimes plays a central role. The modern viewpoints have fostered discussions in the general relativity community because of the need of appropriate definitions of the asymptotics of higher dimensional spacetimes and the associated symmetry groups.

The document is organised as follows: The next section is both a lightning review of the concept of asymptotic structure of a four-dimensional spacetime and the BMS group, and the occasion to start the discussion concerning the difficulties of the higher dimensional definitions of these concepts. This is intended to put on the stage all the ingredients that are discussed in more detail in Sect. 4 and in the Appendix "Einstein's Equations". Section 4 shows how asymptotic flatness can be defined in four and higher dimensions using the Bondi–Sachs procedure and points out at the problems of the higher dimensional definition which were highlighted by holographic considerations in Sect. 3. In this section we have tried to be as pedagogical as possible. Section 3 discusses three relevant ways in which the holographic principle can be implemented in AF spacetimes. The main focus is on holography from the celestial sphere and the connection of BMS symmetries to infrared properties of gravity, but also the isomorphism between the BMS group and the Carroll group is pointed out, as well as the realisation of flat holography via the fluid/gravity correspondence. Section 5 gathers well-known facts from the general relativistic literature, the Wald–Zoupas charges are reviewed somewhat heuristically along with the insights concerning holography form the celestial sphere. In these sections, which are more a showcase of ideas, the relevant references are given. The material is presented at a very basic but broad level and the aim is that this document may be useful to readers who want to approach this field and get an idea of at least some of the various possible ramifications. Several reviews and notes are available today, most of them focussed on the interplay of the BMS symmetries and the infrared structure of gravity [4–7]; we hope this short note can add some other complementary bits to that landscape. Unfortunately the vastity of the field cannot be captured due to space constraints and some other interesting topics (such as the role of BMS group in string theory) have not been covered here.

2 Asymptotic Structure of Spacetime

Given a four-dimensional spacetime (M, g) based on a manifold M without boundary, Penrose conformal methods allow the study of Einstein's equations in an unphysical spacetime $(\bar{M}, \bar{g})$, where $\bar{M}$ is a four-dimensional manifold with a boundary $\mathscr{I}$ and $\bar{g} = \Omega^2 g$ with Ω a smooth function on $\bar{M}$ such that $\Omega = 0$ on $\mathscr{I}$ and $\nabla_\mu \Omega \neq 0$ on $\mathscr{I}$. The boundary represents points at infinity in the physical spacetime. The defining requirement on Ω makes $\mathscr{I}$ a well-defined three-dimensional hypersurface. The causal structure of $\mathscr{I}$ depends on the value of the cosmological constant Λ. When $\Lambda = 0$ the boundary $\mathscr{I}$ is null and composed of two disconnected pieces $\mathscr{I}^+$ and $\mathscr{I}^-$, called respectively future and past null infinity, both of them topologically equivalent to $\mathbb{R} \times S^2$. The two components are joined by the singular point i^0 representing spacelike infinity.[1] The null (and disconnected) nature of the boundary is a defining feature of asymptotically flat spacetimes since also the Minkowski spacetime is characterised by such boundary. However, a crucial difference between Minkowski spacetime and a more general AF spacetime is that in the first case the generators of $\mathscr{I}^-$ can be smoothly identified with those of $\mathscr{I}^+$ across i^0, while for more general spacetimes the identification is made harder by the lacking of a smooth differential structure at i^0 [8]. This point will come back in the next section.

As well-known,[2] studies on the asymptotics of spacetimes are extremely useful because a generally curved spacetime, where no isometries are present, can inherit symmetries from those of the asymptotic structure. Indeed the idea is simple: if the spacetime behaves asymptotically as another spacetime with some set of isometries, then it may inherit at least those symmetries in the asymptotic region. Several ways have been developed to define the asymptotic symmetry group in general relativity (i.e. [1, 8–10]). Four-dimensional asymptotically flat spacetimes at null infinity enjoy an infinite dimensional asymptotic symmetry group known as the BMS group, first studied by Sachs in [10]

$$BMS^{\text{glob}} = L_+^\uparrow \ltimes ST . \tag{2.1}$$

It is the semidirect product of the (proper orthocronous) Lorentz group $L_+^\uparrow$ and an Abelian (normal sub-)group of transformations ST called *supertranslations*. The ST normal subgroup is infinite dimensional and contains translations T, which constitute a four-dimensional normal subgroup of (the proper[3]) BMS^{glob}. Considering null infinity as a hypersurface with coordinates (u, x^1, x^2), where u

---

[1]Points such as i^0 (and $i^\pm$) where $\nabla_\mu \Omega = 0$ (i.e. $i^0, i^\pm$) do not properly belong to $\mathscr{I}$ but are anyway considered part of the boundary of the spacetime even though they have to be treated separately.

[2]See, for example, [2].

[3]See [10] for details; a recent review of Penrose and Bondi–Sachs definition of the BMS group is given in [7].

spans the $\mathbb{R}$ direction and (x^1, x^2) the spherical directions, the group is made of the coordinate transformations

$$u \to \bar{u} = K^{-1}(x^A)(u + \alpha(x^A)), \quad x^A \to \bar{x}^A(x^1, x^2), \qquad (2.2)$$

where the transformation of the angular coordinates is conformal with conformal factor K and α is a totally arbitrary function of the angles. This explains the name supertranslation given to the transformation of u: u translates in an arbitrary way for each given point on the celestial sphere.

There is no way to reduce this group to the Poincaré group preserving the interesting dynamical effects of gravity waves reaching null infinity [11].[4] Rather, holographic considerations hint towards a further extension of the Lorentz subalgebra of the BMS algebra either to the full conformal algebra on S^2 or to the algebra of general diffeomorphisms of S^2, as we discuss in the next section.

As stated in the introduction, modern points of view on gravity and, among them, holography naturally lead to wonder about higher dimensional spacetimes. Hence it is necessary to understand how much of these structures remain in higher dimensions. Conformal methods have been generalised for the purposes of defining higher dimensional asymptotically flat spacetimes provided that the spacetime dimension is even [11, 16, 17], because in other cases the smoothness of the conformal factor cannot be guaranteed. For odd-dimensional spacetimes one has to revert to the metric-based approach [18–22] first systematically developed in the early 1960s by Bondi et al. [9, 23] and Sachs [10, 24] in their studies on gravitational radiation in four-dimensional spacetimes (from which the name

---

[4]This enhancement of the symmetry group with respect to the symmetry group of the Minkowski spacetime came somewhat as a surprise to Bondi and the others. Later on a different kind of enhancement was discovered also by Brown and Henneaux in AdS_3 spacetimes [12]. Symmetry enhancement is a phenomenon blessed by holographysts nowadays. The history of group theoretical research around the BMS group is quite interesting (see [13] for a full list of references) because, soon after its discovery, people started thinking about its representation theory (see [14] and the references therein as well as [13]) as a tool to explain the internal isospin symmetry of hadrons and BMS invariant theories as possible candidates to a unification theory of gravity and particle physics [14, 15]. Komar's paper [15] advances the hypothesis that the BMS^{glob} group should replace the role of the Poincaré group in the definition of particle states because it naturally gives rise also to internal symmetries. Just for historical curiosity let us summarise the line of reasoning. The group $G = BMS/T$ is an infinite group and a symmetry that is most naturally expressed in the centre of mass frame of the system, just as the spin in the Poincaré group. The Lorentz group is $L = G/PST$, where $PST = ST/T$. Hence G contains within its structure a preferred way of representing another spin-like quantity, which Komar suggestively identified with the isospin. He argued that since G comes from general relativity without any empirical assumptions, the G group served as a better explanation of internal symmetries than the other groups that at that time were conjectured using semiempirical reasoning. Furthermore it reflects gravitational properties that at the time were ignored in elementary particle physics, hence paving the way to unification. Komar paper was published in 1965, the "other groups conjectured" to which he refers are most probably the symmetry groups of the eightfold way and the quark model developed between 1961 and 1964. The course of history is known.

Bondi–Metzner–Sachs given to the BMS group). The Bondi–Sachs approach, which exploits a nested structure to solve Einstein's equations, is particularly interesting when discussing holographic matters not only because it can be generalised to any number of dimensions, but also because it can be compared to the well-known Fefferman–Graham expansion for AAdS spacetimes [25].

3 Null Infinity and Holography

The different causal structure of the boundary of asymptotically flat (AF), asymptotically Anti de Sitter (AAdS) and asymptotically de Sitter (dS) spacetimes is at the root of the different ways in which holography can be defined for each spacetime and the difficulties encountered in extending holography to non-AAdS spacetimes [26]. AdS/CFT implements the holographic principle as a correspondence between the bulk gravitational theory and the non-gravitational theory living on the conformal boundary of the asymptotically AdS spacetime. This idea may not survive when discussing AF holography, or else we may have to abandon the idea of a relativistic holographic correspondence in AF spacetimes.

Witten noticed [26] that as opposed to holography in AAdS spacetimes—consisting of a dictionary to compute correlation functions on the two sides of the duality—holography in AF gravity should be given by boundary constraints on the structure of the S-matrix because the S-matrix is the only possible gauge invariant operator in such spacetimes.

In QFT symmetries constrain S-matrix elements via Ward identities. Strominger first conjectured [27] that in an appropriate class of four-dimensional spacetimes (Christodoulou-Klainerman (CK) [28]),[5] a single ASG acting at the same time on the two disjoint component of null infinity $\mathscr{I}^+$ and $\mathscr{I}^-$ can be defined upon identification of the null generators of $\mathscr{I}^-$ and $\mathscr{I}^+$. It has been called BMS_0 and is built in the way we show in Sect. 5. This conjecture has great consequences on both the classical and—if extended to the quantum realm—the quantum gravitational scattering. Indeed classically it implies the conservation of all the charges (and fluxes, to be defined later) associated to the BMS_0 supertranslation symmetries. Furthermore, assuming that the conclusions are valid also in a quantum (or semiclassical picture) and that the S-matrix is invariant under the BMS_0 supertranslations results in the well-known Weinberg soft graviton theorem[6] [27, 30], in the sense that the latter is a manifestation of the invariance of S via a Ward identity. Hence this can

---

[5]CK spacetimes are not exactly AF in the sense of the standard definition, see Appendix "Einstein's Equations".

[6]This theorem [29] states that in any Lorentz invariant theory with massless spin 1 or 2 (photons and gravitons) particles, due to the pole structure of the S-matrix for the scattering of k external particles with momenta p_j ($j = 1, \cdot k$) and a massless particle with momentum $q^\mu \to 0$ (this defines the soft limit), the amplitude $A_{\alpha,\beta;q}$ (α in hard particles and β out hard particles) factorises as $A_{\alpha,\beta;q} = S^{(0)} A_{\alpha,\beta}$, where $S^{(0)}$ which in the gravitational case is $\sum_{j=1}^k (q \cdot k_j)^{-1} \epsilon_{\mu\nu} p_j^\mu p_j^\nu$,

be regarded as a first realisation of Witten's idea. However, if this is to be considered an implementation of holography, then a big difference from the standard one is to be noted: the relevant information comes only from the topological 2-spheres of null infinity, called *celestial spheres* (and any possible theory living there), not from a QFT living on the whole 3-dimensional null boundary.

Despite this, the perspective naturally fosters the question about higher dimensional generalisations of this connection between asymptotic symmetries and soft theorems. Indeed the aforementioned QFT theorems do not depend on the spacetime dimensions: what about the BMS group? The way to investigate this question is to define asymptotic flatness and the associated asymptotic symmetry group for higher dimensional spacetimes. As we will see in Sect. 4 the problem is still open and different focuses brings to different conclusions.

Other considerations from the flat limit ($\Lambda \to 0$) of AdS/CFT motivated the idea that the Lorentz part of the BMS algebra should actually be extended to the full infinite dimensional conformal algebra of transformations on S^2 [31]. Such transformations have been called *superrotations* [31–33]. These enlarged symmetries have been employed [34, 35] to conjecture that they also give rise to Ward identities which are equivalent to subleading (Cachazo-Strominger) soft theorems where the amplitude factorises in soft and hard parts not only at the leading order as in the Weinberg theorem, but also at the first subleading. For several reasons, mainly due to the singularities that plague superrotations defined from Witt/Virasoro symmetries, the proof of this conjecture has not been achieved successfully in both ways: the subleading soft theorems imply the Ward identities of the conformal symmetries but the converse may not be true. See, however, [36] for recent discussions and developments. Instead the group Diff(S^2) of diffeomorphisms of S^2 has been shown to be more successful in this [37, 38]. Both of the two possible transformations are called superrotations and to distinguish among them we will call them respectively BT superrotations (from Barnich-Troessaert [32]) and CL superrotations (from Campiglia-Laddha [37, 38]). Clearly the question on the higher dimensional generalisation is valid also for these new conjectured symmetries. Details and comments about BT superrotations and CL superrotations are given in Sect. 4.

Before we conclude this section we mention other two approaches to flat holography that have been developed and which are somewhat more standard in the sense that they aim at a duality between the bulk theory and a theory on a full codimension one hypersurface.

The null boundary of asymptotically flat spacetime has a degenerate metric $(0, +, +)$ so it is intrinsically non-Riemannian. This is one of the reasons for the difficulties in constructing a bulk/boundary map. In [39] null infinity is recognised as a Carrollian structure and the BMS group isomorphic to a conformal extension of the Carroll group,[7] intended as the group of transformations that preserve the

---

with $\epsilon_{\mu\nu}$ the polarisation tensor of the graviton. It has important consequences: among them the Equivalence Principle can be derived from this.

[7] See [40] for a review.

Carroll structure. If one insists on implementing holography as in AdS/CFT, then an immediate important difference arises: whereas the bulk theory lives on a Riemannian manifold, the dual boundary theory—if exists—should live on this Carroll manifold and be invariant under the BMS, namely the Conformal Carroll group. Two interesting things must be noted. First, the Carroll group is an ultrarelativistic contraction of the Poincaré group (this being the reason of the degeneracy of the underlying metric) so the boundary field theory would be really different from the usual relativistic field theories and the holographic correspondence would be non-relativistic. Some results in this direction have been obtained especially in three-dimensional AF spacetimes, in which case the BMS_3 group is isomorphic to the Galilei Conformal group in $n = 1 + 1$ spacetime dimensions (this comes about because the Carroll group is equivalent to the Galilei group in $n = 1 + 1$ [39]). The reader is referred, i.e. to [41–43] and the references therein. Second, the Carroll group can be defined for any n-dimensional Carroll structure and this motivated the authors in [39] to assume this as a higher dimensional generalisation of the BMS group. The open question is whether this would really be the ASG of some meaningful bulk spacetime. As a final remark we point out that several studies on the representation theory of the BMS algebra, both in four and three spacetime dimensions, have been carried out over the years (see note 3, the references [14, 44–46][8] and the thesis [13] for further references and in depth discussions of the BMS_3 case). Developing the representation theory is necessary at least when the aim is to build BMS/Carroll-invariant field theories (see, for example, [47–49]).

One later approach we mention, totally orthogonal to the two above (but in some sense closer to AdS/CFT), is to exploit and generalise the fluid/gravity relation to finite radius[9]; in this way holography can be implemented considering a timelike hypersurface in a Rindler wedge as the holographic screen [52]. The boundary is then at a finite radius and it is not really the boundary of the spacetime, but it can be pushed to the true boundary [53].

Note Added During the preparation of the final form of the volume, the paper [54] appeared in which a rigorous proof of Strominger's conjecture on the matching of generators of $\mathscr{I}^-$ and $\mathscr{I}^+$ is given for a class of AF spacetimes different from CK spacetimes (see also the final comment in Sect. 5.4).

4 The Metric-Based Approach

In principle for any kind of spacetime the general strategy behind the metric-based approach is to find an appropriate coordinate system with one of the coordinates, say r, parameterising the distance from the bulk of the spacetime. Then the metric

---

[8]The last three are also an early attempt to define holography in AF spacetimes.

[9]See, for example, [50, 51] for the fluid/gravity correspondence.

tensor admit an expansion of the form

$$g_{\mu\nu}(x^\rho) = \tilde{g}_{\mu\nu}(x^\rho) + O((x^{\rho_i})^{-m_{\mu\nu}}), \tag{4.1}$$

as long as r is pushed to the boundary. Here $\tilde{g}_{\mu\nu}$ is the asymptotic metric, which in the flat case is $\tilde{g} = \eta$. The coefficients $m_{\mu\nu}$ are numbers that play a central role in defining the class of spacetimes which are asymptotically $\tilde{g}$; they determine their dynamics and the asymptotic symmetry group. We can naturally cast this discussion as a problem of defining a *phase space* $\mathcal{F}$ for solutions of the theory with given boundary conditions $(\tilde{g}, m_{\mu\nu})$ (we will come back later to this).

There are several requirements that must be respected in order to have a well-defined phase space, at least [55]

(1) the fall-off conditions must be not so strong that they exclude the existence of phenomena that could otherwise occur in the deep interior of the spacetime;
(2) the fall-off conditions should be sufficiently strong that useful notions that characterize the system, such as total mass and radiated energy flux, are well-defined.

We will soon recognise the second requirement, in more general terms, as a condition on the asymptotic symmetry group and the associated charges (or Hamiltonian generators).

As a final comment let us stress that both the casual structure of the boundary and the dimension of the spacetime strongly constrain the subleading terms. The simplest example of this is actually provided by AF spacetimes: spacelike infinity i^0 is a place where dynamical phenomena cannot take place, whereas null infinity $\mathscr{I}^{+/-}$ is the natural place where phenomena involving massless excitations occurs. Thus, a natural condition in studying null infinity is to require that the metric there approaches the Minkowski metric plus corrections that encode the wave modes. This is the starting point of the Bondi, Metzner, van der Burg analysis.

4.1 The Bondi–Sachs Gauge in Any Dimension

According to the definition given by Bondi–Sachs [9, 24] and generalised to any dimension by Tanabe et al. [20], any spacetime is AF if a coordinate system $(u, r, x^1, \cdots, x^{n-2})$ can be found such that the four gauge conditions

$$g_{rr} = 0 = g_{rA}, \quad \det(g_{AB}) = r^{2(n-2)}\det(h_{AB}) = r^{2(n-2)}b(x^A), \tag{4.2}$$

hold, so that the line element takes the form $(A, B = 1, \cdots n - 2)$

$$ds^2 = -\mathcal{U}e^{2\beta}du^2 - 2e^{2\beta}dudr + g_{AB}(dx^A - W^A du)(dx^B - W^B du), \tag{4.3}$$

where $\mathcal{U}$, β, W^A and $g_{AB} := r^2 h_{AB}$ are functions of all the coordinates in general (unless other symmetry requirements are imposed). These admit a proper expansion in inverse powers of r, for large r, such that the metric approaches the Minkowski form as $r \to \infty$

$$h_{AB} = r^2 \gamma_{AB} + O(r^{-j}), \quad g_{uu} = -1 + O(r^{-k}), \quad g_{ur} = -1 + O(r^{-m}), \quad g_{uA} = O(r^{-l}),$$
$$(4.4)$$

where γ_{AB} is the metric on the S^{n-2} sphere. These particular coordinates, called Bondi coordinates, are a useful foliation of the spacetime via null hypersurfaces, as can be seen from the construction of the gauge: the coordinate u is a properly defined retarded time coordinate because $u = const$ defines null hypersurfaces.[10] The normal vector $k^\mu = g^{\mu\nu} \partial_\nu u$ satisfies by definition $k^\mu k_\mu = 0$ ($g_{\mu\nu} g^{\nu\rho} = \delta_\mu{}^\rho$), so that $g^{uu} = 0$. The radial coordinate r is defined so that the determinant condition on g_{AB} holds, which gives to r the meaning of luminosity distance. The coordinates x^A are constant on the integral curves of k^μ, namely $k^\mu \partial_\mu x^A = 0$, implying that $g^{uA} = 0$. Finally it is easily seen from this that $g_{rr} = 0 = g_{rA}$. Hence this establishes the form (4.3).

As mentioned in Sect. 2, in this particular setting the functions β, $\mathcal{U}$ and W^A are determined from h_{AB} which is assumed as an initial data in the integration of Einstein's equations. Detailed comments on the integration scheme can be found in Appendix "Einstein's Equations". The metric h_{AB} must encode the gravity waves degrees of freedom; this leads to a constraint on the form of h_{AB} according to the spacetime dimension. One is led to [18, 20]

$$h_{AB} = \gamma_{AB} + \sum_{k \geq 0} \frac{C_{(n/2+k-1)AB}}{r^{n/2+k-1}}, \quad (4.5)$$

where $k \in \mathbb{Z}$ if n is even and $2k \in \mathbb{Z}$ if n is odd. This is motivated by the fact that in n dimensions, the perturbations to the (angular) metric components h_{AB} start at order[11] $r^{-(n/2-1)}$, namely at order r^{-1} in $n = 4$, and at order $r^{-3/2}$ in $n = 5$. Notice that due to the determinant condition imposed on g_{AB} the first correction to γ_{AB} is traceless; hence it has the right number of degrees of freedom to encode the polarisation of the gravity wave.

In four spacetime dimensions, the solution of Einstein's equations implies

$$\beta = \frac{\beta_{(2)}}{r^2} + \frac{\beta_{(3)}}{r^3} + \cdots \quad \mathcal{U} = 1 - \frac{2m}{r} + \frac{\mathcal{U}_2}{r^2} + \cdots \quad \mathcal{W}^A = \frac{W^A_{(2)}}{r^2} + \frac{W^A_{(3)}}{r^3} + \cdots,$$
$$(4.6)$$

---

[10] It plays the role of the $u = t - r$ coordinate in Minkowski spacetimes, but in curved spacetimes it is not just $u = t - r$, consider, for example, the Schwarzschild case.

[11] This is a general result from the solution of the homogeneous wave equation, see [56].

where all the terms in the expansions are functions of (u, x^A). In this way β is completely determined as well as the functions $W^A_{(2)}$, $W^A_{(>3)}$ and $\mathcal{U}_{(\geq 2)}$, whereas both $m = m(u, x^A)$ and a function $N^A(u, x^A)$ in $W^A_{(3)}$ remain undetermined by the integration procedure and acquire a physical meaning captured by their respective names, *Bondi mass aspect* and *Bondi angular momentum aspect*. The mass aspect and the angular momentum aspect satisfy respectively two evolution equations in u:

$$\partial_u m = \frac{1}{4} D_A D_B N^{AB} - \frac{1}{8} N_{AB} N^{AB}. \tag{4.7}$$

$$
\begin{aligned}
\partial_u N_A = D_A m &+ \frac{1}{4}(D_B D_A D_C C^{BC} \\
&- D^2 D^C C_{CA}) + \frac{1}{4} D_B (N^{BC} C_{CA} + 2 D_B N^{BC} C_{CA}),
\end{aligned} \tag{4.8}
$$

where $N_{AB} := \partial_u C_{AB}$, with C_{AB} the first correction to γ_{AB}, is responsible of the non-conservation of the mass of the system. For this reason, N_{AB} is given the name of the *news tensor*. As opposed to the mass aspect, the angular momentum aspect is not constant when $N_{AB} = 0$. D_A denotes the covariant derivative on S^2 compatible with γ_{AB}.

Apart from the fact that in the expansion (4.5) also half-integer powers of $1/r$ enter, the main feature to be noticed is that the term of order $1/r$ is missing for any $n > 4$. This causes the main difference in the structure of the asymptotic symmetry group in higher dimensions and it is the first point to be addressed when one goes in search of the higher dimensional BMS group. However, as far as the solution of Einstein's equations is concerned, similar results hold. Two free functions, interpreted as the mass and the angular momentum aspects, appear [20, 21][12] and satisfy similar evolution equations. We do not need them for the purposes of Sect. 5.

Comment

The determinant condition on g_{AB}, fixed to be a function of the angle with a specific dependence on the radial coordinate, may be too restrictive from the holographic point of view. This condition may be generalised so that the boundary metric γ_{AB} may evolve in time u. The analysis of four-dimensional asymptotically AdS spacetimes with a Bondi-like expansion instead of the usual Fefferman-Graham one [25] may show this. This generalisation was first shown in [32] where, for the four-dimensional spacetime, γ_{AB} was extended to $\bar{\gamma}_{AB} = e^{2\varphi} \gamma_{AB}$ with $\varphi = \varphi(u, x^A)$. In the same way one can modify the higher dimensional boundary metric γ_{AB} as

---

[12] At different orders with respect to the four-dimensional case, according to the number of dimensions. For example, for $n = 5$ they are respectively at order r^{-2} and r^{-4} [18, 19].

$\bar{\gamma}_{AB} = e^{2\varphi}\gamma_{AB}$ with $\varphi = \varphi(u, x^A)$. With this choice

$$\det(g_{AB}) = r^{2(n-2)}b(u, x^A), \quad b(u, x^A) = e^{2(n-2)\varphi}\gamma \tag{4.9}$$

and

$$h_{AB} = \bar{\gamma}_{AB} + \sum_{k \geq 0} \frac{C_{(n/2+k-1)AB}}{r^{n/2+k-1}}. \tag{4.10}$$

4.2 The Asymptotic Symmetry Group

As said, for a general spacetime there are no isometries, namely no Killing vectors. However, when a metric asymptotically tends to another metric which enjoys some symmetries, we are lead to the definition of asymptotic Killing vector fields as the vector fields ξ that satisfy[13]

$$\mathcal{L}_\xi g_{\mu\nu} = O((x^{\rho_i})^{-m_{\mu\nu}}), \tag{4.11}$$

namely the diffeomorphisms which preserve the asymptotic form of the metric up to the same non-trivial order of the metric expansion. Such diffeomorphism are called *allowed diffeomorphisms*. Among them there are also *trivial diffeomorphisms* which are asymptotically identity. The asymptotic symmetry algebra (asa) is then defined as the quotient "allowed infinitesimal symmetries/trivial infinitesimal symmetry" and the associated group as an exponentiation of the algebra. So that one can write

$$\text{ASG} = \frac{\text{allowed symmetries}}{\text{trivial symmetries}}, \tag{4.12}$$

and consists of all the non-trivial asymptotic transformations that move a point g in the phase space to another point of the phase space.

Another way to characterise the trivial symmetries is via the associated charge. In a general covariant theory (as well as in gauge theories) there are several non-trivial issues in associating a conserved charge, namely a canonical generator, to a symmetry transformation. This problem has been overcome with several different procedures, canonical and covariant, mutually consistent. A general framework, valid for any gauge or generally covariant theory, is the *covariant phase space formalism*.[14] Modulo all the difficulties and caveats, we may claim that to each asymptotic Killing vector a charge Q_ξ is associated as the generator of the

---

[13] The Lie derivative is given by $\mathcal{L}_\xi g_{\mu\nu} := \xi^\rho g_{\mu\nu,\rho} + \xi^\sigma_{,\nu} g_{\mu\sigma} + \xi^\sigma_{,\mu} g_{\nu\sigma} = \xi_{\mu;\nu} + \xi_{\nu;\mu}$.

[14] See [6] for a review.

symmetry, namely

$$\{g, Q_\xi\} = \delta_\xi g \,, \tag{4.13}$$

where the brackets are Dirac brackets, since the gravitational theory is a constrained system. Among the allowed diffeomorphisms, those with vanishing associated charge are called *trivial*. These are gauge transformations that act trivially on the phase space.

The take home lesson of this section is that the nature of the boundary conditions presides over the dimensionality of the asymptotic symmetry group: if the asymptotic fall-off conditions are made weaker, then the group of allowed diffeomorphisms is larger and then the ASG may be larger. An example is again provided by the specific case of four-dimensional asymptotically flat spacetimes. At i^0, the principles (1) and (2) are satisfied by a large class of asymptotic conditions so that one would end up with different ASGs, according to which condition is chosen. For sufficiently strong fall-off conditions at spatial infinity i^0, compatible with (1) and (2), the ASG is the Poincaré group, whereas for weaker choices one can get enlargements of the Poincaré group. As stressed in [55], there are sufficiently good reasons to impose at i^0 the strong asymptotic conditions leading to the Poincaré group. However, at null infinity this freedom is missing and the only sensible boundary conditions allowing for gravity waves in the bulk of the spacetime are so weak that the ASG is infinite dimensional. In the next subsection we see how this comes about in four spacetime dimensions and contrasts to the higher dimensional case.[15]

The Asymptotic Isometries of the Bondi–Sachs Metric

The coordinate transformation (2.2) is the exponentiation of the asymptotic Killing vectors[16] found by solving the gauge preserving conditions

$$\mathcal{L}_\xi g_{rr} = 0 \,, \quad \mathcal{L}_\xi g_{rA} = 0 \,, \quad g^{AB} \mathcal{L}_\xi g_{AB} = 0. \tag{4.14}$$

and the boundary preserving conditions

$$\mathcal{L}_\xi g_{uu} = O\left(\frac{1}{r}\right), \quad \mathcal{L}_\xi g_{ur} = O\left(\frac{1}{r^2}\right), \quad \mathcal{L}_\xi g_{uA} = O(1), \quad \mathcal{L}_\xi g_{AB} = O(r) \,. \tag{4.15}$$

---

[15]Concerning i_n^0 ($n > 4$ is the number of dimensions), one is straightforwardly led to the same conclusion valid at i_4^0 [55].

[16]As usual in the literature we also give the bulk component of them, but on $\mathscr{I}$ it is subleading.

For the general n-dimensional case the gauge preserving conditions remain the same, but the boundary preserving equations become

$$\mathcal{L}_\xi g_{uu} = O(r^{-n/2+1}), \quad \mathcal{L}_\xi g_{ur} = O(r^{-n+2}),$$
$$\mathcal{L}_\xi g_{uA} = O(r^{-n/2+2}), \quad \mathcal{L}_\xi g_{AB} = O(r^{-n/2+3}). \tag{4.16}$$

The last condition in (4.14) can also be generalised to allow for rescalings

$$g^{AB} \mathcal{L}_\xi g_{AB} = 2(n-2)\tilde{\omega}. \tag{4.17}$$

The exact Killing equations determine the general form of the asymptotic Killing vector field

$$\xi = \xi^u \partial_u + \xi^r \partial_r + \xi^A \partial_A \tag{4.18}$$

$$\begin{cases} \xi^u = f(u, x^A), \\ \xi^r = -\dfrac{r}{n-2}\left[\bar{D}_A \xi^A - W^C \partial_C f + \left((n-2)\partial_u \varphi + \dfrac{1}{2}\gamma^{-1}\partial_u \gamma \right) - (n-2)\tilde{\omega} \right], \\ \xi^A = Y^A(u, x^B) - \partial_B f \int_r^\infty dR e^{2\beta} g^{AB}, \end{cases}$$

$$\tag{4.19}$$

as a function[17,18] of two integration functions $f(u, x^A)$ and $Y^A(u, x^B)$, whose behaviour is determined by the asymptotic Killing equations (4.15) or (4.16). At null infinity the Killing field reduces to $\xi_{(\alpha,Y)} = \xi^u \partial_u + \xi^A \partial_A$ and in what follows we are going to restrict to this. The notation $\xi_{(\alpha,Y)}$ is useful to remember that there is a different vector field for each choice of the pair (α, Y); sometimes only the pair is used instead of $\xi_{(\alpha,Y)}$.

As said, the functions f, Y^A are determined by the remaining asymptotic Killing equations (4.16). These make the difference between the four-dimensional and the higher-dimensional case. For this discussion we consider the traditional analysis with both φ and $\tilde{\omega}$ zero. On the one hand, such equations determine the function f to be

$$f(u, x^A) = \alpha(x^A) + \frac{u}{n-2} u D_A Y^A(x^B), \tag{4.20}$$

---

[17]Notice that $\gamma = \det \gamma_{AB}$ is independent of u, so that the $\gamma^{-1}\partial_u \gamma = 0$. This expression reduces to the known expressions for $n = 4$ (eq. 5 in [57]) and for n generic (eq. 52 in [20]). Notice also that $\gamma = \det \gamma_{AB}$ is independent of u, so that the $\gamma^{-1}\partial_u \gamma = 0$.

[18]Here the only difference w.r.t [20] is the term $W^C \partial_C f$ which comes with a $+$ there instead that a $-$ here. This sign difference comes from the fact that the uA mixed terms in the metric (4.3) are preceded by a $-$ sign instead of $+$ used in [20].

with α being

- an arbitrary function of the angles in the four-dimensional case. Hence it can be expanded in an arbitrary number of scalar spherical harmonics

$$\alpha = t^\mu n_\mu + \sum_{l=2}^{\infty} \sum_m \alpha_l^m \mathcal{Y}_l^m , \qquad (4.21)$$

where we denote $\mathcal{Y}_l^m$ the scalar spherical harmonics, $t^\mu = (t^0, t^i)$ and n^μ are $l < 2$ harmonics $(1, \sin\theta\cos\phi, \sin\theta\sin\phi, \cos\theta)$,
- a function constrained to be given by just the $l = 0, 1$ spherical harmonics in the higher dimensional case.

The dramatic reduction in the freedom of the scalar function α happening in higher dimensions is a direct consequence of the g_{AB} fall-off condition and the related Killing equation, the last one in (4.16). Let us see how this comes about. Any asymptotic Killing equation must be solved order by order. Let us take $\mathcal{L}_\xi g_{AB}$ as a generic example and consider first the four-dimensional case. Expanding for $r \to \infty$ the Killing vector and the metric tensor and computing $\mathcal{L}_\xi g_{AB}$, one obtains something like

$$\mathcal{L}_\xi g_{AB} = \ldots + F_3(u, x^A) r^3 + F_2(u, x^A) r^2 + F_1(u, x^A) r + \ldots , \qquad (4.22)$$

and in order to meet the condition $\mathcal{L}_\xi g_{AB} = O(r)$ one has to set to zero all the coefficients of the terms r^m, $m > 1$. An algorithmic but long computation shows that $F_2(u, r, x^A)$ is the only non-trivially vanishing coefficient. In higher dimensions we would have an expression similar to (4.22) with possibly terms with half-integers powers of r; however, this time the Killing equations for g_{AB} requires that all the terms in $\mathcal{L}_\xi g_{AB}$ of order greater than $r^{-n/2+3}$ must be put to zero. In particular, since $-n/2 + 3 < 1$ when $n > 4$, $F_1(u, r, x^A)$ must be equated to zero. This results in the further equation $(n - 2) D_A D_B \alpha = \gamma_{AB} D^2 \alpha$ with respect to the four-dimensional case and this turns out to constrain the form of α because its solutions are the $l = 0, 1$ modes of α.

On the other hand, the asymptotic Killing equations constrain the functions Y^As to be independent of u and to constitute a conformal Killing vector field on S^{n-2}, namely a Killing vector generating conformal transformations.[19] From well-known properties of conformal geometry, we know that

---

[19]A vector field Y satisfying $\mathcal{L}_Y g_{\mu\nu} := 2\nabla_{(\mu} Y_{\nu)} = \psi g_{\mu\nu}$ (∇ covariant derivative) is called *conformal Killing vector field*. The so-called conformal factor ψ is related to the dimension of spacetime by taking the trace of this equation, and it is fixed by this condition:

$$g^{\mu\nu} \mathcal{L}_Y g_{\mu\nu} = \psi g^{\mu\nu} g_{\mu\nu} = \psi n \implies 2\nabla_\mu Y^\mu = n\psi . \qquad (4.23)$$

On the 2-sphere $n = 2$, $\nabla \to D$ and then $D_A Y^A = \psi$.

- when Y^A is a vector field on S^2, that is, when the spacetime is four dimensional, we can choose the conformal transformations to be either globally well-defined or only locally well-defined,
- when Y^A is a vector field on S^{n-2}, $n > 4$, the conformal transformations can only be globally well-defined (Liouville theorem).

Once more we see a reduction in the freedom we have when we go to higher dimensions. In this case the reduction is caused by a mathematical obstruction, in the previous case it depended only on the choice of asymptotic conditions.

Let us expand on the difference between global and local well-definiteness of Y^A which is possible in four-dimensional spacetimes. On the Riemann sphere S^2 one can conveniently adopt stereographic coordinates $z = e^{i\phi} \cot\theta$, $\bar{z} = e^{-i\phi} \cot\theta$ so that $\gamma_{AB}dx^A dx^B = 4(1 + z\bar{z})^{-2}dzd\bar{z}$ and the conformal Killing equations split in two equations for the components Y^z, $Y^{\bar{z}}$ stating that they are respectively holomorphic and antiholomorphic: $\partial_{\bar{z}}Y^z = 0$, $\partial_z Y^{\bar{z}} = 0$. Focusing on the holomorphic part, nothing prevent it to be a meromorphic function expressed as a Laurent expansion for which a basis is given by $l_m = -z^{m+1}\partial_z$, $m \in \mathbb{Z}$, which form the infinite dimensional Witt algebra $[l_m, l_n] = (m - n)l_{m+n}$. The same applies to the antiholomorphic part $\bar{l}_m$, so that these conformal Killing close under the semidirect sum of two copies of the Witt algebra

$$[l_m, l_n] = (m - n)l_{m+n}, \quad [\bar{l}_m, \bar{l}_n] = (m - n)l_{m-n}, \quad [l_m, \bar{l}_n] = 0. \quad (4.24)$$

However, only six of these vector fields $\{l_m, \bar{l}_m\}_{(m=0,\pm 1)}$ are globally well-defined and generate an invertible group of transformations, which is the well-known group of Möbius transformations isomorphic to $SL(2, \mathbb{C})/\mathbb{Z}$ and, in turn, to the orthocronous Lorentz group $L_+^\uparrow = SO(1, 3)^\uparrow$.

Bondi and Sachs originally assumed that the Y^A were globally well-defined. With this choice their definition of the BMS algebra is

$$\mathfrak{bms}_4^{glob} = \mathfrak{l}_+^\uparrow \oplus_s \mathfrak{st}, \quad (4.25)$$

where $\oplus_s$ denotes the semidirect sum, $\mathfrak{st}$ the abelian normal subalgebra of super-translations, $\mathfrak{l}_+^\uparrow$ the Lorentz algebra. This algebra is easily exponentiated to the group in (2.1).

The relevance of the Witt algebra (and most importantly its centrally extended version, the Virasoro algebra) in conformal field theories and holography inspired Banks [31], Barnich and Troessaert [32, 33, 58] to suggest that there is no reason to require that Y^A are globally well-defined. In this case the BMS algebra is generalised to

$$\mathfrak{bms}_4^{loc} = \mathfrak{sr} \oplus_s \mathfrak{st}, \quad (4.26)$$

where $\mathfrak{sr} = \mathfrak{witt} \times \mathfrak{witt}$ is the so-called (BT) superrotation subalgebra. Due to the poles singularities in superrotations, there is no way to exponentiate them to get a finite form of the group.[20] For consistency in this case also supertranslations $\alpha(z, \bar{z})$ have to be expanded as Laurent series, rather than spherical harmonics, and their expansion is $\alpha_{m,n} = (1 + z\bar{z})^{-1} z^m \bar{z}^n$ (see [58] for details). The mode expanded form of the $\mathfrak{bms}_4^{loc}$ algebra is given by (4.24) plus the following commutators[21]:

$$[l_j, \alpha_{m,n}] = \left(\frac{j+1}{2} - m\right) \alpha_{m+j,n}, \quad [\bar{l}_j, \alpha_{m,n}] = \left(\frac{j+1}{2} - n\right) \alpha_{m,n+j}.$$

$$(4.27)$$

The subalgebra spanned by $\alpha_{m,n}, l_j, \bar{l}_k$ with the pair m, n acquiring all possible combinations of $(0, 1)$ and $j, k = 0, \pm 1$ is the Poincaré algebra. Possible central extensions of the $\mathfrak{bms}_4^{loc}$ algebra have been studied [13, 42, 58].

In higher dimensions, instead, all the possible freedom is killed either by the peculiar boundary conditions chosen or by general mathematical facts. The ASG reduces to the Poincaré group (and the corresponding algebra)

$$BMS_{n>4}^{glob} = L_+^\uparrow \ltimes T = ISO(1, n-2).$$

$$(4.28)$$

This reduction of the symmetry group prevents Weinberg soft theorem to be found from asymptotic symmetries. In order to obtain supertranslations in higher dimensions one can question the orthodox choice of asymptotic conditions and explore weaker conditions. This has been first considered for even-dimensional spacetimes [22], where the authors propose to consider the same integer-power expansion valid for $n = 4$, but with the further addition of conditions on the asymptotic expansions of the components of the Ricci tensor. With this choice supertranslations exist also in higher dimensions. The reason for the occurrence of supertranslations is evident: it lies in the non-vanishing $1/r$ term in the expansion of h_{AB}.

Concerning superrotations, apparently there is no hope to obtain them in higher dimensions because of unavoidable mathematical obstructions. To this purpose, defining superrotations in terms of $\text{Diff}(S^2)$ gives better possibilities because it can be extended to $\text{Diff}(S^{n-2})$. As we said in Sect. 3, $\text{Diff}(S^2)$ is preferable because of its tighter connections to subleading soft theorems and also because it has been shown to enjoy less issues than BT superrotations at the level of charges, but as opposed to BT superrotations (in four dimensions) they cannot arise as infinitesimal

---

[20]Notice that in the literature (i.e. [59]) the term "exponentiation" is sometimes used with a different meaning as "reconstruction of the spacetime from the boundary symmetry".

[21]For completeness let us state the general non-mode expanded form for the commutator of the Killing vectors $\xi_{(\alpha,Y)} = \xi^u \partial_u + \xi^A \partial_A$ (with $\varphi = \omega = 0$) restricted to null infinity: $[\xi_{(\alpha_1,Y_1)}, \xi_{(\alpha_2,Y_2)}] = \xi_{(\alpha_3,Y_3)}$ with $\alpha_3 = Y_1^A \partial_A \alpha_2 + \frac{1}{2} \alpha_1 D_A Y_2^A - (1 \leftrightarrow 2)$ and $Y_3^A = Y_1^B D_B Y_2^A - (1 \leftrightarrow 2)$. With a slight modification of the brackets the algebra can also be faithfully represented in the whole bulk [32].

symmetries of the Bondi–Sachs metric without changing the boundary conditions [37, 38, 60, 61], hence leading to a different definition of asymptotic flatness.

5 BMS_4, Charges, Phase Space and Holography from the Celestial Sphere

5.1 Charges and Fluxes

The procedure to associate a Hamiltonian function to a $\mathfrak{bms}$ transformation was developed by Wald and Zoupas [62] based on the covariant phase space formalism [63]. For a general Killing $\xi_{(\alpha,Y)}$ in the $\mathfrak{bms}_4^{glob}$ the Wald–Zoupas charge is[22] [62, 64]

$$ Q_\xi = \frac{1}{16\pi} \int_{\partial\Sigma} d^2\Omega \left[4\alpha m + cY^A Z_A \right] , \tag{5.1} $$

where we denote $d^2\Omega$ the integration measure on S^2, c is a constant whose value is not relevant in our discussion and Z_A is a suitable smooth covariant vector (again the details are not relevant). The integral is only on the boundary $\partial\Sigma$ of a spacelike slice Σ crossing $\mathscr{I}$ because the bulk term vanishes. This is a general property of charges in gravity: they are surface charges. Well-defined charges (integrable) do not depend on the slice Σ (see [6] for a general discussion). Well-defined (integrable) charges do not depend on the choice of the slice Σ (see [6] for a nice pedagogical introduction to the topic). The details of the derivation can be found in [62]. An a posteriori heuristic motivation of the formula may go as follows: Consider the Bondi mass aspect m, on a given null hypersurface $u = u^*$. It can be integrated over the angles and the resulting function is the Bondi mass, representing the mass of the spacetime at a given u instant

$$ M := \frac{1}{4\pi} \int_{\partial\Sigma} d^2\Omega m. \tag{5.2} $$

The mass/energy is related to time translations and it is actually the *charge* associated to that symmetry. Pure time translations are generated by the vector field $\xi_{(\alpha,Y)}$, given in Sect. 4, with $\alpha = 1$ and $Y^A = 0$. The other three spatial translations are given by $\alpha = t^i n_i$ (cfr. 4.21). Hence we can define also a Bondi linear momentum as $P^i := \frac{1}{4\pi} \int_{\partial\Sigma} d\Omega^2 m n^i$. Then we are allowed to guess that for

---

[22]The general definition of charge does not depend on the number of spacetime dimensions, but some points in the discussion in this section must be carefully checked against the change of spacetime dimensions.

each α defining a generic supertranslation ($Y^A = 0$) we can associate a charge as

$$Q_\alpha = \frac{1}{4\pi} \int_{\partial\Sigma} d^2\Omega\, \alpha(x) m(u, x)\,, \tag{5.3}$$

and with Q_α the charge associated to $\xi_{(\alpha,0)}$. We may have justified the more general (5.1).

The charges associated to the BMS symmetry are not conserved because of fluxes of them through null infinity. So in general the difference of the charge (5.3) or (5.1) between two cuts of null infinity is a flux. Let us see this considering the Bondi mass. The Bondi mass aspect m satisfies the evolution Eq. (4.7), namely the Bondi mass satisfies

$$\partial_u M = \frac{1}{32\pi} \int_{\partial\Sigma} d^2\Omega \left[2D_A D_B N^{AB} - N_{AB} N^{AB} \right]. \tag{5.4}$$

This equation is the famous Bondi mass loss equation. The equation can be integrated in the u direction between two points u_i and u_f and the resulting object is the flux of energy radiated through this section of null infinity by gravity waves.[23] This example shows that the charges associated to the BMS symmetry are not conserved because of fluxes of them through null infinity.

The Wald–Zoupas formalism prescribes a flux formula for any asymptotic symmetry vector field ξ and its associated charge, formally

$$Q_\xi^{I_+} - Q_\xi^{I_-} := \Delta_I Q_\xi = -\int_I \delta\mathbf{F}\,, \tag{5.5}$$

where I is a certain region of $\mathscr{I}^+$ and $I_\pm$ denote the upper/lower boundary of I along the u direction. This can be understood as a consistency check on the charge, namely computing the difference in charges on two cuts must be equal to the integral of the form $\mathbf{F}$. This works for the supertranslations ξ_α and for the Lorentz part $\xi_{Y(\text{glob})}$ of the global BMS. For example, in the case of pure supertranslations

$$\mathbf{F}_\alpha = -\frac{1}{32\pi} du\, d^2\Omega\, N^{AB} \pounds_\alpha C_{AB}\,. \tag{5.6}$$

However, Wald–Zoupas prescription does not work for generic BT superrotations $\xi_{Y(\text{sup.rot.})}$, in that case there is a discrepancy of the form [64]

$$-\int_I \delta\mathbf{F} = \Delta_I Q_{\xi_{Y(\text{sup.rot.})}} + \Delta_I \mathcal{F}\,. \tag{5.7}$$

---

[23] Also null matter can be considered and a term proportional to T_{uu} appears.

meaning that for superrotations the charge formula (5.1) and the flux formula (5.5) prescribed by Wald and Zoupas are not consistent. A point to be stressed is that in the Wald–Zoupas construction the consistency between the charge and the flux formula arises from the generator ξ being an asymptotic symmetry vector field, that is, a vector field which preserves asymptotic flatness. The discrepancy found in the case of superrotations hints that superrotations are not really asymptotic symmetry vectors of asymptotically flat spacetimes. In [59] they are named *external symmetries*, signalling that BT superrotations are on a different ground with respect to supertranslations (see Sect. 5.3). The charges associated to CL superrotations have been analysed in [65] while the final version of this book was in press.

5.2   The Phase Space

Wald–Zoupas fluxes can be integrated over the whole $\mathscr{I}^+$. This is going to represent the total charge radiated off from null infinity, for example, the total energy. Physically reasonable spacetimes emits only a finite amount of energy or, in general, charge. Hence a crucial requirement is the flux $\Delta_\infty \mathcal{Q}_\alpha$ to be finite. This is true as long as the news tensor satisfies

$$\lim_{|u| \to \infty} N_{AB} \approx |u|^{-\epsilon_N}, \quad \epsilon_N > 0 \quad . \tag{5.8}$$

The same goes for $\mathcal{Q}_Y$ provided Y is the globally well-defined vector. This fall-off condition along the u direction, implies from (4.8), that only the electric parity piece[24] of C_{AB} is non-zero[25] at $\mathscr{I}^+$ ($\mathscr{I}^+_\pm$ is a standard notation to denote the past and future regions of $\mathscr{I}^+$)

$$C^e_{AB} = (D_A D_B - \frac{1}{2}\gamma_{AB} D^2)C , \tag{5.9}$$

where C is a scalar function. The solution of this equation gives the initial value of C at $u \to -\infty$ which we denotes as $C|_{\mathscr{I}^+_-}$. From the integration procedure of Einstein's equations then we see that the whole spacetime is reconstructed from this datum as well as the Bondi mass and the News function. The space of solutions of AF gravity (or phase space) S is given at the first leading orders by the parameters $\{m|_{\mathscr{I}^+_-}, C|_{\mathscr{I}^+_-}, N_{AB}; N^A|_{\mathscr{I}^+_-}\}$.

---

[24]The tensor C_{AB} on S^2, being traceless and symmetric, can be decomposed into electric and magnetic parity pieces $C_{AB} = C^e_{AB} + C^m_{AB}$ where C^e_{AB} is given by (5.9) and $C^m_{AB} = \epsilon_{E(A}D_{B)}D^E C^m$. Notice that the $l = 0, 1$ harmonics of C^m can automatically be set to zero because they are annihilated by $\epsilon_{E(A}D_{B)}D^E$ as they should because of symmetry.

[25]Indeed from (4.8) the magnetic part must satisfy $(D^4 + D^2/2)C^m = 0$ which implies that $C^m = 0$.

The BMS transformations change these fields via the action of the Lie derivative

$$\mathfrak{L}_\alpha N_{AB} = \alpha \partial_u N_{AB}, \quad \mathfrak{L}_\alpha C_{AB} = \alpha N_{AB} - 2D_A D_B \alpha + \gamma_{AB} D^2 \alpha, \quad (5.10)$$
$$4\mathfrak{L}_\alpha m = 4\alpha \partial_u m + N^{AB} D_A D_B \alpha + 2D_A N^{AB} D_B \alpha, \quad \mathfrak{L}_\alpha C = \alpha.$$

The last one is obtained by comparing (5.9) to $\mathfrak{L}_\alpha C_{AB}$. BMS symmetries change the phase space of solutions. However, more insights on the physical consequences of these changes of solutions are obtained from the perspective of the (enlarged) *radiative phase space*. As we have seen from the definition of the Bondi gauge, apart from m and N^A, the function C is related through C_{AB} to the tensor N_{AB} which can be considered as the true physical observable at null infinity because of its role in the Bondi mass-energy. This perspective has been advocated by Ashtekar who proposed the *asymptotic quantisation* program [66–68] in which AF gravity is quantised from boundary degrees of freedom. This formalism is a canonical formalism in which a phase space $\mathcal{R}$ of *radiative modes* is constructed on the whole of $\mathscr{I}^+$ endowed with a symplectic form generating canonical transformations. One can rigorously see that Ashtekar's symplectic form is strictly related to the Wald–Zoupas $\Delta_\infty Q_\alpha$ fluxes, so that the quantities are equivalent.[26]

In [30] an enlarged phase space $\mathcal{R}^*$ has been introduced so that each commutator of the supertranslation charge with the phase space function reproduces the relevant Lie derivative[27]

$$\{Q_\alpha, N_{AB}\} = \mathfrak{L}_\alpha N_{AB}, \quad \{Q_\alpha, C_{AB}\} = \mathfrak{L}_\alpha C_{AB}, \quad \{Q_\alpha, C\} = \mathfrak{L}_\alpha C. \quad (5.11)$$

Here the generating charge in Ashtekar formalism is really the Wald–Zoupas flux over the whole of $\mathscr{I}^+$, hence it can be obtained from (5.3) upon integration over $\mathscr{I}^+$ using (4.7)

$$Q_\alpha = \frac{1}{32\pi} \int_{\mathscr{I}^+} du d^2\Omega \, \alpha(x) \left[2D_A D_B N^{AB} - N_{AB} N^{AB} \right] = Q_\alpha^\partial + Q_\alpha^b. \quad (5.12)$$

It is given by a boundary Q_α^∂ and a bulk Q_α^b term.[28] The first acts trivially on the classical phase space $\{Q_\alpha^\partial, N_{AB}\} = 0$: it does not contribute classically to the change of solution. However, upon quantisation the boundary term can be shown to be the zero frequency limit of a metric perturbation, hence a *soft graviton* [27]. In addition to this, the field C is naturally interpreted as a Goldstone boson for the breaking of the supertranslation symmetry. Indeed suppose having a starting

---

[26]In the discussion we have preferred starting from Wald–Zoupas charges because somewhat less abstract than Ashtekar formalism, whose review would have taken several pages.

[27]The radiative phase space is only comprised of N_{AB} and the commutators only give this first equation, thus importance of boundary terms on $\mathscr{I}^+$ is missed.

[28]Furthermore notice that the flux form (5.6) is exactly the integrand in (5.12) once $\mathfrak{L}_\alpha C_{AB}$ (cfr. (5.10)) is used.

non-radiative configuration $N_{AB} = 0$. From this configuration, with a *BMS* supertranslation C_{AB} can change by the inhomogeneous term in $\mathcal{L}_\alpha C_{AB}$, which vanishes when the supertranslations are actual translations. Since supertranslations commute with time translations, AF gravity is characterised by a manifold of degenerate vacua ($m = N_{AB} = C_{AB} = 0$) each of them characterised by a different value of C. The supertranslation invariance is broken by the fixing of a particular C. From the quantum picture one recognises that the various degenerate vacua differ only by the insertion of soft gravitons. Finally notice that pure translations are not broken because in that case α is automatically annihilated by the operator $D_A D_B - \gamma_{AB} D^2/2$ to which it couples.

5.3 Gravitational Memory Effects and AF↔LAF Transitions

Another interpretation of these actions—and maybe an observable effect—can be given in terms of the *gravitational memory effect* [69]. This is the permanent displacement of two observers after the passage of a finite burst of gravitational waves.[29] From the geodesic deviation equation applied to two observers[30] that experience the passage of a gravity wave only in the interval $u_f - u_i$, it can be easily seen [6, 69] that the difference $\Delta L = |L_f - L_i|$ in their proper distance depends on the difference ΔC_{AB} between the two times, because the relevant component of the Weyl tensor that enters this equation is proportional to C_{AB}. As we have seen, ΔC_{AB} can be generated by a BMS supertranslation and hence the memory can be read off from the supertranslation charges [64]. Similar considerations have led to the conjecture of the existence of a *spin memory* effect which is related to BT superrotations [75] as well as other memories [60]. As a side comment, it is worth mentioning that in the last few years, fostered by these discussions, the gravity wave memory effect has been studied also in the specific setting of plane-wave spacetimes,[31] which have been shown to possess Carroll symmetries [78].

The interpretation of the action of the asymptotic Killing vectors associated to BT superrotations is more subtle [32, 64] and, as we said, due to several problems at the level of the definition of the phase space there is no agreement on their role (see, for example, [36, 60]). In [79] it is shown that finite superrotations change the topology of $\mathscr{I}^+$ inducing transitions from AF spacetimes with a complete $\mathscr{I}$

---

[29]Historically the distinction between linear [70] and nonlinear memory effect [71] was made and they were considered two separate effects: the first due to the emission process of the radiation and the second due to the full nonlinearity of Einstein's gravity, with no reference to the emission process. However, there may be not such a sharp distinction [72–74].

[30]In the BMS setting they are not inertial because they travel close to the boundary of spacetime along at fixed u and at two different angles.

[31]The impulsive limit of these has spread some confusion in the literature [76, 77]. I would like to thank J. Podolsky for having drawn my attention on the paper [77].

to spacetimes which can be called *locally* AF (LAF) because $\mathscr{I}$ is missing some points. An example of such spacetimes is the cosmic string spacetime [80]

$$ds^2 = -dU^2 - 2dU dR + R^2(d\Theta^2 + K^2 \sin^2 \Theta^2 d\Phi^2) \tag{5.13}$$

where K is a constant acquiring a value in $(0, 1)$, related to the string tension. Because of K there is a deficit angle on the polar axis at $\Theta = 0, \pi$, the north and south poles of a would be sphere S^2. Equivalently one can say that the celestial sphere is missing two points [81]. Such spacetime can be written in Bondi–Sachs form [82] and is characterised by a news tensor N_{AB} and hence also by a tensor C_{AB}. The authors of [79] have shown that finite superrotations can be interpreted as sourcing a transition form an AF to a LAF spacetime and vice versa, using as a first example the snapping of a cosmic string. Such a process is portrayed[32] by a line element describing two half-spaces, each of which with a different coordinate system, joined along the wavefront by appropriate (Penrose) junction conditions. The finite superrotation is then the mapping from one side to the other of the cut. In the snapping of the cosmic string case the cut is a light cone opening at the instant $u = 0$ in which the string snaps and hence, upon writing the metric in Bondi form, the change in coordinates affects C_{AB}. The difference ΔC_{AB} before and after the snapping can be seen as due to a BT superrotation. Higher dimensional generalisation of this can be explored [84] as well as the relation between multiple string configurations (or the other possible defects of the S^2 listed in [81]) with BT superrotations. An alternative point of view on the physical realisation of superrotations has been proposed in [85] while this volume was in press.

5.4   *Christodoulou-Klainerman Spacetimes, Matching Conditions and Scattering*

All the discussion on $\mathscr{I}^+$ can be translated in a Bondi gauge appropriate to $\mathscr{I}^-$, which in particular is characterised by an advanced Bondi time v, and a phase space of radiative modes can be built there. Two distinct copies BMS_+ and BMS_- of the BMS group act respectively on $\mathscr{I}^+$ and $\mathscr{I}^-$. To get a unique symmetry group acting at the same time on the whole of $\mathscr{I} = \mathscr{I}^+ \cup \mathscr{I}^-$ the null generators of the two disjoint components of null infinity have to be matched. This is done in practice via an antipodal identification of points of the spheres at past and future null infinity, so as to define a continuity condition between the asymptotic fields near i^0. This matching condition is always possible in Minkowski spacetime [86], but for general AF spacetime the matching is not at all obvious. Indeed the possible values of ϵ_N may allow for logarithmic u evolution of C_{AB}. To avoid this it is sufficient that $\epsilon_N > 1$. Are there physically meaningful spacetimes corresponding to

---

[32]This description, first developed by Penrose, has been called *cut and paste construction*. See [83] for a review and generalisation of the construction applied to the cosmic string case.

this requirement? For sure there is at least a class: Christodoulou-Klainerman (CK) spacetimes, because they are characterised by $\epsilon_N = 3/2$ [27]. CK spacetimes are arbitrarily close to Minkowski spacetime (in a mathematical well-defined sense), are above the threshold for the formation of black holes and are characterised by $m_B|_{\mathscr{I}^+_+} = 0$, $m_B|_{\mathscr{I}^+_-} = M_{ADM}$, so that they radiate all their energy. However, there may be other spacetimes that satisfy $\epsilon_N > 1$ and allow also for black holes in the interior with both massless and massive final state (see [6] and the references therein). More in general it seems that the only condition is that the spacetimes evolve from a non-radiative ($N_{AB} = 0$) to a non-radiative configuration both at early and late advanced/retarded Bondi times.

Once matched $\mathscr{I}^+$ and $\mathscr{I}^-$, the field C_{AB} at x can be given the same value of[33] C^-_{AB} evaluated at an antipodal point[34] $\tilde{x}$ on S^2: $C_{AB}(x)|_{\mathscr{I}^+_-} = C^-_{AB}(\tilde{x})|_{\mathscr{I}^-_+}$ and the generators of BMS_0 can be identified with the subset of generators of $BMS_+ \times BMS_-$ satisfying $\alpha(x) = \alpha^-(\tilde{x})$ and $Y^A(x) = Y^A_-(\tilde{x})$.

We are in the position to state the BMS invariance of the S-matrix conjecture. In particular, restricting to supertranslations this condition is written as

$$\langle \text{out}| \, [Q^+_\alpha S - S Q^-_\alpha] \, |\text{in}\rangle = 0. \tag{5.14}$$

To get this, one observes that supertranslation charges on $\mathscr{I}^+$ commute with those on $\mathscr{I}^-$ because the supertranslations do not affect each other: $\{Q^+_\alpha, Q^-_{\alpha'}\} = 0$. Furthermore, thanks to the matching conditions $Q^+_\alpha = Q^-_{\alpha'} = Q$. The S matrix commutes with Q_α when $\alpha = 1$ because in that case Q_1 is the Hamiltonian of the system. The conjecture is that for any α: $\{Q_\alpha, S\} = 0$ where it is understood that the operators act on initial and final initial data, states as in (5.14). The full equivalence of (5.14) to the Weinberg soft theorems is shown in [27, 30]. Here we just want to justify the claim in Sect. 3 that in this setting the holographic relations involves the celestial spheres rather than the whole boundary. One constructs a Fock representation for *in* and *out* states.[35] They are labelled by the collection of points from which they enter or exit the spacetime and their energies (assume no spin for simplicity): $|in\rangle = |Z, E^Z\rangle$ and $|out\rangle = |W, E^W\rangle$, where $Z = \{z_1, ..., z_k\}$ and $W = \{w_1, ..., w_k\}$ are the collection of points on $\mathscr{I}^-$, $\mathscr{I}^+$ respectively and E^Z, E^W the energies of the corresponding particles. Easy manipulations [27] turn (5.14) into a Ward identity

$$\langle W, E_W|T(FS)|Z, E^Z\rangle = \sum_j (E^Z_j \alpha(z_j) - E^W_j \alpha(w_j)) \langle W, E_W|S|Z, E^Z\rangle, \tag{5.15}$$

---

[33]Quantities on $\mathscr{I}^-$ are denoted as f^-, later on we denote quantities on $\mathscr{I}^+$ with the overscript +, but this is not necessary.

[34]For example, in standard spherical coordinates $x = (\theta, \phi)$ and $\tilde{x} = (\pi - \theta, \pi + \phi)$, and in stereographic coordinates $(z, \bar{z}) \to (\tilde{z}, \bar{\tilde{z}}) = (-\bar{z}^{-1}, -z^{-1})$.

[35]A Fock representation on $\mathscr{I}^+$ was first constructed in the context of asymptotic quantisation, see [68].

$T(FS)$ is the time ordered product of the S operator and the soft graviton operator $F := F^+ - F^-$ constructed from the soft graviton operator (the boundary part of the charges) on $\mathscr{I}^+$ and the one on $\mathscr{I}^-$. Equation (5.15) is interpreted as the supertranslation Ward identity relating elements of the S-matrix with and without insertions of the soft graviton operator [27] and it is proved [30] to be equivalent to the Weinberg soft graviton theorem. In the past few years several papers have appeared on this topic which further explore the connection between scattering amplitudes in gravity and the structures that emerges from the BMS supertranslations and superrotation symmetries, but it is impossible to cover them here.

As a final comment we should mention that CK spacetime may be too restrictive because indeed the charges associated to general supertranslations vanishes in the limit to spatial infinity from the two components of null infinity [87]. However, there may be more general class of spacetimes in which the ideas here summarised can be safe from this constraint.

Appendix: Einstein's Equations

In [9] it is shown that the Bianchi identity hierarchically organises Einstein's equations in Main equations, Supplementary equations and a Trivial equation, as follows

$$
\text{Main equations} \begin{cases} a) & R_{rr} = 0 \quad \text{diff. eq. for } \beta \text{ w.r.t. } r\,, \\ b) & R_{rA} = 0 \quad \text{diff. eq. for } W^A \text{ w.r.t. } r\,, \\ c) & R_{AB} = 0 \quad \begin{cases} \text{trace part} = 0 \text{ diff. eq. for } \mathcal{U} \text{ w.r.t. } r\,, \\ \text{traceless part} = 0 \text{ diff. eq. for } g_{AB} \text{ w.r.t. } u\,. \end{cases} \end{cases}
$$
$$\tag{A.1}$$

$$
\text{Supplementary equations} \begin{cases} a) & R_{uu} = 0\,, \\ b) & R_{Au} = 0\,, \end{cases} \tag{A.2}
$$

$$
\text{Trivial equation} \quad R_{ur} = 0\,. \tag{A.3}
$$

This pattern remains valid both in higher dimensions and when the theory is coupled to matter provided that the Einstein tensor is used instead of the Ricci tensor and suitable asymptotic conditions are imposed on the stress-energy tensor.[36]

The solution is given order by order in the $1/r$ expansion on each null hypersurface $u = $ constant. Once the Main equations are solved, the Trivial equation is automatically satisfied, while the Supplementary equations are automatically

---

[36] See, for example, [64].

solved at all orders except one in each of them. These further equations are u-evolution equations for the integration functions appearing in the solutions of the Main equations.

In particular, the Main equations R_{rr}, R_{rA}, $g^{AB}R_{AB}$ are respectively differential equations w.r.t. r for the functions β, W^A, $\mathcal{U}$ so that they fixes the form of their $1/r$ expansion in terms of g_{AB} on the given $u = $ constant hypersurface up to possible integration functions independent of r. As seen, g_{AB} plays the role of initial data. The traceless part of (A.1c) gives the u-evolution of the subleading terms in the expansion of g_{AB} in terms of β, W^A, $\mathcal{U}$ which are now known. However, it does not determine the u-evolution of C_{AB}, hence $\partial_u C_{AB} =: N_{AB}$ has to be assumed as a further initial data on the initial $u = $ constant surface. The role of g_{AB} as initial data is quite clear since in the Bondi–Sachs integration scheme the data are placed on a null hypersurface from where the equations are evolved.

The relevant integration functions comes from (A.1b) and from the trace part of (A.1c). Such functions acquire a precise physical meaning and are respectively called *angular momentum aspect* $N^A = N^A(u, x^C)$ and *Bondi mass aspect* $m = m(u, x^C)$. As stated at the beginning of the section, their u-evolution is constrained by the Supplementary equations $R_{uA} = 0$ and $R_{uu} = 0$ respectively. In four dimensions they give Eqs. (4.7) and (4.8).

Comment on Logarithmic Terms To conclude let us make an important observation about the four-dimensional case. A non-vanishing $C_{(2)AB}$ term in the asymptotic expansion of h_{AB} produces, via the equation $R_{rA} = 0$ which contains two derivatives with respect to r of W^A, a logarithmic term

$$W^A = \frac{1}{r^2}\left(-\frac{1}{2}D_B C^{AB}\right)$$
$$-\frac{1}{r^3}\left(\frac{2}{3}N^A(u, x^B) + \left(\ln r + \frac{1}{3}\right)D_B C_{(2)}^{BA} - \frac{1}{3}C^{AB}\nabla^C C_{BC}\right) + O(r^{-4}).$$

$$(A.4)$$

Such a term would not be there either if $C_{(2)AB} = 0$ or $D_B C_{(2)}^{BA} = 0$. As observed in [32], the weaker possibility allows for singularities on S^2, which may seem good for superrotations in the light of the interpretation as cosmic strings. Originally [9, 24] the vanishing of $C_{(2)AB}$ was assumed and such a choice was justified as a sort of Sommerfeld radiation condition, namely the requirement that the radiation outgoing from a source does not have incoming components. This condition is implicitly present also in the textbooks definitions of asymptotic flatness [2, 8] and it is related to the peeling property which characterise these definitions of asymptotic flatness. However, that form of radiation condition has been proved wrong [88, 89] and in general the peeling property too restrictive for several important results in general relativity [88] (including the proof of nonlinear stability of the Minkowski spacetime by Christodoulou and Klainermann). The upshot is that a meaningful definition of asymptotic flatness can be given also for spacetimes that admit logarithmic terms in the asymptotic expansions, which are called *polyhomogeneous* [88], and their ASG is again the BMS_4.

Acknowledgements This work was completed with the support of EPSRC. I would like to thank Marika Taylor for her helpful comments on the manuscript.

References

1. R. Geroch, Asymptotic structure of space-time, in *Asymptotic Structure of Space-Time*, eds. by F.P. Esposito, L. Witten (Springer, Heidelberg, 1977)
2. R.M. Wald, *General Relativity* (University of Chicago Press, Chicago, 1984)
3. A. Strominger, The dS/CFT correspondence. JHEP, 034 (2001). arXiv:hep-th/0106113 [hep-th]
4. A. Strominger, Lectures on the infrared structure of Gravity and Gauge theory (2017). arXiv:1703.05448 [hep-th].
5. A. Ashtekar, Geometry and physics of null infinity (2014). arXiv:1409.1800 [gr-qc]
6. G. Compère, A. Fiorucci, Advanced lectures in general relativity (2018). arXiv:1801.07064 [hep-th]
7. F. Alessio, G. Esposito, On the structure and applications of the Bondi-Metzner-Sachs group. Int. J. Geom. Meth. Mod. Phys. **15**, 1830002 (2018)
8. R. Penrose, W. Rindler, *Spinors and Space-Time: Vol. 2, Spinor and Twistor Methods in Space-Time Geometry*. Cambridge Monographs on Mathematical Physics (Cambridge University Press, Cambridge, 1988)
9. H. Bondi, M.G.J. van der Burg, A.W.K Metzner, Gravitational waves in general relativity VII waves from axi-symmetric isolated systems. Proc. R. Soc. Lond. A **269**, 21–52 (1962)
10. R. Sachs, Asymptotic symmetries in gravitational theory. Phys. Rev. **128**, 2851–2864 (1962)
11. S. Hollands, A. Ishibashi, Asymptotic flatness at null infinity in higher dimensional gravity, in *Proceedings, 7th Hungarian Relativity Workshop (RW 2003), Sarospatak, Hungary, August 10-15, 2003* (2004), pp. 51–61
12. J.D. Brown, M. Henneaux, Central charges in the canonical realization of asymptotic symmetries: An example from three dimensional gravity. Commun. Math. Phys. **104**, 207–226 (1986)
13. B. Oblak, BMS particles in three dimensions (2016). arXiv:1610.08526 [hep-th]
14. P.J. McCarthy, Asymptotically flat space-times and elementary particles. Phys. Rev. Lett. **29**, 817–819 (1972)
15. A. Komar, Quantized gravitational theory and internal symmetries. Phys. Rev. Lett. **15**, 76–78 (1965)
16. S. Hollands, R.M. Wald, Conformal null infinity does not exist for radiating solutions in odd spacetime dimensions. Class. Quantum Grav. **21**, 5139–5145 (2004). gr-qc/0407014
17. S. Hollands, A. Ishibashi, Asymptotic flatness and Bondi energy in higher dimensional gravity. J. Math. Phys. **46**, 022503 (2005). gr-qc/0304054
18. K. Tanabe, S. Kinoshita, T. Shiromizu, On asymptotic structure at null infinity in five dimensions. J. Math. Phys. **51**, 062502 (2010). arXiv:0909.0426v2 [gr-qc]
19. K. Tanabe, S. Kinoshita, T. Shiromizu, Angular momentum at null infinity in five dimensions. J. Math. Phys. **52**, 032501 (2011). http://arxiv.org/abs/1010.1664. arXiv:1010.1664 [gr-qc]
20. K. Tanabe, S. Kinoshita, T. Shiromizu, Asymptotic flatness at null infinity in arbitrary dimensions. Phys. Rev. D, **84**, 044055 (2011). arXiv:1104.0303v2 [gr-qc]
21. K. Tanabe, S. Kinoshita, T. Shiromizu, Angular momentum at null infinity in higher dimensions. Phys. Rev. D, **85**, 124058 (2012). arXiv:1203.0452 [gr-qc]
22. D. Kapec, V. Lysov, S. Pasterski, A. Strominger, Higher-dimensional supertranslations and Weinberg's soft Graviton theorem (2015). arXiv:1502.07644 [gr-qc]

23. H. Bondi, Gravitational waves in general relativity. Nature, **186**(4724), 535 (1960)
24. R.K. Sachs, Gravitational waves in general relativity VIII waves in asymptotically flat space-time. Proc. R. Soc. Lond. A **270**, 103–126 (1962)
25. A. Poole, K. Skenderis, M. Taylor, (A)dS$_4$ in Bondi Gauge (2018). arXiv:1812.05369 [hep-th]
26. E. Witten, *Talk at Strings '98* (1998). http://online.kitp.ucsb.edu/online/strings98/witten/
27. A. Strominger, On BMS invariance of gravitational scattering. JHEP **2014**, 152 (2014). arXiv:1312.2229 [hep-th]
28. D. Christodoulou, S. Klainerman, The global nonlinear stability of the Minkowski space. Princeton Math. **41**, 432 (1993)
29. S. Weinberg, The quantum theory of fields, in *The Quantum Theory of Fields 3 Volume Hardback Set*, vol. 1 (Cambridge University Press, Cambridge, 1995)
30. T. He, V. Lysov, P. Mitra, A. Strominger, BMS supertranslations and Weinberg's soft graviton theorem. JHEP **5**(5), 151 (2015). arXiv:1401.7026 [hep-th]
31. T. Banks, A Critique of pure string theory: heterodox opinions of diverse dimensions (2003). arXiv:hep-th/0306074 [hep-th]
32. G. Barnich, C. Troessaert, Aspects of the BMS/CFT correspondence. JHEP **5** (2010). arXiv:1001.1541 [hep-th]
33. G. Barnich, C. Troessaert, Supertranslations call for superrotations. PoS Ann. U. Craiova Phys. **21**, S11 (2011). arXiv:1102.4632 [gr-qc].
34. F. Cachazo, A. Strominger, Evidence for a New Soft Graviton theorem (2014). arXiv:1404.4091 [hep-th]
35. D. Kapec, V. Lysov, S. Pasterski, A. Strominger, Semiclassical Virasoro symmetry of the quantum gravity S-matrix. JHEP **08** (2014). arXiv:1406.3312 [hep-th]
36. J. Distler, R. Flauger, B. Horn, Double-soft graviton amplitudes and the extended BMS charge algebra (2018). arXiv:1808.09965 [hep-th]
37. M. Campiglia, A. Laddha, Asymptotic symmetries and subleading soft graviton theorem. Phys. Rev. D, **90**, 124028 (2014). arXiv:1408.2228v3 [hep-th]
38. M. Campiglia, A. Laddha, New symmetries for the gravitational S-matrix. JHEP **2015**(4), 76 (2015). arXiv:1502.02318v2 [hep-th]
39. C. Duval, G.W. Gibbons, P.A. Horvathy, Conformal Carroll groups and BMS symmetry. Class. Quantum Grav. **31**(9), 092001 (2014). arXiv:1402.5894 [gr-qc]
40. Y. Calò, *Relation Between Symmetry Groups for Asymptotically Flat Spacetimes*. Master's thesis (University of Salento, Lecce, 2018)
41. A. Bagchi, R. Fareghbal, BMS/GCA redux: towards flatspace holography from non-relativistic symmetries. JHEP, **10**, 92 (2012). arXiv:1203.5795 [hep-th]
42. A. Bagchi, R. Basu, A. Kakkar, A. Mehra, Flat holography: aspects of the dual field theory. JHEP **12**, 147 (2016). arXiv:1609.06203 [hep-th]
43. J. Hartong, Gauging the Carroll algebra and ultra-relativistic gravity. JHEP **2015**, 69 (2015)
44. G. Arcioni, C. Dappiaggi, Exploring the holographic principle in asymptotically flat spacetimes via the BMS group. Nucl. Phys. B **674**, 553–592 (2003). arXiv:hep-th/0306142 [hep-th]
45. G. Arcioni, C. Dappiaggi, Holography in asymptotically flat spacetimes and the BMS group. Class. Quantum Grav. **21**, 5655–5674 (2004). arXiv:hep-th/0312186 [hep-th]
46. C. Dappiaggi, BMS field theory and holography in asymptotically flat space-times. JHEP **2004**, 011 (2004). arXiv:hep-th/0410026 [hep-th]
47. E. Bergshoeff, J. Gomis, G. Longhi, Dynamics of Carroll particles. Class. Quantum Grav. **31**, 205009 (2014). arXiv:1405.2264 [hep-th]
48. A. Barducci, R. Casalbuoni, J. Gomis, Confined dynamical systems with Carroll and Galilei symmetries. Phys. Rev. D **98**, 085018 (2018)
49. A. Barducci, R. Casalbuoni, J. Gomis, VSUSY models with Carroll or Galilei invariance (2018). arXiv:1811.12672 [hep-th]
50. S. Bhattacharyya, S. Minwalla, V.E. Hubeny, M. Rangamani, Nonlinear fluid dynamics from gravity. JHEP **2008**, 045 (2008). arXiv:0712.2456 [hep-th]
51. I. Bredberg, C. Keeler, V. Lysov, A. Strominger, From Navier-Stokes to Einstein. JHEP **2012**, 146 (2012). arXiv:1101.2451 [hep-th]

52. G. Compère, P. McFadden, K. Skenderis, M Taylor, The holographic fluid dual to vacuum Einstein gravity. JHEP **2011**, 50 (2011). arXiv:1103.3022 [hep-th]
53. N. Pinzani-Fokeeva, M.M. Taylor, Towards a general fluid/gravity correspondence. Phys. Rev. D, **91**, 044001 (2015). arXiv:1401.5975 [hep-th]
54. P. Kartik, Conservation of asymptotic charges from past to future null infinity: Supermomentum in general relativity (2019). arXiv:1902.08200 [gr-qc]
55. S. Hollands, A. Ishibashi, R.M. Wald, BMS supertranslations and memory in four and higher dimensions. Class. Quantum Grav. **34** (2017). arXiv:1612.03290 [gr-qc]
56. S. Hassani, *Mathematical Physics: A Modern Introduction to Its Foundations*. 1st edn. (Springer, Berlin, 2002)
57. G. Barnich, C. Troessaert, Symmetries of asymptotically flat four-dimensional spacetimes at null infinity revisited. Phys. Rev. Lett. **105**, 111103 (2010). arXiv:0909.2617 [gr-qc]
58. G. Barnich, C. Troessaert, BMS charge algebra. JHEP, **12**, 105 (2011). arXiv:1106.0213 [hep-th]
59. G. Compère, J. Long, Vacua of the gravitational field. JHEP **2016**(7), 137 (2016). arXiv:1601.04958v3 [hep-th]
60. G. Compère, A. Fiorucci, R. Ruzziconi, Superboost transitions, refraction memory and super-Lorentz charge algebra (2018). arXiv:1810.00377 [hep-th]
61. V. Chandrasekaran, É.É. Flanagan, K. Prabhu, Symmetries and charges of general relativity at null boundaries. JHEP **2018**, 125 (2018)
62. R.M. Wald, A. Zoupas, General definition of "conserved quantities" in general relativity and other theories of gravity. Phys. Rev. D **61**, 084027 (2000). arXiv:gr-qc/9911095
63. C. Crnkovic, E. Witten, Covariant description of canonical formalism in geometrical theories, in *Three Hundred Years of Gravitation*, ed. by W. Israel, S.W. Hawking (Cambridge University Press, Cambridge, 1989)
64. É.É. Flanagan, D.A. Nichols, Conserved charges of the extended Bondi-Metzner-Sachs algebra. Phys. Rev. D **95** (2015). arXiv:1510.03386 [hep-th]
65. É.É Flanagan, K. Prabhu, I. Shehzad, Extensions of the asymptotic symmetry algebra of general relativity. arXiv:1910.04557 [gr-qc]
66. A. Ashtekar, Asymptotic quantization of the gravitational field. Phys. Rev. Lett. **46**, 573–576 (1981)
67. A. Ashtekar, M. Streubel, Symplectic geometry of radiative modes and conserved quantities at null infinity. Proc. Roy. Soc. Lond. A **376**, 585–607 (1981)
68. A. Ashtekar, *Asymptotic Quantization: Based on 1984 Naples Lectures* (Napoli, Bibliopolis, 1987)
69. A. Strominger, A. Zhiboedov, Gravitational memory, BMS supertranslations and soft theorems. JHEP **01**(1), 86 (2016). arXiv:1411.5745 [hep-th]
70. Y.B. Zel'dovich, A.G. Polnarev, Radiation of gravitational waves by a cluster of superdense stars. Sov. Astron. **18**, 17 (1974)
71. D. Christodoulou, Nonlinear nature of gravitation and gravitational-wave experiments. Phys. Rev. Lett. **67**, 1486–1489 (1991)
72. K.S. Thorne, Gravitational-wave bursts with memory: the christodoulou effect. Phys. Rev. D **45**, 520–524 (1992)
73. A.G. Wiseman, C.M. Will, Christodoulou's nonlinear gravitational-wave memory: evaluation in the quadrupole approximation. Phys. Rev. D **44**, 2945–2949 (1991)
74. A. Tolish, R.M. Wald, Retarded fields of null particles and the memory effect. Phys. Rev. D **89**, 064008 (2014). arXiv:1401.5831 [gr-qc]
75. S. Pasterski, A. Strominger, A. Zhiboedov, New gravitational memories. JHEP **12**(12), 53 (2016). arXiv:1502.06120 [hep-th]
76. P.M. Zhang, C. Duval, P.A. Horvathy, Memory effect for impulsive gravitational waves. Class. Quantum Grav. **35**, 065011 (2018)
77. R. Steinbauer, The memory effect in impulsive plane waves: comments, corrections, clarifications (2018). arXiv:1811.10940 [gr-qc]

78. C. Duval, G.W. Gibbons, P.A. Horvathy, P.M. Zhang, Carroll symmetry of plane gravitational waves. Class. Quantum Grav. **34**, 175003 (2017)
79. A. Strominger, A. Zhiboedov, Superrotations and black hole pair creation. Class. Quantum Grav. **34**(6), 064002 (2017). arXiv:1610.00639 [hep-th]
80. A. Vilenkin, E.P.S. Shellard, in *Cosmic Strings and Other Topological Defects*. Cambridge Monographs on Mathematical Physics (Cambridge University Press, Cambridge, 2000)
81. J. Bicak, B. Schmidt, On the asymptotic structure of axisymmetric radiative spacetimes. Class. Quantum Grav. **6**, 1547 (1989)
82. J. Bicák, A. Pravdová, Symmetries of asymptotically flat electrovacuum space—times and radiation. J. Math. Phys **39**(11), 6011–6039 (1998). arXiv:gr-qc/9808068 [gr-qc]
83. J. Podolský, J.B. Griffiths, The collision and snapping of cosmic strings generating spherical impulsive gravitational waves. Class. Quantum Grav. 17(6), 1401 (2000). arXiv:gr-qc/0001049 [gr-qc]
84. F. Capone, M.M. Taylor, Cosmic branes and asymptotic structure. JHEP (accepted for publication). arXiv:1904.04265v2 [hep-th]
85. E. Adjei, W. Donnelly, V. Py, A. J. Speranza, Cosmic footballs from superrotations. arXiv:1910.05435
86. R. Penrose, Asymptotic properties of fields and space-times. Phys. Rev. Lett., **10**, 66–68 (1963)
87. A. Ashtekar, *The BMS Group, Conservation Laws, and Soft Gravitons* (Talk at Perimeter Institute, Waterloo, 2016). http://pirsa.org/16080055
88. P.T. Chrusciel, M.A.H. MacCallum, D.B. Singleton, Gravitational waves in general relativity XIV. Bondi expansions and the 'polyhomogeneity' of $\mathscr{I}$. Phil. Trans. R. Soc. Lond. A **350**, 113–141 (1995)
89. J.A.V. Kroon, A comment on the outgoing radiation condition for the gravitational field and the Peeling theorem. Gen. Rel. Grav. **31**, 1219 (1999). gr-qc/9811034

Relativity of Observer Splitting Formalism and Some Astrophysical Applications

Vittorio De Falco

Abstract This study deals with the relativity of observer splitting formalism, powerful technique in General Relativity to meaningfully distinguish the contributions of the gravitational effects and fictitious forces on the relative motion of two non-inertial observers. In the weak field limit, this approach has a direct connection with the classical description. This technique is particularly useful to model several astrophysical situations. We show two particular applications related to radiation processes and accretion disc physics.

Keywords Equivalence principle · Gravitoelectromagnetism ·
Poynting-Robertson effect

1 Introduction

In Classical Mechanics the fictitious forces appear each time we are in a non-inertial reference frame. The tricky part of the general relativistic description relies on mixing the gravitational field with the accelerated motion of the observers. Such topic achieved resounding successes in General Relativity (GR), for two reasons: (1) for the equivalence principle, which shows how the gravitational forces, generated by a massive body, are closely related to the fictitious forces, due to the accelerated motion of a reference frame; (2) for the gravitoelectromagnetism effects on the motion of a non-inertial observer around a general relativistic object [1].

The first attempts to generalise the classical formalism in GR has been performed following the direct spacetime approach, based on the splitting of the test particle motion with respect to a comoving observer and then relating the resulting quantities with respect to an inertial frame [2–4]. This approach revealed to be misleading in

V. De Falco (✉)
Research Centre for Computational Physics and Data Processing, Faculty of Philosophy &
Science, Silesian University in Opava, Opava, Czech Republic
e-mail: vittorio.defalco@physics.cz

© Springer Nature Switzerland AG 2019
S. Cacciatori et al. (eds.), *Einstein Equations: Physical and Mathematical Aspects of General Relativity*, Tutorials, Schools, and Workshops in the Mathematical Sciences, https://doi.org/10.1007/978-3-030-18061-4_7

the interpretation of the dynamical variables, depending then on further quantities. The successful approach became the relativity of observer splitting formalism, based on the full orthogonal splitting of the test particle motion in local rest space and local time direction (3+1) and transverse and longitudinal components of the local rest space (2+1) with respect to a rest observer [5, 6]. Such formalism offers a natural link with respect to the classical case and provides an explicit physical interpretation of the resulting terms.

Such formalism is fundamental in the description of the motion of inflowing matter towards compact objects. The accretion matter is mainly dominated by the gravitational field and, if endowed with enough angular momentum, can form an accretion disc, emitting in the X-ray energy band from its innermost regions [7]. The radiation emitted by an accreting compact object is intercepted by the inflowing matter and is responsible for significant departures from geodetic motion. Indeed, such radiation field, besides exerting a radial force, may also remove angular momentum from the inflowing matter, thus altering its motion. This radiation drag force has been termed Poynting-Robertson (PR) effect [8, 9]. Bini et al., using the relativity of observer splitting formalism, derived the general relativistic equations of motion for the PR effect in the Kerr metric [10, 11].

The article is structured as follows: in Sect. 2 the relativity of observer splitting formalism is introduced. Then two astrophysical applications are presented; in Sect. 3, the general relativistic PR effect in two dimensions and in Sect. 4 a model describing the dynamical evolution of an accretion disc under the influence of a constant radiation field in the strong field regime.

2   Relativity of Observer Splitting Formalism

We describe the relativity of observer splitting formalism aimed to decompose coherently the relative motion of two non-inertial observers in GR. There are two stages to achieve the complete splitting: (1) $3 + 1$, which is straightforward (see Sect. 2.1); (2) after introduced some notions about the calculus on spatial hypersurfaces (see Sect. 2.2), we arrive to $2 + 1$ splitting (see Sect. 2.3). Finally, we compare such formalism with the classical approach (see Sect. 2.4).

2.1   3+1 Splitting

A spacetime is endowed with a Lorentzian metric $g_{\alpha\beta}$ with signature $(-, +, +, +)$, a symmetric Levi-Civita connection $\Gamma^{\alpha}_{\beta\gamma}$, and a covariant derivative ∇_{α}. We consider the observer congruence, that is a family of observers defined by a future-pointing unit timelike vector field u^{α}, whose proper time τ_u parameterizes the world lines, integral curves of u^{α}. The projector operator $P(u)_{\alpha\beta} = g_{\alpha\beta} + u_{\alpha}u_{\beta}$ permits to orthogonally decompose each tangent space into local rest space Σ and local time t in the direction of the observer u^{α}. The quantities lying on Σ are termed *spatial*.

Let $a(u)^\alpha = u^\beta \nabla_\beta u^\alpha$ be the acceleration of the observer u^α, having the propriety to be always spatial. The kinematical decomposition of the observer congruence is given by Misner and Thorne [12]

$$\nabla_\alpha u_\beta = -a(u)_\alpha u_\beta + \theta_{\alpha\beta} + \omega_{\alpha\beta}, \tag{2.1}$$

where $\theta_{\alpha\beta} = P_\alpha^\mu P_\beta^\nu \nabla_{(\nu} u_{\mu)}$ is the expansion tensor, and $\omega_{\alpha\beta} = P_\alpha^\mu P_\beta^\nu \nabla_{[\nu} u_{\mu]}$ is the vorticity tensor, where $(A, B) = 1/2(AB + BA)$ and $[A, B] = 1/2(AB - BA)$.

2.2 Calculus in the Spatial Hypersurfaces

The calculus in the spatial hypersurfaces permits to yield at the 2+1 splitting. Given any spatial vector field, X^α, and a generic vector, v^α, we have [5, 6]:

- the spatial Lie derivative:

$$\mathcal{L}(u)_X v^\alpha = P(u)\mathcal{L}_X v^\alpha = P(u)_\gamma^\alpha \left(X^\beta \nabla_\beta v^\gamma + v^\beta \nabla_\beta X^\gamma \right); \tag{2.2}$$

- the spatial covariant derivative:

$$\nabla(u)_\beta v^\alpha = P(u)_\delta^\alpha P(u)_\beta^\gamma \nabla_\gamma v^\delta; \tag{2.3}$$

- the temporal Lie derivative:

$$\nabla_{(\text{Lie})}(u) v^\alpha = P(u)\mathcal{L}_u v^\alpha; \tag{2.4}$$

- the temporal Fermi–Walker derivative:

$$\nabla_{(\text{fw})}(u) v^\alpha = P(u)_\gamma^\alpha u^\beta \nabla_\beta v^\gamma; \tag{2.5}$$

- the temporal co-rotating Fermi–Walker (or co-expanding Lie) derivative:

$$\nabla_{(\text{cfw})}(u) v^\alpha = \nabla_{(\text{fw})}(u) v^\alpha + \omega(u)_\beta^\alpha v^\beta = \nabla_{(\text{Lie})}(u) v^\alpha + \theta_\beta^\alpha v^\beta. \tag{2.6}$$

As occurs in continuum mechanics with the Lagrangian and Eulerian derivatives, these definitions imply three different approaches to perform the transport of quantities in GR. They are all equivalent, but each one is more suitable than the others depending on the problem to solve. We note, that the last definition is a combination of the previous definitions together with the kinematical decomposition of the observer [5, 6, 13].

Let U^α be the unit timelike four-velocity of a test particle, where τ_U is the particle's proper time along its world line and $\tau_{(U,u)}$, is its proper time relative to the observer u^α such that $d\tau_{(U,u)}/d\tau_U = \gamma(U,u)$ with $\gamma(U,u)$ the Lorentz factor. Therefore, the four-velocity U^α can be decomposed along u^α as follows:

$$U(\tau_U)^\alpha = \gamma(U,u)\left[u^\alpha + v(U,u)^\alpha\right] = E(U,u)\,u^\alpha + p(U,u)^\alpha, \tag{2.7}$$

where $v^\alpha \equiv v^\alpha(U,u)$ is the relative spatial velocity of U^α, $\gamma \equiv \gamma(U,u) = (1 - v^2)^{-1/2}$ with $v \equiv v(U,u) = \sqrt{v_\alpha v^\alpha}$ the module of the relative spatial velocity of U^α, $E \equiv E(U,u) = \gamma$ is the relative energy per unit mass, and $p^\alpha \equiv p(U,u)^\alpha = \gamma(U,u)v(U,u)^\alpha$ is the relative spatial momentum of U^α per unit mass, and $p \equiv p(U,u) = \sqrt{p_\alpha p^\alpha}$ is the module of the relative spatial momentum.

To operate along the test particle curve, we define the intrinsic (or absolute) derivative of a spatial vector field X^α along the test particle trajectory as

$$\frac{DX^\alpha(\tau_U)}{d\tau_U} = \frac{dX^\alpha(\tau_U)}{d\tau_U} + \Gamma^\alpha_{\beta\gamma}\,U(\tau_U)^\beta\,X^\gamma(\tau_U). \tag{2.8}$$

In this way we can extend the notions of Fermi–Walker, Lie, and co-rotating Fermi–Walker transport along the test particle curve as [5, 6, 13]

$$\frac{D_{(\text{tem})}(U,u)\,X^\alpha(\tau_U)}{d\tau_{(U,u)}} = \left[\nabla_{(\text{tem})}(u) + v^\beta\,\nabla(u)_\beta\right]X^\alpha(\tau_U) \quad \text{tem=fw, Lie, cfw.} \tag{2.9}$$

2.3 2+1 Splitting

A system of coordinates is adapted to the observer congruence when given $\{E_a^\alpha\}$, a spatial frame, being a basis for Σ, we have that the frame derivatives of a function f are $u(f) = f_{,0}$, $E_a^\alpha\partial_\alpha f = f_{,a}$. In such a frame we can construct [5, 6, 13]:

$$\nabla_{(\text{tem})}(u)\,E_a^\alpha = C_{(\text{tem})}(u)_a^b\,E_b^\alpha, \quad \text{tem=fw, Lie, cfw,} \tag{2.10}$$

$$\nabla_{E_a}E_b^\alpha = \Gamma(u)_{ab}^c\,E_c^\alpha, \tag{2.11}$$

$$(P(u)\,[E_a, E_b])^\alpha = C(u)_{ab}^c\,E_c^\alpha, \tag{2.12}$$

where $C_{(\text{tem})}(u)_a^b$ are the temporal constant structures, $\Gamma(u)_{ab}^c$ are the spatial connections, and $C(u)_{ab}^c$ are the spatial constant structures.

The spatial projection of the test particle's four-acceleration, $a(U)^\alpha = DU^\alpha/d\tau_U$, measured in the observer frame, $A(U,u)^\alpha$, is given by $A(U,u)^\alpha = 1/\gamma\,P(u)_\beta^\alpha a(U)^\beta$. Therefore, we have [5, 6, 13]:

$$a(U)^\alpha = \gamma\,P(U)_\beta^\alpha\left\{\frac{D_{(\text{tem})}(U,u)}{d\tau_{(U,u)}}\left[\gamma\,u^\alpha + p(u,U)^\alpha\right]\right\}$$
$$= \gamma^2\frac{D_{(\text{tem})}(U,u)}{d\tau_{(U,u)}}u^\alpha + \gamma\frac{D_{(\text{tem})}(U,u)}{d\tau_{(U,u)}}p(u,U)^\alpha. \tag{2.13}$$

We have split the test particle acceleration into temporal and spatial part with respect to the observer congruence. The temporal projection along u^α leads to evolution of the observer under the influence of the general relativistic effects (gravitoelectromagnetism forces); while the spatial projection orthogonal to u^α leads to the general relativistic non-inertial forces.

The first term in Eq. (2.13) can be decomposed into [5, 6, 13]:

$$\gamma^2 \frac{D_{(tem)}(U, u)}{d\tau_{(U,u)}} u^\alpha = \gamma^2 \left[a(u)^\alpha + H_{(tem)}(u)^\alpha_\beta\, v(U, u)^\beta \right], \tag{2.14}$$

where

$$H_{(tem)}(u)^\alpha_\beta = \begin{cases} \theta(u)^\alpha_\beta - \omega(u)^\alpha_\beta, & \text{tem=fw}; \\ 2\theta(u)^\alpha_\beta - 2\omega(u)^\alpha_\beta, & \text{tem=Lie}; \\ \theta(u)^\alpha_\beta - 2\omega(u)^\alpha_\beta, & \text{tem=cfw}. \end{cases} \tag{2.15}$$

In Eq. (2.14) the first term leads to the gravitoelectric force, instead the second term to the gravitomagnetic force.

The second term in Eq. (2.13) can be decomposed into longitudinal and transversal part with respect to the observer congruence as [5, 6, 13]

$$\gamma \frac{D_{(tem)}(U, u)}{d\tau_{(U,u)}} p(u, U)^\alpha$$

$$= \gamma \frac{D_{(tem)}(U, u)\, p(U, u)}{d\tau_{(U,u)}} \hat{v}(U, u)^\alpha + \gamma\, p(U, u) \frac{D_{(tem)}(U, u)\, \hat{v}(U, u)^\alpha}{d\tau_{(U,u)}}$$

$$= \frac{dp(U, u)^\alpha}{d\tau_U} + \gamma^2\, \Gamma(u)^\alpha_{\beta\gamma}\, v(U, u)^\beta\, v(U, u)^\gamma + \gamma^2\, C_{(tem)}(u)^\alpha_\beta\, v(U, u)^\beta,$$

$$\tag{2.16}$$

where the second and third terms are the relative centrifugal force and the relative centripetal force, respectively.

2.4 Comparison with the Classical Formalism

Now we move to the classical formalism for describing the relative motions of two non-inertial observers. We consider two reference frames $\mathbb{R} \equiv \{O, x, y, z\}$ and $\mathbb{R}' \equiv \{O', x', y', z'\}$ in relative motion to each other, observing the dynamical evolution of a point P in a classical framework. We call $\mathbf{r}(t) = P(t) - O$ and $\mathbf{r}'(t) = P(t) - O'$ the radius vectors in the reference frames $\mathbb{R}$ and $\mathbb{R}'$, respectively. The relationship between the accelerations is [14, 15]:

$$\mathbf{a} = \mathbf{a}' + \mathbf{a}_{O'} + \boldsymbol{\omega} \times \boldsymbol{\omega} \times \mathbf{r}' + 2\boldsymbol{\omega} \times \mathbf{v}', \tag{2.17}$$

where $\mathbf{a_t} = \mathbf{a}' + \mathbf{a}_{O'}$ is the translatory acceleration, $\boldsymbol{\omega} \times \boldsymbol{\omega} \times \mathbf{r}'$ is the centrifugal force, and $2\boldsymbol{\omega} \times \mathbf{v}'$ is the Coriolis force. All the terms figuring in Eq. (2.17) have a precise and clear physical meaning.

Exploiting the relativity of observer formalism in GR we have found that the acceleration of the test particle with respect to the observer u^α assumes the following form (see Eqs. (2.14) and (2.16)):

$$\begin{aligned}
a(U)^\alpha &= \frac{dp(U, u)^\alpha}{d\tau_U} + \gamma^2 \left\{ \left[a(u)^\alpha + H_{(\text{tem})}(u)^\alpha_\beta \, v(U, u)^\beta \right] \right. \\
&\left. + \Gamma(u)^\alpha_{\beta\gamma} \, v(U, u)^\beta \, v(U, u)^\gamma + C_{(\text{tem})}(u)^\alpha_\beta \, v(U, u)^\beta \right\}.
\end{aligned} \tag{2.18}$$

As in the classical case also this formalism leads to a precise splitting, where all the terms have a transparent meaning. In addition, we note the following one-to-one correspondence between Eqs. (2.17) and (2.18):

$$\begin{aligned}
a(U)^\alpha &\longrightarrow \mathbf{a}, \\
\frac{dp(U, u)^\alpha}{d\tau_U} &\longrightarrow \mathbf{a}', \\
\gamma^2 \left[a(u)^\alpha + H_{(\text{tem})}(u)^\alpha_\beta \, v(U, u)^\beta \right] &\longrightarrow \mathbf{a}_{O'}, \\
\gamma^2 \Gamma(u)^\alpha_{\beta\gamma} \, v(U, u)^\beta \, v(U, u)^\gamma &\longrightarrow \boldsymbol{\omega} \times \boldsymbol{\omega} \times \mathbf{r}', \\
\gamma^2 C_{(\text{tem})}(u)^\alpha_\beta \, v(U, u)^\beta &\longrightarrow 2\boldsymbol{\omega} \times \mathbf{v}'.
\end{aligned} \tag{2.19}$$

Here, it becomes more evident that the role played by the general relativistic effects on the motion of the observer (gravitoelectromagnetism force) and on the fictitious non-inertial forces, and the exact link with the classical formalism.

3   The General Relativistic PR Effect in Two Dimensions

We study the motion of a test particle in the Kerr metric (see Sect. 3.1) under the influence of a radiation field, constituted by photons travelling along null geodesics on the background spacetime, and the general relativistic PR effect (see Sect. 3.2). We then derive the test particle equations of motion (see Sect. 3.3) and investigate the test particle trajectories in Schwarzschild and Kerr metric (see Sect. 3.4).

3.1   Geometrical Setting

We consider the spacetime outside a compact object of mass M and spin a, described by the Kerr metric. Using Boyer–Lindquist coordinates (t, r, θ, φ) adapted to the spacetime symmetries, the line element of the Kerr metric, in the equatorial plane $\theta = \pi/2$ and in geometrical units $(G = c = 1)$, reads as [12]

$$ds^2 \equiv g_{\mu\nu}dx^\mu dx^\nu = -\left(1 - \frac{2M}{r}\right)dt^2 - \frac{4Ma}{r}dt\,d\varphi + \frac{r^2}{\Delta}dr^2 + \rho\,d\varphi^2, \quad (3.1)$$

where $\Delta \equiv r^2 - 2Mr + a^2$, and $\rho \equiv r^2 + a^2 + 2Ma^2/r$. We consider two family of observers: the distant static observer, whose adapted frame is given by the holonomic basic $\{\partial_t, \partial_r, \partial_\theta, \partial_\varphi\}$; the Zero Angular Momentum Observers (ZAMOs), whose adapted frame is the anholonomic (tetraed) frame $\{e_{\hat{t}}, e_{\hat{r}}, e_{\hat{\theta}}, e_{\hat{\varphi}}\}$. The two reference frames are connected through the following transformation law [10–12]:

$$e_{\hat{t}} = n, \quad e_{\hat{r}} = \frac{1}{\sqrt{g_{rr}}}\partial_r, \quad e_{\hat{\varphi}} = \frac{1}{\sqrt{g_{\varphi\varphi}}}\partial_\varphi, \quad e_{\hat{\theta}} = \frac{1}{\sqrt{g_{\theta\theta}}}\partial_\theta, \quad (3.2)$$

where $n = N^{-1}(\partial_t - N^\varphi \partial_\varphi)$ is the ZAMO four velocity, $N = (-g^{tt})^{-1/2}$ is the time lapse function, and $N^\varphi = g_{t\varphi}/g_{\varphi\varphi}$ is the spatial shift vector field.

3.2   Radiation Field

We consider the presence of a pure electromagnetic field coming around the compact object, modelled as a coherent flux of photons travelling along null geodesics of the background spacetime. The related energy-momentum tensor is [10, 11]:

$$T^{\mu\nu} = \Phi^2 k^\mu k^\nu, \quad k^\mu k_\mu = 0, \quad k^\mu \nabla_\mu k^\nu = 0, \quad (3.3)$$

where Φ is the intensity of the radiation field and k is the photon four-momentum field. The photon four-momentum can be split in the ZAMO frame as $k = E(n)[n + \hat{\nu}(k, n)]$ and $\hat{\nu}(k, n) = \sin\beta\, e_{\hat{r}} + \cos\beta\, e_{\hat{\varphi}}$ [10, 11]. We have that $E(n) = E/N(1 + bN^\varphi)$ is the relative energy of the photon, $E = -k_t$ is the constant conserved energy, $L_z = k_\varphi$ is the constant conserved angular momentum along the z axis, $b \equiv L_z/E$ is the impact parameter, and $\beta = \arccos[bE/\sqrt{g_{\varphi\varphi}}E(n)]$ is the photon angle in the ZAMO $\hat{r} - \hat{\varphi}$ plane. The quantity Φ can be completely determined, imposing the conserved equations $\nabla_\nu T^{\mu\nu} = 0$, which gives $\Phi^2 = \Phi_0^2/(\sqrt{g_{\varphi\varphi}}N|b\tan\beta|)$, where Φ_0 is the intensity at the emission surface [10, 11].

3.3 Equations of Motion

Let us consider a test particle moving in the equatorial plane with velocity U^α, that can be split in the ZAMO as $U = \gamma(U, u)[n + v(U, n)]$ and $v(U, n) = v \sin \alpha \, e_{\hat{r}} + \cos \alpha \, e_{\hat{\varphi}}$, where $\gamma(U, n) = (1 - ||v(U, n)||^2)^{-1/2}$ and $v = ||v(U, n)||$. The radiation field interacts with the test particle through the Thomson scattering cross section σ, independent from the direction and frequency of the radiation. The associated force is $\mathcal{F}_{(\text{rad})}(U)^\mu = -\sigma P(U)^\mu{}_\nu T^\nu{}_\delta U^\delta$, where $P(U)^\mu{}_\nu$ is the project operator with respect to U. The equation of motions are given by $ma(U)^\mu = \mathcal{F}_{(\text{rad})}(U)^\mu$, where m is the test particle mass and $a(U)^\mu$ is the test particle four acceleration. Decomposing the photon four-momentum with respect to U^α as $k = E(U)[U + \hat{\mathcal{V}}(k, U)]$, we arrive to the following equation $a(U) = \tilde{\sigma} \Phi^2 E(U)^2 \hat{\mathcal{V}}(k, U)$, where $\tilde{\sigma} = \sigma/m$. We defined $A = \tilde{\sigma} \Phi_0^2 E^2 \equiv ML_\infty/L_{\text{EDD}}$, where L_∞ is the luminosity of the central source at infinity and L_{EDD} is the Eddington luminosity at infinity, that is termed luminosity parameter and it ranges in the interval $[0, 1]$. Using the relativity of observer splitting formalism for the test particle acceleration, Eq. (2.18),[1] and for the radiation field (see [10, 11] for further details) we obtain the following coupled set of first order differential equations [10, 11]:

$$\frac{dv}{d\tau_U} = -\frac{\sin \alpha}{\gamma}[a(n)^{\hat{r}} + 2v \cos \alpha \, \theta(n)^{\hat{r}}{}_{\hat{\varphi}}]$$

$$+ \frac{A(1 + bN^\varphi)}{rN^2\sqrt{g_{\varphi\varphi}}|\sin \beta|}[\cos(\alpha - \beta) - v][1 - v\cos(\alpha - \beta)], \qquad (3.4)$$

$$\frac{d\alpha}{d\tau_U} = -\frac{\gamma \cos \alpha}{v}\left[a(n)^{\hat{r}} + 2v\cos \alpha \, \theta(n)^{\hat{r}}{}_{\hat{\varphi}} - v^2\frac{\sqrt{\Delta}(r^3 - Ma^2)}{r^3\rho}\right]$$

$$+ \frac{A\,(1 + bN^\varphi)[1 - v\cos(\alpha - \beta)]}{v} \sin(\beta - \alpha), \qquad (3.5)$$

$$\frac{dr}{d\tau_U} = \frac{\gamma v \sin \alpha}{\sqrt{g_{rr}}}, \qquad (3.6)$$

$$\frac{d\phi}{d\tau_U} = \frac{\gamma}{\sqrt{g_{\varphi\varphi}}}\left(v \cos \alpha + \frac{2aM}{r\sqrt{\Delta}}\right), \qquad (3.7)$$

where

$$a(n)^{\hat{r}} = \frac{M[(r^2 + a^2)^2 - 4a^2 Mr]}{r^3\rho\sqrt{\Delta}}, \qquad \theta(n)^{\hat{r}}{}_{\hat{\varphi}} = \frac{-aM(3r^2 + a^2)}{r^3\rho}. \qquad (3.8)$$

---

[1]Using the Lie transport the terms $C_{(\text{Lie})}(n)^\alpha_{\beta\gamma}$, $C_{(\text{Lie})}(n)^\alpha_\beta$, vanish in the Kerr metric (see [5, 6, 13], for further details).

3.4 Test Particle Trajectories

We study through the test particle trajectories this dynamical system, which admits a critical radius $r_{(\text{crit})}$, region where the radiation force balances the gravitational pull and the PR drag force, which is a function of the luminosity parameter A [10, 11]. The final destiny of the test particle strongly depends on its initial conditions (position and velocity fields), which is either at infinity or on the critical region. It is important to note that the test particle can never cross the event horizon, because the radiation pressure is so strong to push the particle away.

Schwarzschild Metric

We distinguish two families of orbits depending on the initial position of the test particle with respect to the critical radius: inside and outside. In both cases we analyse the behaviour of the test particle under different initial conditions on the velocity field and the radiation field, regulated by the impact parameter b, see Fig. 1. We note how the radial photons reach faster the test particle than the other photons, giving thus more time to the PR effect to remove angular momentum from the test particle. In addition, for radial photon the test particle stops on a point of the critical region, while in the other case it moves on this region with constant velocity [10, 11].

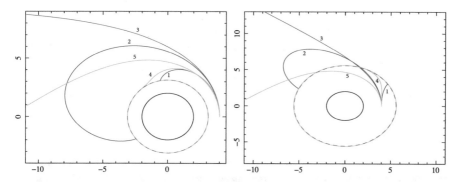

Fig. 1 Orbits of test particles in the Schwarzschild spacetime plotted in units of $M = 1$. The continuous black line is the Schwarzschild radius, the dashed red line is the critical orbit, the blue lines are for radial photon with $b = 0$, instead the green lines for photon with $b \neq 0$. **Left panel.** The luminosity parameter is at $A/M = 0.6$ and the critical orbit is at $r_{(\text{crit})} = 3.125M$. All the test particles have the same initial positions $(r_0, \varphi_0, \alpha_0) = (4M, 0, 0)$, while the initial velocities are: (1) $v_0 = 0.5$, (2) $v_0 = 0.64$, (3) $v_0 = 0.9$, (4) $(v_0, b) = (0.2, 0.2)$, and (5) $(v_0, b) = (0.64, 5.2)$. **Right panel.** The luminosity parameter is at $A/M = 0.8$ and the critical orbit is at $r_{(\text{crit})} = 5.5M$. All the test particles have the same initial positions $(r_0, \varphi_0, \alpha_0) = (4M, 0, 0)$, while the initial velocities are: (1) $v_0 = 0.5$, (2) $v_0 = 0.71$, (3) $v_0 = 0.8$, (4) $(v_0, b) = (0.5, 0.2)$, and (5) $(v_0, b) = (0.71, 5.2)$

Kerr Metric

In this case we have to distinguish four cases depending not only on the position of the test particle with respect to the critical radius, but also if it co-rotates ($a > 0$) or counter-rotates ($a < 0$) with the central object, see Fig. 2. The same observations drawn for the Schwarzschild spacetime can be similarly applied to this case. However, here it is also present the frame dragging effect, responsible to incredibly alterate the shape of the test particle orbits and, to force them to move with constant velocity on the critical region [10, 11].

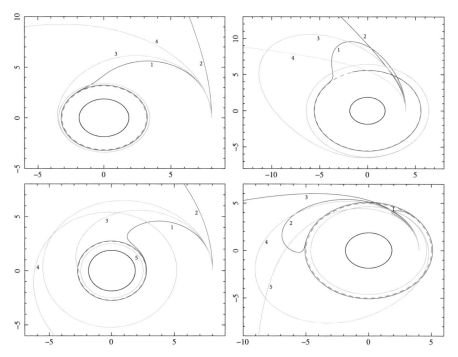

Fig. 2 Orbits of test particles in the Kerr spacetime plotted in units of $M = 1$ in all panels. It is valid the same notation of Fig. 1. **Left upper panel.** The event horizon is at $1.866M$, the spin is $a = 0.5$, the luminosity parameter is $A/M = 0.6$, and the critical radius is at $r_{(\text{crit})} = 3.154M$. All the test particles have the same initial positions $(r_0, \varphi_0, \alpha_0) = (8M, 0, 0)$, while the initial velocities are: (1) $v_0 = 0.5$, (2) $v_0 = 0.8$, (3) $(v_0, b) = (0.2, 0.5)$, and (4) $(v_0, b) = (0.2, 5.2)$. **Right upper panel.** The event horizon is at $1.866M$, the spin is $a = 0.5$, the luminosity is $A/M = 0.8$, and the critical radius is at $r_{(\text{crit})} = 5.551M$. All the test particles have the same initial positions $(r_0, \varphi_0, \alpha_0) = (4M, 0, 0)$, while the initial velocities are: (1) $v_0 = 0.5$, (2) $v_0 = 0.8$, (3) $(v_0, b) = (0.5, 0.5)$, and (4) $(v_0, b) = (0.5, 5.2)$. **Left lower panel.** The event horizon is at $1.866M$, the spin is $a = -0.5$, the luminosity parameter is $A/M = 0.6$, and the critical radius is at $r_{(\text{crit})} = 2.71M$. All the test particles have the same initial positions $(r_0, \varphi_0, \alpha_0) = (8M, 0, 0)$, while the initial velocities are: (1) $v_0 = 0.2$, (2) $v_0 = 0.5$, (3) $(v_0, b) = (0.2, 1.5)$, and (4) $(v_0, b) = (0.2, 3.5)$. **Right lower panel.** The event horizon is at $1.866M$, the spin is $a = -0.5$, the luminosity parameter is $A/M = 0.8$, and the critical radius is at $r_{(\text{crit})} = 5M$. All the test particles have the same initial positions $(r_0, \varphi_0, \alpha_0) = (4M, 0, 0)$, while the initial velocities are: (1) $v_0 = 0.5$, (2) $v_0 = 0.8$, (3) $v_0 = 0.9$, (4) $(v_0, b) = (0.8, 0.5)$, and (5) $(v_0, b) = (0.8, 3.5)$

4   Accretion Disc Model Influenced by a Radiation Field in the Strong Field Regime

We develop a numerical model of a geometrically thin accretion disc in the radiation field of a central source, by using the Kerr metric and the relativistic PR effect (see Sects. 4.1 and 4.2). Our numerical simulations determine the PR radius, zone located between the unperturbed region of the standard disc and the inner region dominated by the relativistic PR effect (see Sect. 4.3). In some of our simulations we explain the formation of a denser ring-like structure slightly inwards of the PR radius, being a consequence of the PR effect (see Sect. 4.4).

4.1   Accretion Disc Model

A geometrically thin accretion disc, placed in the equatorial plane of a rotating compact object of mass M and spin a, described by the Kerr metric, Eq. (3.1), is invested by a constant radiation field. The accretion disc is filled by many non-interacting test particles coming continuously and with constant accretion rate from the outer boundary. The radiation field propagates radially and interacts with each test particle through Thomson scattering, where Eqs. (3.4)–(3.7) govern the equatorial motion of the test particles (see Sect. 3.3 and [16]).

   The radiation field illuminates the disc, penetrating in its inner regions in accordance with the radial optical depth given by $\tau_r(r)$. Therefore, the effective luminosity inside the disc is $\mathcal{A}_{\text{eff}}(r) = A\,e^{-\tau_r(r)}$, where A is the unshielded luminosity, see Sect. 3.3. The proprieties of the matter inflow in the disc, where radiation has negligible effects, are modelled either on the standard α-viscosity [17] or on the α_g-viscosity prescription (viscosity proportional to gas-pressure) [18]. Only part of the radiation intersects the accretion disc and after each test particle–photon interaction the photon is lost to infinity without being scattered again.

   In the inner regions the disc quantities gradually depart from the standard model, because the radiation-drag effects dominate over viscous stresses. A steady state is attained when the inner disc is eroded up to a radius at which matter is dragged in at the same rate at which it drifts inwards from the outer disc. This transition zone is termed PR radius, R_{PR} and in the PR region, between R_{PR} and R_{in}, the radial velocity increases rapidly, the number density steeply reduces, and the optical depth remains low (see [16] for further details).

4.2   Numerical Implementation

The C++ code PRTRAJECTORIES uses the OPENMP library for massive parallelization in the integration of up to millions of test particle trajectories, performed by a high-precision 8-order Runge–Kutta method [16]. The matter inflow is modelled by injecting test particles with a random azimuthal coordinate on the outer boundary of the disc. The numerical calculations can be divided in two stages. The key outputs of the first stage are the values of the standard disc model to simulate the initial propriety of the accretion flow (see [16] for technical details). In the second stage the resulting disc is exposed to the influence of the radiation. Using constant values of A, each simulation is terminated when the flow configuration reaches a quasi-steady state, corresponding to a constant value of R_{PR}, at the same precision of the radial discretization extension, during 300 integration steps [16]. These simulations require thus a considerable computational power.

4.3 Results of the Numerical Simulations

We run different simulations for α- and α_g-viscosity discs by inserting different input parameters (luminosity, spin, viscosity, mass, and outer radius). A steady state regime is attained in all cases after an integration time that corresponds to about $\sim$100 Keplerian orbits at the PR radius. Initially, the inner boundary of the viscosity-dominated disc is set at the Innermost Stable Circular Orbit (ISCO) radius, but it gradually moves to larger radii during the simulations until it stabilises at R_{PR}, see Fig. 3. We checked that the optical depth in the vertical direction remains $\ll 1$ throughout the PR dominated region in all cases. Therefore, at first-order, photons undergo a single interaction with matter which scatters them away from the equatorial plane, without returning back to the matter. The PR radius increases few gravitational radii from the ISCO radius as the luminosity increases from 10^{-3}

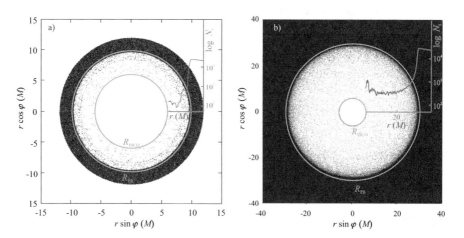

Fig. 3 Test particle distribution for the final state of simulations, where the parameters are: (**a**) $A = 0.1$, $a = 0$, $\alpha = 0.01$, $M = 1.5M_\odot$, $R_{out} = 12M$ and (**b**) $A = 0.8$, $a = 0$, $\alpha = 0.01$, $M = 1.5M_\odot$, $R_{out} = 50M$. The inset shows the radial profile of the particle density (arbitrary units)

to 0.8, with a gradually steeper dependence as A approaches unity. The PR radius correlates also with the viscosity parameter, as a result of the anticorrelation between viscosity and disc density [16].

We note that the density does not decrease monotonically inwards of the PR radius and displays a peak inside R_{PR}, as a slightly darker inner annuli, see Fig. 3. No clear feature is seen at the corresponding radius in the optical depth, testifying that the narrow peak is not due to radiation shielding. In Sect. 4.4 we show that the way in which angular momentum is extracted through the PR effect generates an oscillatory behaviour in the radial velocity which, in combination with the increasing effective luminosity at smaller radii, leads in some cases to the formation of a single narrow annulus of higher density.

4.4 Rings of High Density

The origin of such rings of higher density, seen in Fig. 3, depends significantly on the initial conditions and the interplay between angular momentum losses driven by the PR effect and changes of eccentricity of the originally circular Keplerian trajectories caused by radiation pressure [16]. In the simplified case of an isolated test particle, spiralling towards the central radiating object in a Keplerian circular orbit, the profile of the radial four-velocity, $U^r(r)$, displays characteristic loops and peaks (see left and middle panel of Fig. 4). We note that the position of the loops and peaks of the four-velocity profile coincides with that of the rings of higher density in the disc. In order to quantify the distance between the loops and/or peaks we follow a perturbation analysis with a constant $A \ll 1$ in the Schwarzschild metric. We linearise the radial coordinate of the test particle, spiralling initially in a Keplerian circular orbit at the radius r_0, with respect to A [11, 16]:

$$r(t) = r_0 + A\frac{B}{\omega_r}\left[2B\sin(\omega_r t) - \cos(\omega_r t) - \frac{2M}{\sqrt{r_0^3(r_0 - 3M)}}t + 1\right],$$

(4.1)

where t is the proper time, $B = \sqrt{\frac{M}{r_0 - 6M}}$, and $\omega_r = \sqrt{M\frac{r_0 - 6M}{r_0^3(r_0 - 3M)}}$ is the radial epicyclic frequency. The corresponding linearized formula for U^r is:

$$U^r = A\,B\,[\sin(\omega_r t) + 2B\cos(\omega_r t) - 2B].$$

(4.2)

Eqs. (4.1) and (4.2) represent the parametric expressions of a quasi-cycloidal curve in the $r - U^r$ plane. The distance between the nodes of the loops (or between the peaks) of the cycloid is $D = 2B^2A/\omega_r$ [16]. Instead, $D_0 = D/2$ determines the shift of the node of the first loop with respect to the position of the perturbed

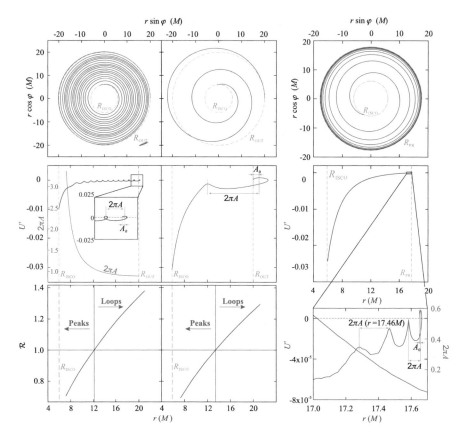

Fig. 4 Spiral infall of an isolated test particle from an initial Keplerian circular orbit under the influence of a central radiation field in the Schwarzschild geometry. The left and the middle column correspond to the case of an isolated particle spiralling under the influence of a constant luminosity $A = 0.01$ and $A = 0.1$, respectively, starting from an initial Keplerian circular orbit at $r_0 = 20M$. The right column corresponds the case of case of a test particle spiralling under the influence of variable effective luminosity $\mathcal{A}_{\mathrm{eff}}(r)$, where the initial Keplerian circular orbit is located at $R_{\mathrm{PR}} = 17.65M$ (credit to [16])

orbit. These predictions are in a very good agreement with the results of numerical integration of the trajectory (see left column of Fig. 4). Also in the more complex case of variable effective luminosity $\mathcal{A}_{\mathrm{eff}}(r)$, this approximation works successfully (see the bottom panel of the right column of Fig. 4).

5   Conclusions

In this work we have underlined the pedagogical function of the relativity of observer splitting formalism in better understanding the GR, in the immediate connection with the classical description, and in the useful modelling of astrophysical problems. In particular, we have presented two possible astrophysical applications: the general relativistic motion of a test particle around a rotating compact object influenced by a radiation field together with the PR effect; an accretion disc model around a rotating compact object, made by many non-interacting test particles and affected by a constant radiation field together with the general relativistic PR effect (based on the first application).

The potentiality of this formalism is wide, because it can be validly exploited to describe the motion of a test particle under a generic force in the GR frame. Regarding the first application, the general relativistic PR effect in two dimensions can be further improved in many directions: extending it in three dimensions, considering an extended source of emission, improving a non-constant radiation field-test particle interaction, and considering more realistic radiation field models. Instead for the accretion disc model under the influence of a radiation field can be further ameliorate considering the evolution of an accretion disc under the influence of a type-I X-ray burst, variable radiation field.

Acknowledgements I thank the International Space Science Institute (ISSI) in Bern for the support to carry out this work and the Silesian University in Opava for the partial support in participating at the Domoschool Summer School 2018. I am grateful to Pavel Bakala for the kind help in providing me the Figs. 3 and 4 in this article.

References

1. F. de Felice, On the gravitational field acting as an optical medium. Gen. Relativ. Gravit. **2**, 347–357 (1971)
2. M.A. Abramowicz, B. Carter, J.P. Lasota, Optical reference geometry for stationary and static dynamics. Gen. Relativ. Gravit. **20**, 1173–1183 (1988)
3. M.A. Abramowicz, G.F.R. Ellis, A. Lanza, Relativistic effects in superluminal jets and neutron star winds. Astrophys. J. **361**, 470–482 (1990)
4. M.A. Abramowicz, P. Nurowski, N. Wex, Covariant definition of inertial forces. Class. Quantum Grav. **10**, L183–L186 (1993)
5. D. Bini, P. Carini, R.T. Jantzen, The intrinsic derivative and centrifugal forces in general relativity: I. Theoretical foundations. Int. J. Mod. Phys. D **6**, 1–38 (1997)
6. F. de Felice, D. Bini, *Classical Measurements in Curved Space-Times* (Cambridge University Press, Cambridge, 2010)
7. J. Frank, A. King, D.J. Raine, *Accretion Power in Astrophysics.* (Cambridge University Press, Cambridge, 2002)
8. J. H. Poynting, Radiation in the solar system: its effect on temperature and its pressure on small bodies. Mon. Not. R. Astron. Soc. **64**, 1 (1903)
9. H.P. Robertson, Dynamical effects of radiation in the solar system. Mon. Not. R. Astron. Soc. **97**, 423 (1937)
10. D. Bini, R.T. Jantzen, L. Stella, The general relativistic Poynting Robertson effect. Class. Quantum Grav. **26**, 5 (2009)
11. D. Bini, A. Geralico, R.T. Jantzen, O. Semerák, L. Stella, The general relativistic Poynting-Robertson effect: II. A photon flux with nonzero angular momentum. Class. Quantum Grav. **28**, 3 (2011)

12. C.W. Misner, K.S. Thorne, J.A. Wheeler, *Gravitation* (W.H.Freeman and Co., San Francisco, 1973)
13. V. De Falco, E. Battista, M. Falanga, Lagrangian formulation of the general relativistic Poynting-Robertson effect. Phys. Rev. D **97**, 8 (2019)
14. H. Goldstein, C.P. Poole, J.L. Safko, *Classical Mechanics* (Addison Wesley, Boston, 2002)
15. K. Vogtmann, A. Weinstein, V.I. Arnol'd, *Mathematical Methods of Classical Mechanics* (Springer, New York, 2013)
16. P. Bakala, et al., The relativistic Poynting-Robertson effect: numerical models of accretion discs in the strong field regime. Astron. Astrophys. (2018, submitted)
17. N.I. Shakura, R.A. Sunyaev, Black holes in binary systems. Observational appearance. Astron. Astrophys. **24**, 337–355 (1973)
18. M. Nannurelli, L. Stella, The X-ray spectrum of modified alpha-viscosity accretion disks. Astron. Astrophys. **226**, 343–346 (1989)

Some Remarks on a New Exotic Spacetime for Time Travel by Free Fall

Davide Fermi

Abstract This work is essentially a review of a new spacetime model with closed causal curves, recently presented in another paper. The spacetime at issue is topologically trivial, free of curvature singularities, and even time and space orientable. Besides summarizing previous results on causal geodesics, tidal accelerations, and violations of the energy conditions, here redshift/blueshift effects and the Hawking–Ellis classification of the stress–energy tensor are examined.

Keywords General relativity · Closed causal curves · Time machines · Energy conditions · Hawking–Ellis classification

Mathematics Subject Classification (2000) Primary 83C99; Secondary 83-06

1 Introduction

In spite of being inherently locally causal, the classical theory of general relativity is well-known to encompass violations of the most intuitive notions of chronological order and causality. Such violations are generically ascribable to the fact that "faster than light" motions and "time travels" (namely, closed causal curves) are not ruled out by first principles. Several spacetime models sharing the above exotic features are known nowadays; see [22–24, 39] for comprehensive reviews. In [10] a four-fold classification of the existing literature was suggested:

First Class Exact solutions of the Einstein's equations, where closed timelike curves (CTCs) are produced by strong angular momenta or naked singularities.

D. Fermi (✉)
Dipartimento di Matematica, Università degli Studi di Milano, Milano, Italy

Istituto Nazionale di Fisica Nucleare, Sezione di Milano, Italy
e-mail: davide.fermi@unimi.it

These include the Gödel metric [14], the Van-Stockum dust [41, 44], the Taub-NUT space [27, 28, 31, 38, 40], the Kerr black hole [18, 19, 43], the radiation models of Ahmed et al. [1, 36], as well as certain cosmic strings models [4, 15, 16, 25].

Second Class Specifically designed Lorentzian geometries, describing CTCs at the price of violating the usual energy conditions. Examples of such geometries are those analysed by Ori and Soen[5, 32, 33, 35, 37], and by Tippett and Tsang[42].

Third Class Ad hoc spacetimes for "faster than light" motions, violating the standard energy conditions. These include the Alcubierre warp drive [2], the Krasnikov–Everett–Roman tube [8, 9, 20] and the Ellis–Morris–Thorne wormhole [7, 29, 30].

Fourth Class A single model by Ori [34] which presents CTCs, still fulfilling all the classical energy conditions. The corresponding metric is obtained evolving via Einstein's equations some suitable Cauchy data, corresponding to dust-type matter content; unluckily, this characterization of the metric is not explicit enough to rule out the appearance of pathologies (e.g., black holes).

Many problematic aspects are known to affect spacetimes with CTCs. For example, the well-posedness plus existence and uniqueness of solutions for various kinds of Cauchy problems is a rather delicate issue [3, 6, 11–13]. Besides, divergences often arise in semiclassical and quantized field theories living on the background of spacetimes with CTCs [17, 21, 45]; arguments of this type led Hawking to formulate the renowned "chronology protection conjecture" [17].

This work reviews a spacetime of the second class, recently proposed in [10]. In the spirit of Alcubierre's and Krasnikov's interpolation strategies [2, 20], the said spacetime is intentionally designed as a smooth gluing of an outer Minkowskian region, and an inner flat region with CTCs; the topology is trivial and no curvature singularity occur. The geometry of the model is established in Sect. 2; natural time and space orientations are described in Sect. 3. Some manifest symmetries of the model are pointed out in Sect. 4 and then employed in Sect. 5 for reducing to quadratures the equations of motion for a specific class of causal geodesics. Section 6 discusses the existence of causal geodesics connecting a point in the outer Minkowskian region to an event in its causal past. Tidal accelerations experienced by an observer moving along a timelike geodesic of the latter type are discussed in Sect. 7, while frequency shifts of light signals propagating along null geodesics are analysed in Sect. 8. The matter content of the spacetime is derived in Sect. 9; there it is shown that the classical energy conditions are violated and the Hawking–Ellis classification of the stress–energy tensor is determined. Finally, Sect. 10 mentions some possible developments, to be analysed in future works.

2 Postulating the Geometry

Assume *ab initio* that the geometry of the spacetime is described by the Lorentzian manifold

$$\mathfrak{T} := (\mathbb{R}^4, g), \tag{2.1}$$

where g is the metric defined by the line element

$$ds^2 = -\big[(1-\mathcal{X})\,dt + \mathcal{X}aR\,d\varphi\big]^2 + \big[(1-\mathcal{X})\,\rho\,d\varphi - \mathcal{X}\,b\,dt\big]^2 + d\rho^2 + dz^2. \tag{2.2}$$

The above expression for ds^2 is characterized by the following building blocks:

1. a set of cylindrical-type coordinates $(t, \varphi, \rho, z) \in \mathbb{R} \times \mathbb{R}/(2\pi\mathbb{Z}) \times (0, +\infty) \times \mathbb{R}$ on $\mathbb{R}^4$ (notice that units are fixed so that the speed of light is $c = 1$);
2. a pair of dimensionless parameters $a, b \in (0, +\infty)$, playing the role of scale factors (cf. Eq. (2.7), below);
3. two concentric tori $\mathbf{T}_\lambda, \mathbf{T}_\Lambda$ in $\mathbb{R}^3$ (for fixed $t \in \mathbb{R}$), with common major radius R and minor radii λ, Λ such that $\lambda < \Lambda < R$, namely, (see Fig. 1)

$$\mathbf{T}_\ell := \big\{ \sqrt{(\rho - R)^2 + z^2} = \ell \big\} \qquad (\ell = \lambda, \Lambda); \tag{2.3}$$

4. an at least twice continuously differentiable function $\mathcal{X} \equiv \mathcal{X}(\rho, z)$, such that

$$\mathcal{X} = \begin{cases} 1 \text{ inside } \mathbf{T}_\lambda, \\ 0 \text{ outside } \mathbf{T}_\Lambda. \end{cases} \tag{2.4}$$

For later convenience, we fix

$$\mathcal{X}(\rho, z) := \mathcal{H}\big(\sqrt{(\rho/R - 1)^2 + (z/R)^2}\big), \tag{2.5}$$

where $\mathcal{H}$ is a shape function on $[0, +\infty)$ of class C^k for some $k \in \{2, 3, \ldots, \infty\}$, such that $\mathcal{H}(y) = 1$ for $0 \leqslant y \leqslant \lambda/R$, $\mathcal{H}(y) = 0$ for $y \geqslant \Lambda/R$ and $\mathcal{H}'(y) < 0$ for $\lambda/R < y < \Lambda/R$ (piecewise polynomial choices of $\mathcal{H}$ fulfilling these requirements are described in [10, App. A]; see also the forthcoming Eq. (2.8) and Fig. 2).

Fig. 1 Representation (for fixed t) of the tori $\mathbf{T}_\lambda$ (in blue) and $\mathbf{T}_\Lambda$ (in orange)

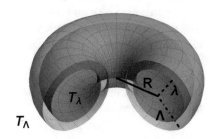

Fig. 2 Plot of an admissible shape function $\mathcal{H}$ (cf. Eq. (2.8))

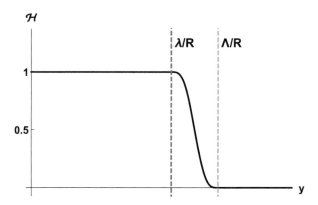

Paying due attention to the usual coordinate singularity at $\rho = 0$, it can be checked that the line element (2.2) does indeed define a non-degenerate, symmetric bilinear form of constant signature (3, 1), i.e., a Lorentzian metric g on the whole spacetime manifold $\mathbb{R}^4$ (see [10] for more details). Furthermore, on account of the presumed regularity features of $\mathcal{X}$, the said metric g is granted by construction to be of class C^k, with $k \geqslant 2$; this suffices to infer that the corresponding Riemann curvature tensor Riem_g is of class C^{k-2} (whence, at least continuous) and, in particular, free of singularities, alike all the associated curvature invariants.

Next, notice that outside the larger torus $\mathbf{T}_\Lambda$ (where $\mathcal{X} \equiv 0$) the line element ds^2 of Eq. (2.2) reproduces the usual Minkowskian analogue

$$ds_0^2 := -dt^2 + \rho^2 \, d\varphi^2 + d\rho^2 + dz^2 \,. \tag{2.6}$$

On the other hand, inside the smaller torus $\mathbf{T}_\lambda$ (where $\mathcal{X} \equiv 1$), ds^2 matches the flat line element

$$ds_1^2 := -a^2 R^2 d\varphi^2 + b^2 \, dt^2 + d\rho^2 + dz^2 \,. \tag{2.7}$$

This shows that inside $\mathbf{T}_\lambda$ the periodic variable $\varphi \in \mathbb{R}/(2\pi\mathbb{Z})$ is a time coordinate, meanwhile $t \in \mathbb{R}$ is a space coordinate: the opposite of what happens in the Minkowskian region outside $\mathbf{T}_\Lambda$. In consequence of this, within $\mathbf{T}_\lambda$ the curves naturally parametrized by φ (with t, ρ, z fixed) are indeed geodesic CTCs.

As a matter of fact, the line element (2.2) is designed on purpose as a sort of interpolation (parametrized by the shape function $\mathcal{X}$) between the two flat line elements (2.6), (2.7); this is achieved by means of a strategy similar to those originally devised by Alcubierre [2] and Krasnikov [20], and later successfully adopted by other authors for the construction of exotic spacetimes with CTCs.

Summing up, the spacetime $\mathfrak{T}$ defined in Eq. (2.1) possesses no curvature singularity and contains CTCs, besides being manifestly topologically trivial. The forthcoming sections describe other noteworthy features of $\mathfrak{T}$, mostly retracing the analysis detailed in [10].

A Case Study

Following [10], as a prototype for numerical computations in this work we systematically refer to the setting specified hereafter:

$$\lambda = \frac{3}{5} R, \quad \Lambda = \frac{4}{5} R, \quad a = \frac{9}{100}, \quad b = 10, \quad \mathcal{H}(y) := \mathfrak{H}\left(\frac{\Lambda/R - y}{\Lambda/R - \lambda/R}\right);$$

$$(2.8)$$

$$\mathfrak{H}(w) := \begin{cases} 0 & \text{for } w < 0, \\ 35 \, w^4 (1-w)^3 + 21 \, w^5 (1-w)^2 + 7 \, w^6 (1-w) + w^7 & \text{for } 0 \leqslant w \leqslant 1, \\ 1 & \text{for } w > 1. \end{cases}$$

For $\mathcal{H}$ as above, the shape function $\mathcal{X}$ is of class C^3. The choices indicated in Eq. (2.8) are not completely arbitrary, but partly motivated by some requirements to be discussed in Sect. 6 (see Eq. (6.3)). The parameter R, intentionally left unspecified, plays the role of a natural length scale for the model under analysis.

3 Natural Choices of Time and Space Orientations

Consider the following vector fields:

$$E_{(0)} := \frac{(1 - \mathcal{X}) \rho \, \partial_t + \mathcal{X} \, b \, \partial_\varphi}{\rho \, (1 - \mathcal{X})^2 + a \, b \, R \, \mathcal{X}^2}, \qquad E_{(1)} := \frac{(1 - \mathcal{X}) \partial_\varphi - \mathcal{X} a R \, \partial_t}{\rho \, (1 - \mathcal{X})^2 + a \, b \, R \, \mathcal{X}^2},$$

$$E_{(2)} := \partial_\rho, \qquad E_{(3)} := \partial_z.$$

$$(3.1)$$

These are everywhere defined on $\mathbb{R}^4$ and have the same regularity of $\mathcal{X}$ (i.e., are of class C^k); furthermore, it can be checked by explicit computations (see [10]) that

$$g(E_{(\alpha)}, E_{(\beta)}) = \eta_{\alpha\beta} \qquad (\alpha, \beta \in \{0, 1, 2, 3\}), \qquad (3.2)$$

where $(\eta_{\alpha\beta}) := \mathrm{diag}(-1, 1, 1, 1)$. In other words, the set $(E_{(\alpha)})_{\alpha \in \{0,1,2,3\}}$ forms a pseudo-orthonormal tetrad; this can be used to induce natural choices of time and space orientations for $\mathfrak{T}$, following the prescriptions outlined in the sequel.

Notably, $E_{(0)}$ is everywhere timelike (see Eq. (3.2)). Taking this into account, we fix the time orientation of $\mathfrak{T}$ assuming that $E_{(0)}$ is future-directed at all points. This time orientation reproduces the familiar Minkowskian one in the spacetime region outside $\mathbf{T}_\Lambda$, where $E_{(0)} = \partial_t$. At the same time, the established convention implies that ∂_φ is timelike and future-directed inside $\mathbf{T}_\lambda$, since $E_{(0)} = 1/(aR) \, \partial_\varphi$ therein. Of course, the integral curves of $E_{(0)}$ can be interpreted as the worldlines of certain *fundamental observers*; any event $p \in \mathfrak{T}$ uniquely identifies one of these observers and $E_{(0)}$ coincides with the 4-velocity of the latter.

Next, consider the vector fields $(E_{(i)})_{i\in\{1,2,3\}}$; these are everywhere spacelike (see Eq. (3.2)) and further fulfil $E_{(1)} = (1/\rho)\,\partial_\varphi$, $E_{(2)} = \partial_\rho$, $E_{(3)} = \partial_z$ in the Minkowskian region outside $\mathbf{T}_\Lambda$. In view of these facts, we choose to equip $\mathfrak{T}$ with the space orientation induced by the left-handed ordered triplet $(E_{(1)}, E_{(2)}, E_{(3)})$.

Finally, let us mention two additional facts regarding the tetrad $(E_{(\alpha)})_{\alpha\in\{0,1,2,3\}}$. On the one hand, the family $(E_{(i)})_{i\in\{1,2,3\}}$ is not involutive; by Frobenius theorem, this means that there does not exists any foliation of $\mathfrak{T}$ into spacelike hypersurfaces orthogonal to $E_{(0)}$. On the other hand, none of the vector fields $(E_{(\alpha)})_{\alpha\in\{0,1,2,3\}}$ fulfils the Killing equation $\mathcal{L}_{E_{(\alpha)}}g = 0$ ($\mathcal{L}$ indicates the Lie derivative); so, none of them generates an isometry of g.

4 Manifest Symmetries and Stationary Limit Surfaces

The spacetime under analysis exhibits a number of symmetries; hereafter these symmetries are discussed, together with some of their most relevant implications (see [10, Sec. 4] for more details on this theme).

First of all, notice that the line element (2.2) is invariant under each of the following coordinate transformations, describing *discrete symmetries* of the spacetime $\mathfrak{T}$:

$$(t, \varphi, \rho, z) \rightarrow (-t, -\varphi, \rho, z)\,; \tag{4.1}$$

$$(t, \varphi, \rho, z) \rightarrow (t, \varphi, \rho, -z)\,. \tag{4.2}$$

Incidentally, it should be observed that the symmetry by reflection across the plane $\{z = 0\}$ represented by Eq. (4.2) depends crucially on the distinguished choice (2.5) of the shape function $\mathcal{X}$ (granting that $\mathcal{X}(\rho, z) = \mathcal{X}(\rho, -z)$).

In terms of the orthonormal tetrad introduced in the previous Sect. 3 (see Eq. (3.1)), the above transformations (4.1) (4.2) induce, respectively, the mappings $(E_{(0)}, E_{(1)}, E_{(2)}, E_{(3)}) \rightarrow (-E_{(0)}, -E_{(1)}, E_{(2)}, E_{(3)})$ and $(E_{(0)}, E_{(1)}, E_{(2)}, E_{(3)}) \rightarrow (E_{(0)}, E_{(1)}, E_{(2)}, -E_{(3)})$. The facts mentioned above and the considerations of Sect. 3 indicate, in particular, that the spacetime $\mathfrak{T}$ is invariant under the (separate) inversion of time and space orientations.

To proceed, let us examine the presence of (global) *continuous symmetries*. In this connection, it should be noticed that the coefficients of the metric g do not depend explicitly on t, φ (recall that $\mathcal{X} \equiv \mathcal{X}(\rho, z)$); this suffices to infer that the following are Killing vector fields, generating each a one-parameter group of isometries of g:

$$K_{(0)} := \partial_t\,, \qquad K_{(1)} := \partial_\varphi\,. \tag{4.3}$$

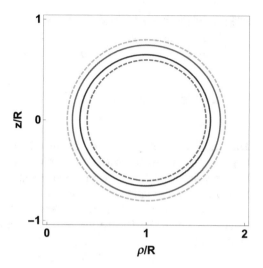

Fig. 3 Representation (for fixed t, φ) of the stationary limit surfaces $\Sigma_{(0)}$ (in red), $\Sigma_{(1)}$ (in purple), along with the tori $\mathbf{T}_\lambda$ (in blue), $\mathbf{T}_\Lambda$ (in orange) (cf. Eqs. (2.8)–(4.5))

Next, consider the related set of events

$$\Sigma_{(i)} := \{p \in \mathfrak{T} \mid g_p(K_{(i)}, K_{(i)}) = 0\} \qquad (i = 0, 1), \qquad (4.4)$$

admitting the explicit coordinate representations

$$\Sigma_{(0)} = \{\mathcal{X}(\rho, z) = (1+b)^{-1}\}, \qquad \Sigma_{(1)} = \{\mathcal{X}(\rho, z) = (1+aR/\rho)^{-1}\}. \qquad (4.5)$$

Both $\Sigma_{(0)}$ and $\Sigma_{(1)}$ are timelike hypersurfaces contained in the spacetime region delimited by the tori $\mathbf{T}_\lambda, \mathbf{T}_\Lambda$ (see Fig. 3); in particular, neither of them is a Killing horizon. As a matter of fact, $\Sigma_{(0)}$ and $\Sigma_{(1)}$ are *stationary limit surfaces* [18] for $K_{(0)}$ and $K_{(1)}$, respectively. This means that the integral curves of $K_{(0)}$ can be interpreted as the worldlines of stationary observers outside $\Sigma_{(0)}$ (where $\mathcal{X}(\rho, z) > (1 + b)^{-1}$ and $K_{(0)}$ is timelike), but no particle travelling along a timelike curve inside $\Sigma_{(0)}$ (where $\mathcal{X}(\rho, z) < (1 + b)^{-1}$ and $K_{(0)}$ is spacelike) can remain at rest with respect to such observers. Similarly, no particle outside $\Sigma_{(1)}$ can remain at rest with respect to observers moving along the orbits of $K_{(1)}$ inside $\Sigma_{(1)}$.

5 Quadratures for a Special Class of Geodesics

Consider the Lagrangian function

$$L(U) := \frac{1}{2} g(U, U) \qquad (U \in T\mathfrak{T}) \qquad (5.1)$$

and recall that the solutions of the associated Euler–Lagrange equations describe affinely parametrized geodesics. It should also be kept in mind that for any geodesic ξ with tangent vector field $\dot{\xi}$, modulo possible re-parametrizations, there holds

$$L(\dot{\xi}) = \text{const.} = E\,, \qquad E = \begin{cases} -1/2 & \text{if } \xi \text{ is timelike}\,, \\ 0 & \text{if } \xi \text{ is null}\,, \\ +1/2 & \text{if } \xi \text{ is spacelike}\,. \end{cases} \tag{5.2}$$

This fact and the symmetries discussed in Sect. 4 allow to reduce the Euler–Lagrange equations to quadratures, at least for a special class of geodesics. More precisely, in the following we restrict the attention to geodesics lying in the plane $\{z = 0\}$, and essentially retrace the analysis of [10, Sec. 5].

On account of the reflection symmetry (4.2), the geodesics mentioned above can be analysed considering just the restriction of the Lagrangian (5.1) to the tangent bundle of the plane $\{z=0\}$; this has the following coordinate representation, where $\mathcal{X} \equiv \mathcal{X}(\rho, 0) \equiv \mathcal{H}(|\rho/R - 1|)$ (see Eq. (2.5)):

$$L(\dot{\xi})\big|_{z=0,\dot{z}=0} = -\frac{1}{2}\big[(1-\mathcal{X})\,\dot{t} + aR\,\mathcal{X}\dot{\varphi}\big]^2 + \frac{1}{2}\big[\rho\,(1-\mathcal{X})\dot{\varphi} - b\,\mathcal{X}\dot{t}\,\big]^2 + \frac{1}{2}\,\dot{\rho}^2\,. \tag{5.3}$$

The above reduced Lagrangian possesses a maximal number of first integrals: namely the canonical momenta associated to the cyclic coordinates t, φ (coinciding with the Noether charges related to the Killing vector fields $K_{(0)}$, $K_{(1)}$ of Eq. (4.3)), and the energy of the Lagrangian system.

In place of the above-mentioned canonical momenta, it is convenient to consider the related dimensionless parameters

$$\gamma := -\frac{\partial L(\dot{\xi})|_{z=0,\dot{z}=0}}{\partial \dot{t}}\,, \qquad \omega := -\frac{1}{\gamma\,R}\frac{\partial L(\dot{\xi})|_{z=0,\dot{z}=0}}{\partial \dot{\varphi}}\,. \tag{5.4}$$

It can be checked by explicit computations that the above definitions yield

$$\dot{t} = \gamma\,, \qquad \dot{\varphi} = -\frac{\gamma\,\omega\,R}{\rho^2} \qquad \text{outside } \mathbf{T}_\Lambda\,, \tag{5.5}$$

$$\dot{t} = -\frac{\gamma}{b^2}\,, \qquad \dot{\varphi} = \frac{\gamma\,\omega}{a^2 R} \qquad \text{inside } \mathbf{T}_\lambda\,. \tag{5.6}$$

Equations (5.5) (5.6) allow to infer a number of notable facts.

1. Causal geodesics traversing the Minkowskian region outside $\mathbf{T}_\Lambda$ are future-directed (recall the time orientation established in Sect. 3) only for $\gamma > 0$. In particular, for timelike curves of this kind, $\gamma \equiv dt/d\tau$ is the usual Lorentz factor of special relativity; so, $\gamma \geqslant 1$ and the limits $\gamma \to 1^+$, $\gamma \to +\infty$ correspond, respectively, to non-relativistic and ultra-relativistic regimes.
2. Causal geodesics travelling across the flat region inside $\mathbf{T}_\lambda$ are future-directed only if $\gamma \cdot \omega > 0$.

3. In view of the previous remarks (1)(2), causal geodesics crossing both the region outside $\mathbf{T}_\Lambda$ and the one inside $\mathbf{T}_\lambda$ are future-directed only if

$$\gamma > 0 \qquad \text{and} \qquad \omega > 0. \tag{5.7}$$

Remarkably, for $\gamma > 0$ the first identity in Eq. (5.6) gives $\dot{t} < 0$ inside $\mathbf{T}_\lambda$, indicating that the coordinate t decreases while moving towards the future. One can make sense of the latter (apparently paradoxical) feature keeping in mind that t is a spatial coordinate inside $\mathbf{T}_\lambda$; this fact plays a crucial role in the construction of time travels, to be discussed in the upcoming Sect. 6.

To proceed, notice that the conservation law for the energy associated to the reduced Lagrangian (5.3) can be expressed, in terms the parameters γ, ω via the relation

$$\frac{1}{2}\dot{\rho}^2 + V_{\gamma,\omega}(\rho) = E, \tag{5.8}$$

where E is a constant fixed as in Eq. (5.2) and ($\mathcal{X} \equiv \mathcal{X}(\rho, 0) \equiv \mathcal{H}(|\rho/R - 1|)$)

$$V_{\gamma,\omega}(\rho) := \gamma^2 \frac{[a\,\mathcal{X} - (1-\mathcal{X})\,\omega]^2 - [(\rho/R)\,(1-\mathcal{X}) + b\,\mathcal{X}\,\omega]^2}{2\,[(\rho/R)\,(1-\mathcal{X})^2 + a\,b\,\mathcal{X}^2]^2}. \tag{5.9}$$

In particular, there holds

$$V_{\gamma,\omega}(\rho) = \frac{\gamma^2}{2}\left(\frac{R^2\omega^2}{\rho^2} - 1\right) \qquad \text{outside } \mathbf{T}_\Lambda, \tag{5.10}$$

$$V_{\gamma,\omega}(\rho) = \text{const.} = \frac{\gamma^2}{2a^2}\left(\frac{a^2}{b^2} - \omega^2\right) \qquad \text{inside } \mathbf{T}_\lambda. \tag{5.11}$$

At the same time, $V_{\gamma,\omega}$ depends sensibly on the sign of ω in the intermediate region delimited by $\mathbf{T}_\lambda$ and $\mathbf{T}_\Lambda$. Figure 4 shows a plot of $V_{\gamma,\omega}$ in the setting (2.8), for a particular choice of γ, ω (specified in the caption).

Fig. 4 Plot of the potential $V_{\gamma,\omega}$ for $\gamma = 1.1$, $\omega = 0.08$ (cf. Eqs. (2.8)–(5.9)). The grey line represents the energy level $E = -1/2$ (cf. Eqs. (5.2)–(5.8))

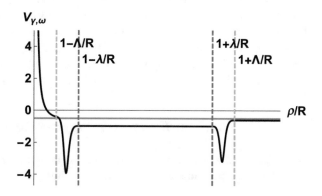

Equation (5.8) and the definitions in Eq. (5.4) allow to derive, by standard methods, explicit quadrature formulas for the solutions $t(\tau), \varphi(\tau), \rho(\tau)$ ($\tau \in \mathbb{R}$) of the Euler–Lagrange equations associated to the reduced Lagrangian (5.3); for brevity, we do not report these formulas here and refer to [10] for more details.

6   Exotic Causal Geodesics and Time Travel by Free Fall

Building on the results described in the previous section, it can be shown that there exist future-directed causal geodesics starting at a point in the Minkowskian region outside $\mathbf{T}_\Lambda$, reaching the flat region inside $\mathbf{T}_\lambda$ and finally returning outside $\mathbf{T}_\Lambda$, arriving at an event in the causal past of the initial point. Such geodesics can be unequivocally interpreted as the trajectories of (massive or massless) test particles performing a free fall time travel into the past, with respect to observers at rest in the outer Minkowskian region. Using a suggestive terminology, the latter observers could refer to the spacetime region delimited by $\mathbf{T}_\Lambda$ as a "time machine".

As anticipated in item (3) of the preceding section, it should be kept in mind that the existence of geodesics such as those described above relies significantly on the possibility to implement the condition $\dot{t} < 0$ inside $\mathbf{T}_\lambda$. Understandably, this region can be reached only if the parameters $a, b, \lambda, \Lambda, R$ and the shape function $\mathcal{X}$ characterizing the spacetime $\mathfrak{T}$, as well as the initial data related to the geodesic coefficients γ and ω, are chosen suitably (in this connection, see Eq. (6.2) below).

Let ξ be a causal geodesic with initial data $\xi(0), \dot{\xi}(0)$, described in coordinates by

$$
\begin{aligned}
&t(0) = 0, && \varphi(0) = 0, && \rho(0) = \rho_0 > R + \Lambda, \; z(0) = 0; \\
&\dot{t}(0) = \gamma > 0, && \dot{\varphi}(0) = -\gamma\,\omega R/\rho_0^2 < 0, && \dot{\rho}(0) < 0, && \dot{z}(0) = 0.
\end{aligned}
\tag{6.1}
$$

In the above positions, $t(0)$ and $\varphi(0)$ are fixed arbitrarily, while the choices of $\rho(0)$ and $z(0)$ indicate that the initial event lies outside $\mathbf{T}_\Lambda$, on the plane $\{z = 0\}$. On the other hand, $\dot{t}(0) = \gamma > 0$ grants that ξ is future-directed (see item (1) of Sect. 5), whereas $\dot{\rho}(0) < 0$ implies that ξ is initially moving towards the torus $\mathbf{T}_\Lambda$. The specific expression for $\dot{\varphi}(0)$ follows essentially from the definition of ω given in Eq. (5.4); note that it understands the assumption $\omega > 0$, ensuring that ξ remains future-directed when it enters the region delimited by $\mathbf{T}_\lambda$. Finally, notice that the results of Sect. 5 can certainly be employed, since $z(0) = 0, \dot{z}(0) = 0$.

A sufficient condition granting that ξ crosses the region inside $\mathbf{T}_\lambda$ is the following:

$$
\exists\,\rho_1 \in (0, R - \lambda) \;\; \text{s.t.} \;\; V_{\gamma,\omega}(\rho) \leqslant E \;\; \text{for all } \rho \geqslant \rho_1,
\tag{6.2}
$$

where $V_{\gamma,\omega}$ and E are defined, respectively, as in Eqs. (5.9) and (5.2). Especially, it should be kept in mind that $E = -1/2$ if ξ is timelike, and $E = 0$ if ξ is null.

In view of Eqs. (5.10) (5.11), to fulfil the condition (6.2) it is necessary to comply at least with the requirements

$$
\frac{a}{b} < 1 - \frac{\Lambda}{R}, \qquad \gamma > \sqrt{|2E|}\sqrt{\frac{(1-\Lambda/R)^2 + a^2}{(1-\Lambda/R)^2 - a^2/b^2}},
$$

$$
\frac{a}{b}\sqrt{1 - \frac{2E\,b^2}{\gamma^2}} < \omega < \left(1 - \frac{\Lambda}{R}\right)\sqrt{1 + \frac{2E}{\gamma^2}}. \tag{6.3}
$$

This shows that a carefully devised choice of parameters and initial data is required. For example, radial motions starting outside $\mathbf{T}_\Lambda$ with $\dot\varphi = 0$ cannot enter the spacetime region delimited by $\mathbf{T}_\lambda$, since the corresponding momentum $\omega = 0$ (see Eq. (5.5)) does not fulfil the above Eq. (6.3).

Incidentally, the specific choices reported in Eq. (2.8) do satisfy the condition (6.2) for any γ, ω fulfilling the requirements in Eq. (6.3), for both $E = -1/2$ (timelike geodesics) and $E = 0$ (null geodesics).

Whenever the condition (6.2) is fulfilled, the radial variable $\rho(\tau)$ associated to the causal geodesic ξ starts at $\rho(0) = \rho_0 > R + \Lambda$ (see Eq. (6.1)), decreases until it reaches the value ρ_1 (defined implicitly by Eq. (6.2)), and then begins to increase endlessly. Especially, $\rho(\tau)$ returns to its initial value at some proper time τ_2, i.e.,

$$
\rho(\tau_2) = \rho_0; \tag{6.4}
$$

in addition, it is possible to arrange things so that

$$
t(\tau_2) < 0, \qquad \varphi(\tau_2) = 0 \,(\mathrm{mod}\, 2\pi). \tag{6.5}
$$

The above relations indicate that the geodesic ξ returns at exactly the same spatial position from which it started off, but at an earlier coordinate time. Thus, a test particle freely falling along ξ would be able to reach an event lying in the causal past of its departure, even in the outer Minkowskian region.

Using the quadrature formulas mentioned in Sect. 5, it is possible to characterize quantitatively the geodesic motion described above, both for timelike and null curves. As an example, consider a timelike geodesic ξ with

$$
E = -1/2, \tag{6.6}
$$

fulfilling the conditions (6.1) (6.4) (6.5). An exhaustive analysis of this case is provided in [10, Sec. 6]; in particular, therein it is shown that in the limit (cf. Eq. (6.3))

$$
\varpi := \sqrt{\gamma^2 - \frac{a^2}{b^2}\left(1 + \frac{b^2}{\gamma^2}\right)} \to 0^+ \tag{6.7}
$$

there holds (at least for suitable choices of the shape function $\mathcal{H}$; see [10])

$$\frac{t(\tau_2)}{R} = -\left(\frac{4a}{b^2}\frac{\lambda}{R}\right)\frac{1}{\varpi}\left(1+o(1)\right), \qquad \frac{\tau_2}{R} = \left(\frac{4a}{\gamma}\frac{\lambda}{R}\right)\frac{1}{\varpi}\left(1+o(1)\right),$$

$$\varphi(\tau_2) = \left(\frac{4}{b}\frac{\lambda}{R}\sqrt{1+\frac{b^2}{\gamma^2}}\right)\frac{1}{\varpi}\left(1+o(1)\right) \pmod{2\pi}. \tag{6.8}$$

A few remarks regarding the above asymptotic expansions are in order:

1. $t(\tau_2)$ can be large and negative at will, showing that the amount of coordinate time travelled into the past can be made as large as desired.
2. τ_2 becomes awfully large as well for $\varpi \to 0^+$. However, since $\tau_2/|t(\tau_2)| \sim b^2/\gamma$, one has that τ_2 remains comparatively small with respect to $|t(\tau_2)|$ at least for ultra-relativistic motions (i.e., for large γ); this shows that the amount of proper time spent during the time travel can be kept small by moving at sufficiently high speeds, exactly as in special relativity.
3. $\varphi(\tau_2)$ varies quickly for $\varpi \to 0^+$, indicating that the identity $\varphi(\tau_2)=0 \pmod{2\pi}$ is always allowed in principle, yet difficult to achieve in practice since it relies on a fine tuning of the parameters. Of course, this fine tuning cannot be described using just the above asymptotic expansion; on the other hand, let us suggest that it could be avoided by considering a piecewise geodesic motion (see Sect. 10 for further comments of this topic).

To conclude this section let us report a few numerical results extracted from Tables 1–3 of [10], where many more examples can be found. For definiteness, we refer to the specific setting (2.8) and consider a massive particle starting in the outer Minkowskian region with a speed comparable to those attained at the Large Electron–Positron collider; namely we set

$$\gamma = 10^5. \tag{6.9}$$

As an example, we require the said particle to perform a time travel by free fall so as to reach an event lying roughly a 1000 years in the past, while experiencing an interval of proper time of only about 1 year:

$$t(\tau_2) \sim -10^3\,\mathrm{y}, \qquad \tau_2 \sim 1\,\mathrm{y}. \tag{6.10}$$

To this purpose it is necessary to fix suitably the parameter ϖ of Eq. (6.7), depending on the size R of the apparatus. The forthcoming Eqs. (6.11)–(6.13) indicate the required choices of ϖ for three prototypical values of R, corresponding, respectively, to a nominal human-sized length (10^2m), the Sun-Earth distance (10^{11}m), and to an interstellar distance of a hundred light-years (10^{18}m):

$$\varpi \sim 2\cdot 10^{-20} \qquad \text{for } R = 10^2\mathrm{m}; \tag{6.11}$$

$$\varpi \sim 2 \cdot 10^{-11} \qquad \text{for } R = 10^{11} \text{m} ; \tag{6.12}$$

$$\varpi \sim 2 \cdot 10^{-4} \qquad \text{for } R = 10^{18} \text{m} . \tag{6.13}$$

In all the above cases it is possible to fine tune ϖ so as to implement the condition $\varphi(\tau_2) = 0 \,(\text{mod } 2\pi)$, but we omit the discussion of this issue for brevity (see [10]).

7 Tidal Forces During Time Travel by Free Fall

Consider a small extended body performing a time travel by free fall into the past. More precisely, assume that the body consists of nearby massive point particles whose worldlines form a beam of timelike geodesics of the type described in Sect. 6; this understands, in particular, that the interaction between different particles is being neglected. Hereafter the tidal forces experienced by the said extended body (i.e., the relative accelerations of its constituent particles) are investigated.

Let ξ be the geodesic followed by one of the particles in the body and let $\delta\xi$ be the deviation vector from it [46]. As well-known, tidal accelerations between ξ and nearby geodesics are described by the Jacobi deviation equation

$$\frac{\nabla^2 \delta\xi}{d\tau^2}(\tau) = \mathcal{A}_\tau \, \delta\xi(\tau), \qquad \mathcal{A}_\tau X := -\, \text{Riem}\big(X, \dot{\xi}(\tau)\big) \dot{\xi}(\tau), \tag{7.1}$$

where Riem is the Riemann tensor for the metric g. For any fixed τ, the orthogonal complement of $\dot{\xi}(\tau)$ is a Euclidean subspace of $T_{\xi(\tau)}\mathfrak{T}$ and $\mathcal{A}_\tau$ determines a self-adjoint operator on it. The *maximal tidal acceleration per unit length* at $\xi(\tau)$ is (see [10, Sec. 7] for further details)

$$\alpha(\tau) := \sup_{X \in T_{\xi(\tau)}\mathfrak{T}\setminus\{0\}, \, g(\dot{\xi}(\tau), X)=0} \frac{\sqrt{g(\mathcal{A}_\tau X, \mathcal{A}_\tau X)}}{\sqrt{g(X, X)}} . \tag{7.2}$$

It can be shown by direct inspection that

$$\alpha(\tau) = \frac{\gamma^2}{R^2} \, \text{a}\big(\rho(\tau)/R\big), \tag{7.3}$$

where $\text{a}(\rho/R)$ is a dimensionless function, depending on the spacetime parameters $\lambda/R, \Lambda/R, a, b$ and on the geodesic coefficients γ, ϖ. Of course, $\text{a}(\rho/R) = 0$ for $\rho \in (0, R - \Lambda) \cup (R - \lambda, R + \lambda) \cup (R + \Lambda, +\infty)$, correctly indicating that tidal accelerations vanish in the flat spacetime regions outside $\mathbf{T}_\Lambda$ and inside $\mathbf{T}_\lambda$.

As an example, refer again to the setting (2.8) and consider a geodesic motion with $\gamma = 10^5$ (see Eq. (6.9)). Figure 5 shows the plot of $\text{a}(\rho/R)$ in this case, for $\varpi = 2 \cdot 10^{-20}$ (see Eq. (6.11)). The following numerical results are extracted from Tables 1–3 of [10] ($g_\oplus = 9.8 \, \text{m/s}^2$ is a nominal value for the Earth's gravitational acceleration):

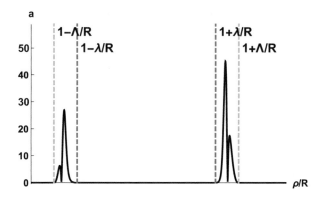

Fig. 5 Plot of the tidal acceleration function $a(\rho/R)$ (cf. Eqs. (2.8)–(7.3)) for $\gamma = 10^5$, $\varpi = 2 \cdot 10^{-20}$

$$\max_{\tau \in [0, \tau_2]} \alpha(\tau) \sim 4 \cdot 10^{23} \, g_{\pm}/m \qquad \text{for } R = 10^2 m, \varpi = 10^{-20}; \qquad (7.4)$$

$$\max_{\tau \in [0, \tau_2]} \alpha(\tau) \sim 4 \cdot 10^{5} \, g_{\pm}/m \qquad \text{for } R = 10^{11} m, \varpi = 10^{-11}; \qquad (7.5)$$

$$\max_{\tau \in [0, \tau_2]} \alpha(\tau) \sim 4 \cdot 10^{-9} \, g_{\pm}/m \qquad \text{for } R = 10^{18} m, \varpi = 10^{-4}. \qquad (7.6)$$

The above numerics show that, for the chosen value of γ, tidal accelerations are intolerable for a human being if the radius R is smaller than the Sun-Earth distance, while they become essentially negligible for a radius R of galactic size.

8   Redshift and Blueshift of Light Signals

The existence of future-directed causal geodesics lying in the plane $\{z = 0\}$ and connecting the spacetime region outside $\mathbf{T}_\Lambda$ to the one inside $\mathbf{T}_\lambda$ was discussed in Sect. 6. In the following, working in the geometric optics approximation, it is analysed the frequency of light signals propagating along null geodesics of the above type, as measured by suitable observers.

More precisely, consider the fundamental observers described in Sect. 3 and recall that their worldlines coincide with the integral curves of the timelike vector field $E_{(0)}$, belonging to the tetrad (3.1). The frequency of a light signal propagating along a null geodesic ξ (with tangent $\dot{\xi}$), measured at $p \in \mathfrak{T}$ by a fundamental observer is

$$\Omega := - g_p(\dot{\xi}, E_{(0)}) . \qquad (8.1)$$

Let a light signal of frequency Ω_i be emitted at a spacetime point $p_i \in \mathfrak{T}$ by an observer of the above family and assume it to reach an event p_f where another fundamental observer measures the frequency Ω_f; the conventional *redshift factor* associated to this scenario is

$$\zeta_{i \to f} := \frac{\Omega_i}{\Omega_f} - 1 . \tag{8.2}$$

As usual, the terms *redshift* and *blueshift* refer, respectively, to the cases $\zeta_{i \to f} > 0$ and $\zeta_{i \to f} < 0$.

Keeping in mind the facts mentioned above, as an example we proceed to determine the redshift factor for a light signal which is emitted by a fundamental observer at a generic point p_i in the plane $\{z=0\}$, propagates along a null geodesic ξ of the type analysed in Sect. 6 (with $E = 0$), and finally reaches an event p_f in the Minkowskian region outside $\mathbf{T}_\Lambda$ where its frequency is measured by another observer of the same family.

Firstly, it should be noticed that, in the case under analysis, the frequency at any event p can be expressed as

$$\Omega = \gamma \, \Theta(\rho/R) , \tag{8.3}$$

where γ is the geodesic momentum defined in Eq. (5.4), ρ is the radial coordinate of p, and Θ is a dimensionless function depending on the spacetime parameters $\lambda/R, \Lambda/R, a, b$ and on the geodesic coefficient ω (but not on γ). In particular, taking into account Eqs. (3.1) (5.5) (5.6) and (8.1) (8.3), one has

$$\Theta(\rho/R) = \text{const.} = 1 , \qquad \text{outside } \mathbf{T}_\Lambda , \tag{8.4}$$

$$\Theta(\rho/R) = \text{const.} = \omega/a > 0 , \qquad \text{inside } \mathbf{T}_\lambda . \tag{8.5}$$

Equations (8.3) (8.4) show that the frequency measured at any event p_f outside $\mathbf{T}_\Lambda$ is

$$\Omega_f = \text{const.} = \gamma > 0 ; \tag{8.6}$$

this result, together with Eqs. (8.2) (8.3), allows to infer that the redshift factor relative to the said emission p_i and measurement p_f reads

$$\zeta_{i \to f} = \Theta(\rho_i/R) - 1 , \tag{8.7}$$

where ρ_i is the radial coordinate of p_i. As a direct consequence of Eqs. (8.3) (8.6), $\zeta_{i \to f}$ does not depend on the momentum γ associated to ξ. As an example, Fig. 6 shows the plot of the redshift parameter $\zeta_{i \to f}$ as a function of the coordinate ρ_i/R corresponding to the emission p_i, in the setting of Eq. (2.8) for $\omega = 0.08$.

Consider now a light signal emitted in the Minkowskian region outside $\mathbf{T}_\Lambda$, performing a time travel along the null geodesic ξ, and finally returning outside

Fig. 6 Plot of the redshift factor $\zeta_{i \to f}$ (cf. Eqs. (2.8)–(8.7)) for $\omega = 0.08$, as a function of the emission coordinate ρ_i/R

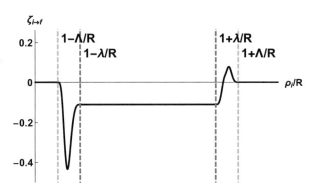

$\mathbf{T}_\Lambda$. In this case the previous results (8.4) and (8.7) yield

$$\zeta_{i \to f} = \text{const.} = 0 \qquad (p_i \text{ outside } \mathbf{T}_\Lambda), \tag{8.8}$$

indicating that the light signal is received in the past with no frequency shift with respect to its emission.

Conversely, for a light signal emitted inside $\mathbf{T}_\lambda$ and observed outside $\mathbf{T}_\Lambda$, Eqs. (8.5) and (8.7) imply

$$\zeta_{i \to f} = \text{const.} = \frac{\omega}{a} - 1 \qquad (p_i \text{ inside } \mathbf{T}_\lambda). \tag{8.9}$$

The above relation (8.9) makes patent that, for fixed values of the spacetime parameter a, either redshift or blueshift can occur, depending on the specific value of the geodesic coefficient ω (which is related, in turn, to the direction of emission in the plane $\{z = 0\}$; see Eq. (5.5)). In this connection, note that the constraints in Eq. (6.3) (necessary for the null geodesic ξ to exist) yield

$$\frac{1}{b} - 1 < \zeta_{i \to f} < \frac{1}{a}\left(1 - \frac{\Lambda}{R}\right) - 1. \tag{8.10}$$

A few remarks are in order:

1. if $b < 1$, then $\zeta_{i \to f} > 0$ for all ω as in Eq. (6.3). This means that the light signal under analysis gets unavoidably redshifted;
2. if $a > 1 - \Lambda/R$, then $\zeta_{i \to f} < 0$ for all ω as in Eq. (6.3). In this case, the light signal gets certainly blueshifted;
3. if $a < 1 - \Lambda/R$ and $b > 1$, then both redshifts and blueshifts are allowed. Actually, for ω as in Eq. (6.3), the light signal gets redshifted (resp., blueshifted) if $\omega < a$ (resp., $\omega > a$). The prototypical setting (2.8) belongs to this case.

9   The Exotic Matter Content of the Spacetime

Recall that the geometry of the spacetime $\mathfrak{T}$ was established *a priori* in Sect. 2. The corresponding matter content is hereafter deduced *a posteriori* by implementing a "reversed" approach to Einstein's equation, originally proposed in [2, 20] for the construction of exotic spacetimes. In this sense, Einstein's equations (with zero gravitational constant) are deliberately enforced by hand, defining the stress–energy tensor for $\mathfrak{T}$ as the symmetric bilinear form

$$T := \frac{1}{8\pi G} \left(\text{Ric} - \frac{1}{2} R g \right), \tag{9.1}$$

where G, Ric, and R denote, respectively, the universal gravitational constant, the Ricci tensor, and the scalar curvature of the metric g. Of course, T vanishes identically outside $\mathbf{T}_\Lambda$ and inside $\mathbf{T}_\lambda$, where the spacetime is flat.

As usual, the *energy density* measured at any event $p \in \mathfrak{T}$ by an observer with normalized 4-velocity $U \in T_p\mathfrak{T}$ (such that $g(U, U) = -1$) is given by

$$\mathcal{E}(U) := T(U, U). \tag{9.2}$$

In the following we first compute the energy densities measured by two families of observers. Next, we determine the Hawking–Ellis class [18, 45] of the stress–energy tensor T.

The Energy Density Measured by the Fundamental Observers   Consider the fundamental observers of Sect. 3, with 4-velocity coinciding with the timelike element $E_{(0)}$ of the tetrad (3). In [10] it is shown that the energy density measured by such observers at a point $p \in \mathfrak{T}$ of coordinate (t, φ, ρ, z) can be expressed as

$$\mathcal{E}\left(E_{(0)}|_p\right) = \frac{1}{8\pi G R^2} \, \mathfrak{E}_f(\rho/R, z/R), \tag{9.3}$$

where $\mathfrak{E}_f$ is a dimensionless function, depending on the parameters $\lambda/R, \Lambda/R, a, b$.

As an example, consider the setting (2.8). Figure 7 shows the graph of the map $\rho/R \mapsto \mathfrak{E}_f(\rho/R, 0)$. In the same setting one has $\min \mathfrak{E}_f = \mathfrak{E}_f(1.69 \ldots, 0) \sim -2 \cdot 10^4$, yielding

$$\min_{p \in \mathfrak{T}} \mathcal{E}\left(E_{(0)}|_p\right) \sim -\frac{10^{27}}{(R/\text{m})^2} \, \text{gr/cm}^3 . \tag{9.4}$$

This makes patent that $\mathcal{E}(E_{(0)}|_p) < 0$ for some $p \in \mathfrak{T}$, which proves that the weak (and, hence, the dominant) energy condition is violated; by similar computations it can be inferred that the strong condition fails as well. Moreover, the numerics in Eq. (9.4) reveal that (the absolute value of) the energy density, while considerably

Fig. 7 Plot of the function $\mathfrak{E}_f(\rho/R, 0)$ (cf. Eqs. (2.8)–(9.3))

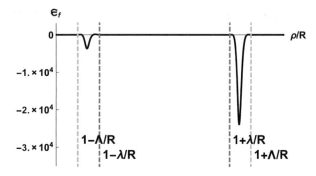

Fig. 8 Plot of the function $\mathfrak{E}_\xi(\rho/R)$ (cf. Eqs. (2.8)–(9.5)) for $\varpi = 2 \cdot 10^{-20}$

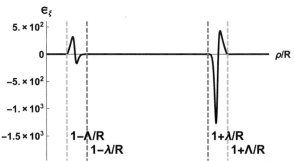

smaller than the Planck density $\rho_P := c^5/(\hbar\, G^2) \sim 10^{93}$gr/cm^3, is huge on human scales unless $R \gtrsim 10^{13}$m.

The Energy Density Measured During a Time Travel by Free Fall Let ξ be a timelike geodesic (with tangent $\dot{\xi}$) of the type described in Sect. 6. The energy density measured at proper time τ by an observer in free fall along ξ is given by (see [10])

$$\mathcal{E}\big(\dot{\xi}(\tau)\big) = \frac{\gamma^2}{8\pi G\, R^2}\, \mathfrak{E}_\xi\big(\rho(\tau)/R\big)\,, \tag{9.5}$$

where $\rho(\tau)$ is the radial coordinate associated to ξ and $\mathfrak{E}_\xi$ is a dimensionless function, depending on λ/R, Λ/R, a, b and on γ, ω.

As an example, refer again to the setting (2.8) and consider a geodesic with $\gamma = 10^5$ (see Eq. (6.9)). Figure 8 shows the graph of the function $\rho/R \mapsto \mathfrak{E}_\xi(\rho/R)$ for $\varpi = 2 \cdot 10^{-20}$ (see Eq. (6.11)). The following numerical results are extracted from Tables 1–3 of [10] (cf. also the preceding Eqs. (6.11)–(6.13) and (7.4)–(7.6)):

$$\min_{\tau \in [0, \tau_2]} \mathcal{E}\big(\dot{\xi}(\tau)\big) \sim -7 \cdot 10^{31}\, \text{gr/cm}^3 \qquad \text{for } R = 10^2\, \text{m}\,,\ \varpi = 10^{-20}; \tag{9.6}$$

$$\min_{\tau \in [0, \tau_2]} \mathcal{E}\big(\dot{\xi}(\tau)\big) \sim -7 \cdot 10^{13}\, \text{gr/cm}^3 \qquad \text{for } R = 10^{11}\, \text{m}\,,\ \varpi = 10^{-11}; \tag{9.7}$$

$$\min_{\tau \in [0, \tau_2]} \mathcal{E}\big(\dot{\xi}(\tau)\big) \sim -0.7\, \text{gr/cm}^3 \qquad \text{for } R = 10^{18}\, \text{m}\,,\ \varpi = 10^{-4}. \tag{9.8}$$

Comments analogous to those reported below Eq. (9.4) can be made even in the present situation. In particular, the above numerics show that, for $\gamma = 10^5$, the energy densities measured during a time travel by free fall are comparable to those of everyday experience only for enormous values of the radius R.

Hawking–Ellis Classification of the Stress–Energy Tensor Let $(E_{(\alpha)})_{\alpha\in\{0,1,2,3\}}$ be the pseudo-orthonormal tetrad of Sect. 3 and consider the stress–energy components

$$T_{\alpha\beta} := T\big(E_{(\alpha)}, E_{(\beta)}\big) \qquad (\alpha, \beta \in \{0, 1, 2, 3\}) . \tag{9.9}$$

Incidentally, notice that T_{00} coincides by definition with the previously discussed energy density $\mathcal{E}(E_{(0)})$ measured by a fundamental observer.

It can be checked by direct inspection that $(T_{\alpha\beta})$ is a block matrix of the form

$$(T_{\alpha\beta}) = \begin{pmatrix} \mathbf{P} & \mathbf{0} \\ \mathbf{0} & \mathbf{Q} \end{pmatrix}, \qquad \mathbf{P} = \mathbf{P}^T, \ \mathbf{Q} = \mathbf{Q}^T, \tag{9.10}$$

where $\mathbf{0}$ is the zero 2×2 matrix and $\mathbf{P}, \mathbf{Q}$ are 2×2 real symmetric matrices. The (Segre-Plebański-)Hawking–Ellis classification [18, 26, 45] relies on the analysis of the eigenvalues μ determined by the characteristic equation

$$0 = \det\big(T_{\alpha\beta} - \mu\, \eta_{\alpha\beta}\big) = \det(\mathbf{P} - \mu\, \mathbf{J}) \cdot \det(\mathbf{Q} - \mu\, \mathbf{1}) , \tag{9.11}$$

where $(\eta_{\alpha\beta}) := \operatorname{diag}(-1, 1, 1, 1)$, $\mathbf{J} := \operatorname{diag}(-1, 1)$ and $\mathbf{1} := \operatorname{diag}(1, 1)$.

The study of the $\mathbf{Q}$-block is elementary. In fact, $\mathbf{Q}$ is symmetric with respect to the standard Euclidean product on $\mathbb{R}^2$, represented by $\mathbf{1}$; so, by the spectral theorem it follows that the equation $\det(\mathbf{Q} - \mu\, \mathbf{1}) = 0$ has two real (possibly coinciding) solutions, which correspond to orthogonal spacelike eigenvectors of $(T_{\alpha\beta})$.

Regarding the $\mathbf{P}$-block, consider the related discriminant

$$\Delta := (T_{00} + T_{11})^2 - 4T_{01}^2 , \tag{9.12}$$

where the components T_{00}, T_{01}, T_{11} are as in Eq. (9.9). It can be checked that the eigenvalues determined by $\det(\mathbf{P} - \mu\, \mathbf{J}) = 0$ are both real if $\Delta \geqslant 0$ (coinciding if $\Delta = 0$) and complex conjugate if $\Delta < 0$. As a matter of fact, the sign of Δ varies, depending on the parameters λ/R, Λ/R, a, b, and on the shape function $\mathcal{X}$. As an example, consider the setting (2.8); Fig. 9 shows the plot of $(\operatorname{sgn}\Delta)(\rho/R, z/R)$ and reveals that Δ can be either positive or negative.

More precisely, the equation $\Delta = 0$ implicitly defines two surfaces, yielding a three-layered partition of the curved spacetime region delimited by $\mathbf{T}_\lambda$ and $\mathbf{T}_\Lambda$:

1. in the *intermediate layer* it is $\Delta > 0$, so $\det(\mathbf{P} - \mu\, \mathbf{J}) = 0$ has two distinct real solutions. In view of the previous results for the $\mathbf{Q}$-block, this means that

Fig. 9 Representation of the regions where $\mathrm{sgn}\,\Delta = 1$ (in red) and $\mathrm{sgn}\,\Delta = -1$ (in purple) (cf. Eqs. (2.8)–(9.12))

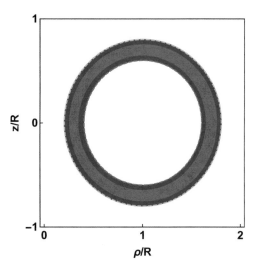

in this layer the stress–energy tensor T fits into *type I* of the Hawking–Ellis classification, namely the one corresponding to most common matter forms.

2. in the *inner* and *outer layers* there holds $\Delta < 0$, so $\det(\mathbf{P} - \mu\,\mathbf{J}) = 0$ has two complex-conjugate solutions. Recalling again the remarks about the **Q**-block, this implies that in these layers T fits into *type IV* of the Hawking–Ellis classification. Stress–energy tensors of this type are known to suffer serious interpretation problems in terms of classical matter sources; nonetheless, the renormalized expectation values of the stress–energy tensor are of this type in a number of semiclassical models for quantum fields on curved backgrounds (see [45] and references therein).

3. at the *transition surfaces* delimiting the previous layers (where $\Delta = 0$), the equation $\det(\mathbf{P} - \mu\,\mathbf{J}) = 0$ has two coinciding real solutions. Thus, T fits into *type II* of the Hawking–Ellis classification, which typically admits a classical interpretation in terms of radiation.

10   Future Investigations

This conclusive section hints at some possible developments related to the spacetime $\mathfrak{T}$, whose investigation is deferred to future works.

Accelerated Trajectories   The alleged time travels considered in [10] and in Sect. 6 of this work exclusively refer to observers in free fall along specific timelike geodesics. Some troublesome features related to such trajectories could be partly overcome by means of accelerated motions. For example, tiny accelerations could allow to avoid the fine tuning issue for the parameter ϖ of Eq. (6.7), required at the present stage to return to the same spatial position. Moreover, accelerating observers could travel into the past of significantly larger amounts of time, possibly averting the need to produce enormous (but still negative) energy densities. To this purpose, their journey should be scheduled as follows: (1) freely fall inside $\mathbf{T}_\lambda$, following a timelike geodesic like those described in Sect. 6; (2) once inside $\mathbf{T}_\lambda$, produce a small acceleration so as to end upon a nearby geodesic along which the decrease of the coordinate time t is somehow optimal; (3) follow the said geodesic for an arbitrarily large interval of proper time, thus moving into the past with respect to t; (4) produce again a small acceleration, in order to return on a nearby geodesic connecting the interior of $\mathbf{T}_\lambda$ to the Minkowskian region outside $\mathbf{T}_\Lambda$.

Causal Structure and Creation of Closed Causal Curves The results of Sect. 6 indicate that there exist (non-necessarily geodetic) closed causal curves self-intersecting at any spacetime point $p \in \mathfrak{T}$. Thus, $\mathfrak{T}$ is an "eternal time machine" (in the sense of [6, 22]) where causality is violated at any point. A clearer understanding of the causal structure of $\mathfrak{T}$ could be achieved characterizing its past and future null infinities. In this connection, it would also be of interest to consider a more realistic variant of $\mathfrak{T}$, evolving from an initial non-exotic chronal region; such a variant could be described in rather simple terms assuming the shape function $\mathcal{X}$ of Eq. (2.5) to depend also on the coordinate t and, possibly, on φ (in addition to ρ, z). Conceivably, this alternative model would present a chronology horizon, delimiting the spacetime region with closed causal curves.

Classical and Quantum Fields Propagating on the Background of $\mathfrak{T}$ A fundamental issue typically affecting spacetimes with CTCs is the stability under back-reaction effects related to (classical and quantum) fields propagating on the said curved backgrounds; namely such effects are expected to drastically alter the structure of the spacetime under analysis, intrinsically preventing the formation of closed causal curves. This type of arguments led Hawking to formulate the so-called Chronology Protection Conjecture [17]. A first step towards the investigation of the stability issue for $\mathfrak{T}$ would be the analysis of the Cauchy problem for the wave equation on it, describing the propagation of a scalar field. In this connection, the construction of the causal propagator comprises non-trivial problems of micro-local analysis, which could be partly eased by the specific symmetries of $\mathfrak{T}$.

Acknowledgements I wish to thank Livio Pizzocchero for valuable comments and suggestions. This work was supported by: INdAM, Gruppo Nazionale per la Fisica Matematica; "Progetto Giovani GNFM 2017 - Dinamica quasi classica per il modello di polarone" fostered by GNFM-INdAM; INFN, Istituto Nazionale di Fisica Nucleare.

References

1. F. Ahmed, A type N radiation field solution with $\Lambda < 0$ in a curved space-time and closed time-like curves. Eur. Phys. J. C **78**, 385 (6pp.) (2018)
2. M. Alcubierre, The warp drive: hyper-fast travel within general relativity. Class. Quant. Grav. **11**, L73-L77 (1994)

3. A. Bachelot, Global properties of the wave equation on non-globally hyperbolic manifolds. J. Math. Pures Appl. **81**, 35–65 (2002)
4. S. Deser, R. Jackiw, G. 't Hooft, Physical cosmic strings do not generate closed timelike curves. Phys. Rev. Lett. **68**(3), 267–269 (1992)
5. J. Dietz, A. Dirmeier, M. Scherfner, Geometric analysis of Ori-type spacetimes, 24pp. (2012). arXiv:1201.1929 [gr-qc]
6. F. Echeverria, G. Klinkhammer, K.S. Thorne, Billiard balls in wormhole spacetimes with closed timelike curves: classical theory. Phys. Rev. D **44**(4), 1077–1099 (1991)
7. H. Ellis, Ether flow through a drainhole: a particle model in general relativity. J. Math. Phys. **14**, 104–118 (1973)
8. A.E. Everett, Warp drive and causality. Phys. Rev. D **53**(12), 7365–7368 (1996)
9. A.E. Everett, T.A. Roman, Superluminal subway: the Krasnikov tube. Phys. Rev. D **56**(4), 2100–2108 (1997)
10. D. Fermi, L. Pizzocchero, A time machine for free fall into the past. Class. Quant. Grav. **35**(16), 165003 (42pp.) (2018)
11. J.L. Friedman, The Cauchy problem on spacetimes that are not globally hyperbolic, in *The Einstein Equations and the Large Scale Behavior of Gravitational Fields*, ed. by P.T. Chrusciel, H. Friedrich (Springer/Birkhäuser, Basel, 2004)
12. J. Friedman, M.S. Morris, I.D. Novikov, F. Echeverria, G. Klinkhammer, K.S. Thorne, U. Yurtsever, Cauchy problem in spacetimes with closed timelike curves. Phys. Rev. D **42**(6), 1915–1930 (1990)
13. V. Frolov, I. Novikov, *Black Hole Physics. Basic Concepts and New Developments* (Kluwer Academic Publisher, Dordrecht, 1998)
14. K. Gödel, An example of a new type of cosmological solutions of Einstein's field equations of gravitation. Rev. Mod. Phys. **21**(3), 447–450 (1949)
15. J.R. Gott III, Closed timelike curves produced by pairs of moving cosmic strings: exact solutions. Phys. Rev. Lett. **66**(9), 1126–1129 (1991)
16. J.D.E. Grant, Cosmic strings and chronology protection. Phys. Rev. D **47**(6), 2388–2394 (1993)
17. S.W. Hawking, Chronology protection conjecture. Phys. Rev. D **46**(2), 603–611 (1992)
18. S.W. Hawking, G.F.R. Ellis, *The Large Scale Structure of Space-Time* (Cambridge University Press, Cambridge, 1975)
19. R.P. Kerr, Gravitational field of a spinning mass as an example of algebraically special metrics. Phys. Rev. Lett. **11**(5), 237–238 (1963)
20. S.V. Krasnikov, Hyperfast travel in general relativity. Phys. Rev. D **57**(8), 4760–4766 (1998)
21. S.V. Krasnikov, Time machines with the compactly determined Cauchy horizon. Phys. Rev. D **90**(2), 024067 (8pp.) (2014)
22. S.V. Krasnikov, *Back-in-Time and Faster-Than-Light Travel in General Relativity* (Springer, Berlin, 2018)
23. F.S.N. Lobo, Closed timelike curves and causality violation, in *Classical and Quantum Gravity: Theory, Analysis and Applications*, ed. by V.R. Frignanni (Nova Science Publishers, New York, 2008)
24. F.S.N. Lobo, *Wormholes, Warp Drives and Energy Conditions* (Springer, Berlin, 2017)
25. C. Mallary, G. Khanna, R.H. Price, Closed timelike curves and 'effective' superluminal travel with naked line singularities. Class. Quant. Grav. **35**, 175020 (18pp.) (2018)
26. P. Martín-Moruno1, M. Visser, Essential core of the Hawking–Ellis types. Class. Quant. Grav. **35**, 125003 (12pp.) (2018)
27. C.W. Misner, The flatter regions of Newman, Unti, and Tamburino's generalized Schwarzschild space. J. Math. Phys. **4**(7), 924–937 (1963)
28. C.W. Misner, Taub-NUT space as a counterexample to almost anything, in *Relativity Theory and Astrophysics*, ed. by J. Ehlers. Relativity and Cosmology, vol. 1 (AMS, Providence, 1967)
29. M.S. Morris, K.S. Thorne, Wormholes in spacetime and their use for interstellar travel: a tool for teaching general relativity. Am. J. Phys. **56**(5), 395–412 (1988)
30. M.S. Morris, K.S. Thorne, U. Yurtsever, Wormholes, time machines, and the weak energy condition. Phys. Rev. Lett. **61**(13), 1446–1449 (1988)

31. E. Newman, L. Tamburino, T. Unti, Empty space generalization of the Schwarzschild metric. J. Math. Phys. **4**(7), 915–923 (1963)
32. A. Ori, Must time-machine construction violate the weak energy condition? Phys. Rev. Lett. **71**, 2517–2520 (1993)
33. A. Ori, A class of time-machine solutions with a compact vacuum core. Phys. Rev. Lett. **95**, 021101 (4pp.) (2005)
34. A. Ori, Formation of closed timelike curves in a composite vacuum/dust asymptotically flat spacetime. Phys. Rev. D **76**(4), 044002 (14pp.) (2007)
35. A. Ori, Y. Soen, Causality violation and the weak energy condition. Phys. Rev. D **49**(8), 3990–3997 (1994)
36. D. Sarma, M. Patgiri, F.U. Ahmed, Pure radiation metric with stable closed timelike curves. Gen. Relativ. Gravit. **46**, 1633 (9pp.) (2014)
37. Y. Soen, A. Ori, Improved time-machine model. Phys. Rev. D **54**(8), 4858–4861 (1996)
38. A.H. Taub, Empty space-times admitting a three parameter group of motions. Ann. Math. **53**(3), 472–490 (1951)
39. K.S. Thorne, Closed timelike curves, in *General Relativity and Gravitation*, ed. by R.J. Gleiser, C.N. Kozameh, O.M. Moreschi, 1992. *Proceedings of the 13th International Conference on General Relativity and Gravitation* (Institute of Physics Publishing, Bristol, 1993)
40. K.S. Thorne, Misner space as a counterexample to almost any pathology, in *Directions in General Relativity*, ed. by B.L. Hu, M.P. Ryan Jr., C.V. Vishveshwara. *Proceedings of the 1993 International Symposium, Maryland*, vol. 1. Papers in Honor of Charles Misner (Cambridge University Press, Cambridge, 1993)
41. F.J. Tipler, Rotating cylinders and the possibility of global causality violation. Phys. Rev. D **9**(8), 2203–2206 (1974)
42. B.K. Tippett, D. Tsang, Traversable acausal retrograde domains in spacetime. Class. Quantum Grav. **34**, 095006 (12pp.) (2017)
43. A. Tomimatsu, H. Sato, New series of exact solutions for gravitational fields of spinning masses. Prog. Theor. Phys. **50**(1) (1973), 95–110
44. W.J. van Stockum, The gravitational field of a distribution of particles rotating about an axis of symmetry. Proc. Royal Soc. Edinburgh **57**, 135–154 (1938)
45. M. Visser, The quantum physics of chronology protection, in *Workshop on Conference on the Future of Theoretical Physics and Cosmology in Honor of Steven Hawking's 60th Birthday. Proceedings*, ed. by G.W. Gibbons, E.P.S. Shellard, S.J. Rankin (Cambridge University Press, Cambridge, 2002)
46. R.M. Wald, *General Relativity* (The University of Chicago Press, Chicago, 1984)

Schwarzschild Spacetime Under Generalised Gullstrand–Painlevé Slicing

Colin MacLaurin

Abstract We investigate a foliation of Schwarzschild spacetime determined by observers freely falling in the radial direction. This is described using a generalisation of Gullstrand–Painlevé coordinates which allows for any possible radial velocity. This foliation provides a contrast with the usual static foliation implied by Schwarzschild coordinates. The 3-dimensional spaces are distinct for the static and falling observers, so the embedding diagrams, spatial measurement, simultaneity, and time at infinity are also distinct, though the 4-dimensional spacetime is unchanged. Our motivation is conceptual understanding, to counter Newton-like viewpoints. In future work, this alternate foliation may shed light on open questions regarding quantum fields, analogue gravity, entropy, energy, and other quantities. This article is aimed at experienced relativists, whereas a forthcoming series is intended for a general audience of physicists, mathematicians, and philosophers.

Keywords Gullstrand–Painlevé · Falling observer · Coordinates

1 Introduction

Schwarzschild-Droste spacetime is most commonly expressed in terms of Schwarzschild-Droste coordinates:

$$ds^2 = -\left(1 - \frac{2M}{r}\right)dt^2 + \left(1 - \frac{2M}{r}\right)^{-1}dr^2 + r^2 d\Omega^2, \tag{1}$$

where $d\Omega^2 := d\theta^2 + \sin^2\theta\, d\phi^2$. This is a natural choice because of their simplicity and intuition, also r is the curvature coordinate and ∂_t is the preferred Killing vector field. However, in pedagogical settings, many presentations express the properties

C. MacLaurin (✉)
University of Queensland, Brisbane, QLD, Australia
e-mail: colin.maclaurin@uqconnect.edu.au

© Springer Nature Switzerland AG 2019
S. Cacciatori et al. (eds.), *Einstein Equations: Physical and Mathematical Aspects of General Relativity*, Tutorials, Schools, and Workshops in the Mathematical Sciences, https://doi.org/10.1007/978-3-030-18061-4_9

of space and time in terms of the foliation $t = \text{const}$, but without a clear qualifier about this choice. This downplays the lesson that space and time are relative to the observer. This is particularly true in the case of the radial proper distance, and also for the "time at infinity" when interpreted as a global simultaneity convention. We seek to improve conceptual understanding by taking a complementary description based on freely falling observers, and re-examining familiar descriptions from this alternate $3 + 1$-splitting. These observers and their space (the hypersurfaces orthogonal to them) are conveniently described using a generalisation of Gullstrand–Painlevé coordinates which allows for all possible radial velocities.

While most textbook material and research calculations correctly account for coordinates or frames, some conceptual confusion remains. For example the mathematics of arbitrary foliations is well understood and clearly taught, including the intrinsic and extrinsic curvature of hypersurfaces, and the ADM formalism along with lapse-shift notation. However, this technical knowledge has not always been utilised in some expositions of Schwarzschild coordinates. Another topic which is generally handled well is the careful extraction of measurable observables, for instance the decomposition of the velocity gradient of a congruence into shear, expansion, and vorticity. Yet while for proper distance the general definition $\int ds$ is standard, it is rarely applied to give anything beyond the static radial measurement $(1 - 2M/r)^{-1/2}dr$. The book by Taylor & Wheeler [53] is a notable exception. Yet it seems "neo-Newtonian" [10] interpretations of relativity have not completely died out.

The generalised Gullstrand–Painlevé coordinates are well suited to the falling observers, in that the x^0-coordinate is their proper time. These coordinates are convenient for computation and intuition. One might protest that calculations can be made in any coordinate system, however, in history new coordinates have often helped to advance understanding [27]. In particular, the popularisation of Eddington–Finkelstein and Kruskal–Szekeres coordinates extended understanding across the horizon(s). However, these null coordinates are less insightful for timelike observers [13]. There are probably hundreds of coordinate systems for Schwarzschild spacetime used in the literature [54] [52, §7] [37, §2.2] [49] [17, §10], but the coordinates studied here are arguably the most intuitive for radial timelike geodesics. Another obvious "natural" choice is circular orbits, however, these only exist for $r > 3M$ so cannot probe the horizon.

In future work, the generalised Gullstrand–Painlevé coordinates could have applications to open research questions about any quantities which depend on foliation, such as entropy or the decomposition of quantum fields on curved spacetime into positive and negative frequency modes [56]. A special case has given alternate descriptions for Hawking radiation [24, §3] [40] [1, §4.3]. Also related coordinates are considered in laboratory analogues of gravity based on fluid flow [27, 44].

Section 4 presents isometric embedding diagrams for the new spatial slices which are orthogonal to the observer congruence. This is a way of visualising the curvature of 3-dimensional space. Section 5 discusses measurement of the radial proper distance, for different observers including inside the horizon. Sec-

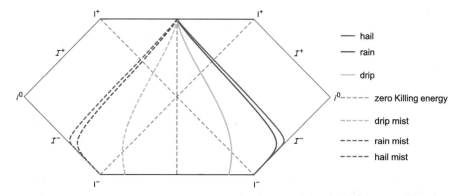

Fig. 1 Some maximally extended geodesics in a Penrose diagram, picked to coincide at the "event" $(t, r) \to (0, 0)$. Rain and hail start from past timelike infinity, and fall inwards to the black hole singularity; their mist variants are similar. Drip and drip mist start from the past singularity in the white hole interior, fall out of the past horizon into an exterior region, then reach a maximum height before falling through the future horizon into the black hole interior. Zero Killing energy geodesics pass from the white hole interior through the bifurcate horizon directly into the black hole interior. From our universe, negative energy worldlines may be achieved by passing inside the horizon and then accelerating "outwards"

tion 6 clarifies some potential misconceptions about coordinate basis vectors and coordinate gradients. While this section contains fairly straightforward material, it has rarely been explained in print. Section 7 explains the concept of "time at infinity" contains an implicit simultaneity convention, if interpreted globally. The falling observers offer a different simultaneity convention, which helps to avoid recurring misconceptions. The following Sects. 2 and 3 give a careful analysis of the worldlines and coordinate charts; for a quicker read, glance at Fig. 1 and Eq. 4 before skipping to Sect. 4.

2   Radial Timelike Geodesics

The geodesics of Schwarzschild (and Kerr) have been thoroughly researched. Still, it is helpful to clarify the allowable radial timelike geodesics, in particular the existence of negative Killing energy orbits within the extended manifold. Textbooks justifiably focus on orbits with angular momentum. Of our four classes of radial geodesics given below, Chandrasekhar [6, §19] covers just one in detail (but clearly implies there are others), Frolov & Novikov [15, §2.3.2] briefly allude to all four, and O'Neill [38, §4.10] has a detailed study for Kerr; however, the older classic Hagihara [19, ff.] may exclude the $r < 2M$ black hole region and hence the negative Killing energy geodesics, and the language in Taylor & Wheeler [53] denies their existence.

Table 1 Radial timelike geodesics in terms of e

Nickname	Traditional name	Killing energy per mass	Range
Hail	Hyperbolic	$e > 1$	All
Rain	Parabolic	$e = 1$	All
Drip	Elliptic	$0 < e < 1$	$r \leq r_{\max} = \frac{2M}{1-e^2}$
Mist	—	$e \leq 0$	$r < 2M$

This is restricted to regions I and II. Note the energy measured by any *local* observer is always positive

The quantity e derived from the "static" Killing vector field ξ which is timelike at infinity, is a natural parameter choice:

$$e := -\mathbf{u} \cdot \xi, \tag{2}$$

where $\mathbf{u}$ is the 4-velocity field, and the dot represents the metric. This Killing energy per mass or "energy per mass at infinity" is preserved along geodesics.[1] ξ has the components $(1, 0, 0, 0)$, where all components are expressed in Schwarzschild charts unless otherwise stated. (This is familiar and simple, but internal computations are made in Gullstrand–Painlevé or Kruskal–Szekeres coordinates where necessary.) The 4-velocity is

$$u^\mu = \left(e\left(1 - \frac{2M}{r}\right)^{-1}, \pm\sqrt{e^2 - 1 + \frac{2M}{r}}, 0, 0 \right). \tag{3}$$

These have zero angular momentum, as determined from the Killing vector fields orthogonal to ξ. Taylor & Wheeler [53, §B2] use the metaphors *hail, rain*, and *drips* for classes of ingoing radial motion, see Table 1. Rain are the quintessential radial geodesics, and "fell from rest at infinity" so to speak, or have $e = 1$ to be precise. Hail has $e > 1$, and fell from infinity with initial inward velocity having Lorentz factor $\gamma = e$ relative to a static observer there. Drips have $0 < e < 1$ and fell from rest at some finite $r_{\max} > 2M$. Taylor & Wheeler state these "cover all possible radially moving free-float frames" [53, §B2], however, there are trajectories with $e \leq 0$ which exist only inside the horizon. We dub these *mist* because they fall more slowly than drips (in the sense their relative 3-velocity points outwards), and seem ethereal to observers in region I. See Table 1. To achieve these worldlines, pass into the horizon, accelerate sufficiently "outwards", then return to freefall. This possibility of negative Killing energy observers is best known in the context of the Penrose process or superradiant scattering in Kerr spacetime, or from the heuristic

---

[1]Infinity can be made rigorous by conformal compactification, which produces a new manifold with a boundary consisting of timelike, null, and spacelike infinities. However, for our purposes a simple limit $r \to \infty$ is often sufficient, or at least an approximation $r \gg 2M$. Physically, infinity is loosely analogous with Solar System observers far from a black hole.

Table 2 Allowed regions, this generalises Table 1

Energy	Direction	Regions
$e > 0$	Ingoing	I and II
$e < 0$	Ingoing	III and II
$e > 0$	Outgoing	IV and I
$e < 0$	Outgoing	IV and III
$e = 0$	–	IV and II

Drip and drip mist are further subject to $r \leq r_{max} = 2M/(1 - e^2)$. There is no ingoing/outgoing freedom for $e = 0$

description of Hawking radiation as particle pairs [21, §1]. It can occur when the relevant timelike Killing vector field becomes spacelike.

So far this discussion is limited to inward motion in the "physical" spacetime consisting of black hole interior and one exterior region. In the maximal analytic extension of the manifold, the $e < 0$ worldlines emerge from the parallel exterior region, if continued backwards as geodesics, as Fig. 1 shows. We can also subdivide "mist" into rain mist ($e = -1$), hail mist ($e < -1$), drip mist ($-1 < e < 0$), and zero Killing energy observers ($e = 0$). The sign of $dr/d\tau$ in Eq. 3 is an additional parameter which specifies ingoing or outgoing motion (lower and upper signs respectively), which extends the classes to outgoing variants, where allowed. Table 2 shows the allowed parameter combinations for all four regions. Hence the extended) timelike radial geodesics are classified uniquely, modulo translation in "time" (ξ).

3   Generalised Gullstrand–Painlevé Coordinates

For $e \neq 0$ we use a coordinate $T \equiv T_e$ which is proper time along the worldlines. While one possibility is to extend a single local frame outwards by geodesics, we consider an entire congruence of worldlines [31, §2.6] [15, §3.2.2]. The metric becomes

$$ds^2 = -\frac{1}{e^2}\left(1 - \frac{2M}{r}\right)dT^2 \mp \frac{2}{e^2}\sqrt{e^2 - 1 + \frac{2M}{r}}\,dT\,dr + \frac{1}{e^2}dr^2 + r^2 d\Omega^2. \quad (4)$$

This line element is suited to the congruence, and aids intuition and computation. (We avoid the term "adapted" coordinates which implies $g_{0i} = 0$.) The transformation from Schwarzschild coordinates is

$$dT = e\,dt \mp \left(1 - \frac{2M}{r}\right)^{-1}\sqrt{e^2 - 1 + \frac{2M}{r}}\,dr. \quad (5)$$

Note this must match $\partial T / \partial t \cdot dt + \partial T / \partial r \cdot dr$. The rain case $e = 1$ is due to Gullstrand [18] and Painlevé [39]. It was generalised by Gautreau & Hoffmann [16] using the parameter $r_{\max}$ described previously, hence limited to the drip case $0 < e < 1$. Martel & Poisson [33] derived a similar coordinate eT, using instead a parameter $p := 1/e^2$ they limited to the rain and hail cases, and under which Eddington–Finkelstein null coordinates are a limiting case. Finch [13] clarified the unity for $e > 0$. Bini et al. [2] have the most general treatment, using e as parameter, including outgoing motions, and applied to various spacetimes. There are numerous other related works. My contribution is to clarify the allowable worldlines in this context (specifically $e \leq 0$), and to explore the resulting properties of the $3 + 1$-splitting.

Various related derivations have been given, which can be generalised if necessary. One uses local Lorentz boosts from the static to falling orthonormal frames [53, §B-4] [28]. Another considers the proper time on freely falling clocks [16] [36, §15]. We follow instead a mathematically elegant approach which defines the time gradient from the co-velocity:

$$dT := -\mathbf{u}^\flat, \tag{6}$$

where $\mathbf{u}^\flat$ is dual to the 4-velocity $\mathbf{u}$, and the minus sign compensates for our signature convention -+++. This definition is the unique choice which is both proper time along the worldlines: $dT/d\tau \equiv dT(\mathbf{u}) = 1$, and constant along the 3-space orthogonal to them—that is, Einstein simultaneous: $dT(\mathbf{v}) = 0$ for $\mathbf{u} \cdot \mathbf{v} = 0$.[2] (Eq. 6 is a valid definition locally if and only if a timelike congruence is *geodesic* and *vorticity-free*. With the inclusion of a scalar integrating factor $1/N$, where N is the lapse, the geodesic requirement is dropped.) This co-velocity approach was applied in our context by Martel & Poisson [33], Finch [13], and Bini et al. [2]. Amongst general treatments, [11, §4.6.2] and [41, §2.3.3] are especially clear and relevant here; see also early sources [51] [9, §2.2] or numerical relativity textbooks. The vorticity-free requirement follows from Frobenius' theorem, see also [45, §2.3] [55, §B-3] [8, §2.12] or the differential geometry literature.

The coordinates are regular at $r = 2M$. The cross-term indicates motion relative to r, as seen from the inverse metric component

$$g^{Tr} = dT \cdot dr = dr(dT^\sharp) = dr(-\mathbf{u}) = -u^r = -\frac{dr}{d\tau} \neq 0 \tag{7}$$

in general, using Eqs. 3 and 6. The coordinates have lapse $N = 1$ since T is proper time, also a shift of $(\mp\sqrt{e^2 - 1 + 2M/r}, 0, 0)$. Indeed, because of the unit lapse, the inverse metric components $(g^{Ti})_{i=1,2,3}$ are precisely the shift vector [2, §2] [35, §21.4]. The line element is independent of T, so unchanged by translation in ∂_T

---

[2]Recall the intuition for combining a vector with a 1-form dT to produce a scalar, as the number of level sets $T = $ const crossed by the vector [47, §3.3] [35, §2.5].

Fig. 2 Standard Penrose diagram for Schwarzschild spacetime, with regions dubbed our universe exterior (I), black hole interior (II), parallel universe exterior (III), and white hole interior (IV). The orange line borders our default generalised Gullstrand–Painlevé chart, for an ingoing variant with $e \geq 1$

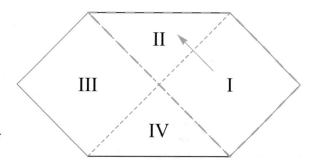

which is the Killing vector field $\boldsymbol{\xi}/e$. Even though the normal $\mathbf{u}$ to the hypersurfaces has a nonzero r-component, the shift 4-vector compensates for this, so the sum of these (normal times lapse plus shift) is the coordinate vector ∂_T. The $T = \text{const}$ slices are identical, it does not matter the symmetry underlying this is spacelike inside the horizon.

A single coordinate chart covers two adjacent regions of the extended manifold, including the horizon segment between them. Figure 2 shows an ingoing chart with $e > 0$. The outgoing variant with $e > 0$ covers regions IV and I. As pointed out for the $e = +1$ coordinates by [24, §2], all regions can be covered by reversing time for these charts: this means nothing more exotic than reinterpreting T as *decreasing* towards the future. Hence in the reversed charts, *minus* $\mathbf{u}$ is future-pointing. Our inclusion of $e < 0$ coordinates allows a more intuitive interpretation: time-reversing a chart is equivalent to reversing both the sign of e and ingoing / outgoing, as seen from Eq. 3. Table 2 summarises the variants. (Note the line element is formally the same under $\pm e$, but the meaning of T is not.)

It must be qualified the drip and drip mist charts are bounded by $r < r_{\text{max}}$. Martel & Poisson [33] discount them for this reason, however, the worldlines which motivate these charts only exist in the same sub-region, and we can certainly supplement with different coordinate charts. The boundary at r_{max} is excluded because technically charts are open subsets; in fact the Christoffel symbol Γ^T_{rr} diverges at the boundary. Any two complementary Gullstrand–Painlevé charts still omit one branch of the horizon. All four chart combinations together still omit the bifurcate horizon, as pointed out for Eddington–Finkelstein coordinates [5]. One can use Kruskal–Szekeres coordinates there. To a physicist these details seem pedantic, but they are important for the $e = 0$ trajectories.

We briefly mention some properties useful in various $3 + 1$-applications. Firstly for the worldlines themselves, the kinematic decomposition shows shear and expansion, but zero vorticity and acceleration. The 3-spaces $T = \text{const}$ have metric

$$^{(3)}ds^2 = \frac{1}{e^2}dr^2 + r^2 d\Omega^2 \tag{8}$$

in terms of r. (In fact this 3-metric occurs within any static, spherically symmetric, vacuum spacetime [2, §3].) The Riemann tensor has (at most) one non-trivial component up to symmetry: $^{(3)}R^{\theta\phi}{}_{\theta\phi} = (1 - e^2)/r^2$, the Ricci tensor components $^{(3)}R^\theta{}_\theta$ and $^{(3)}R^\phi{}_\phi$ have the same value, and the Ricci scalar is double this [33, §3] [2, §3]. Hence the curvature of 3-space has the same sign everywhere, but not the same magnitude. The Kretschmann scalar is $4(1 - e^2)^2/r^4$. The extrinsic curvature has trace K behaving like $r^{-3/2}$ for $r \to 0$. This is significant for quantum fields on curved spacetime, where authors require "nice slices" with small intrinsic and extrinsic curvature [29, 34]. Another consideration is asymptotic flatness, but even the $e = \pm 1$ coordinates fail to meet some technical conditions here, so their ADM mass is not valid [17, §8.3.1]. Hence our coordinates, though nice, are not nice enough at $r \to 0$ and $r \to \infty$ by some requirements. But they are physically motivated, unlike the contrived singularity-avoiding slices in the quantum field theory references, so might still prove useful even in these ongoing fields of research.

An additional coordinate system suited to $e \neq 0$ observers will also prove useful. Keep the new T-coordinate, but replace r with $\rho \equiv \rho_e$ defined such that the metric becomes diagonal:

$$ds^2 = -dT^2 + \frac{1}{e^2}\left(e^2 - 1 + \frac{2M}{r}\right)d\rho^2 + r^2 d\Omega^2 \tag{9}$$

A falling observer with the same e as the coordinate system is comoving: $\rho = $ const. Hence these coordinates are even better adapted to the congruence, at the expense of a less intuitive radial coordinate. These coordinates were given by Gautreau & Hoffmann [16, §4] for $0 < e < 1$. They are a generalisation of the $e = 1$ "Lemaître coordinates", but a special case of the general Lemaître[-Tolman-Bondi] coordinates for a spherically symmetric dust spacetime [26]. The transformation from Schwarzschild coordinates is

$$d\rho = e\,dt \mp e^2\left(1 - \frac{2M}{r}\right)^{-1}\left(e^2 - 1 + \frac{2M}{r}\right)^{-1/2}dr. \tag{10}$$

While the line element Eq. 9 is formally the same for $\pm e$ and ingoing/outgoing variants, the coordinates T and ρ depend intrinsically on these.

Finally for $e = 0$ observers (at $r \neq 2M$), ordinary Schwarzschild coordinates are adapted, because r is *purely* timelike to them ($dr \propto \mathbf{u}^b$) and t purely spacelike, so each worldline has $t = $ const. At the bifurcate horizon, Kruskal–Szekeres coordinates will suffice.

4   Embedding Diagrams: Cones and Funnels

One use of isometric embeddings is as a visual representation of curvature. For any foliation respecting the "time" and spherical symmetries, a 2-dimensional equatorial surface $\theta = \pi/2$ of a constant time slice is representative. Under the usual static foliation $t = $ const, the surface has metric

$$^{(2)}ds^2 = \left(1 - \frac{2M}{r}\right)^{-1}dr^2 + r^2d\phi^2. \tag{11}$$

This can be embedded in Euclidean $\mathbb{R}^3$: under cylindrical coordinates (r, ϕ, z), the surface $z = z(r)$ has metric

$$ds^2 = dr^2 + dz^2 + r^2d\phi^2 = \left(1 + \left(\frac{dz}{dr}\right)^2\right)dr^2 + r^2d\phi^2. \tag{12}$$

The metrics match when $z = \sqrt{8M(r - 2M)}$, a popular funnel-shaped surface of revolution known as "Flamm's paraboloid" [14]. However, this is not the only possible depiction of the curvature of space. The $T = $ const hypersurfaces have equatorial slice

$$^{(2)}ds^2 = \frac{1}{e^2}dr^2 + r^2d\phi^2. \tag{13}$$

For drips and drip mist $0 < |e| < 1$, the embedded surface $z = \sqrt{1/e^2 - 1}\,r$ yields matching metrics, defined up to $r = r_{\max}$. For $|e| = 1$ the plane $z = 0$ suffices, indeed many authors have noticed 3-space is flat for $e = +1$ [26, 39]. For $|e| > 1$ the above embedding approach fails, as the Euclidean surface must have $g_{rr} \geq 1$. However, an alternate choice is to embed in Minkowski spacetime:

$$ds^2 = dr^2 - dz^2 + r^2d\phi^2 = \left(1 - \left(\frac{dz}{dr}\right)^2\right)dr^2 + r^2d\phi^2 \tag{14}$$

This leads to $z = \sqrt{1 - 1/e^2}\,r$, as drawn in Fig. 3.

The $e = 0$ case is unique. These worldlines have $t = $ const, and the hyperplanes orthogonal to them in the tangent space have $r = $ const. In fact under this congruence, all events at a given $r = r_0$ in the same region (II or IV) mesh together

Fig. 3 Embedding diagrams in flat space, restricted to regions I and II only. The left is Flamm's paraboloid, which shows the curvature of static slices. The right diagram shows the curvature of the falling observer space $T = $ const. The latter is a cone with $z \propto r$, and extends inside the horizon (with the tip excluded since $r = 0$ is not part of the 4-manifold). While a 2-cone is intrinsically flat, the "3-cone" it represents is not

to form a hypersurface $r = $ const on the manifold. Its metric follows from Eq. 1:

$$^{(3)}ds^2 = \left(\frac{2M}{r_0} - 1\right)dt^2 + r_0{}^2(d\theta^2 + \sin^2\theta\, d\phi^2) \tag{15}$$

which is the geometry $\mathbb{R} \times S^2$. The 2-surface with $\theta = \pi/2$ embeds in Euclidean space as the cylindrical surface with $r = r_0$ and the z-axis replaced by $\sqrt{2M/r_0 - 1}\, t$. There is a single exception: at the bifurcate horizon the orthogonal hypersurfaces immediately exit the congruence, but if continued as spatial geodesics they extend into regions I and III as the static slices $t = $ const. This is an Einstein–Rosen bridge, so the embedding diagram is two of Flamm's surfaces stuck together. Figure 5 shows these hypersurfaces in a Penrose diagram.

Flamm's paraboloid is not the only depiction of the curvature of space. While the static foliation is a particularly natural choice, the $e = 1$ choice is also natural and we argue more physically reasonable. Pedagogically, sources could more clearly qualify that Flamm's paraboloid is not the only possible embedding diagram, but represents the curvature of space as measured by people at a fixed location. Having said that, many sources do provide embeddings from alternate slices [35, §21.8, §31.6] [12], via $x^0 = $ const under various coordinates, or other creative choices [32] [20, §7.13]. Perhaps the Gauss–Codazzi equations for a congruence would sharpen the arguments about spatial geometry.

To push the point, we can make a "fake black hole" via an unusual foliation of Minkowski spacetime [30]. From spherical coordinates $-dt^2 + dr^2 + r^2 d\Omega^2$, take an observer field moving radially as a function of r only: $u^\mu = (u^t, u^r, 0, 0)$ where $u^t = \sqrt{1 + (u^r)^2}$. A new time coordinate $dT := -\mathbf{u}^b/u^t$ has level sets $T = $ const orthogonal to the observers, even though a proper time coordinate does not exist in general. The metric becomes

$$ds^2 = -dT^2 - 2\frac{u^r}{u^t}dT\, dr + \frac{1}{1 + (u^r)^2}dr^2 + r^2 d\Omega^2. \tag{16}$$

Various different choices of u^r lead to 3-slices which imitate selected aspects of static and falling observers in Schwarzschild spacetime, but here we duplicate Flamm's paraboloid by taking $u^r := \pm\sqrt{2M/(r - 4M)}$. Since $g_{rr} \leq 1$, (re-)embed in Minkowski spacetime as before in Eq. 14. The surface is valid for $r \geq 4M$, and formally matches Flamm's paraboloid! Admittedly this is contrived, as the worldlines are accelerated and seem unnatural, and the underlying 3-geometries are distinct. Nevertheless, it cautions against overinterpretation of the funnel picture. Perhaps this foliation of Minkowski spacetime would make a useful test case for quantum effects in analogy with black holes, just as Rindler coordinates are; presumably here a null-result is expected.

5   Spatial Measurements

In this section we discuss the spatial distance measured by observers of various velocities near a black hole. The familiar textbook "radial proper distance" interval

$$ds = \left(1 - \frac{2M}{r}\right)^{-1/2} dr \tag{17}$$

follows from setting $dt = d\theta = d\phi = 0$ in Eq. 1. But since the general definition of proper distance is $\int ds$ over any spacelike curve, we can apply the analogous procedure in the new coordinates, to measure along the slice $dT = 0$:

$$ds = \frac{1}{|e|} dr \tag{18}$$

Painlevé [39] contrasted the above results in the $e = 1$ case, and concluded general relativity is self-contradictory. Instead, these describe measurements by different observers: *static* and *falling* respectively.[3] Equation 18 is valid even inside the horizon! While the common interpretation of Eq. 17 as measurement by an observer at infinity is not completely without merit, measurement by local observers is more directly meaningful. Any generalisations of this quantity are surprisingly little known. Gautreau & Hoffmann [16] showed Eq. 18 for the drip case $0 < e < 1$, and Taylor & Wheeler [53, §B-3] justify the rain case $e = 1$. Similarly, the 3-volume inside the event horizon for our congruence is $1/|e|$ times the Euclidean ball volume $\frac{4}{3}\pi(2M)^3$, which Finch [13] showed for $e > 0$.

There are other derivations of these results. The most fashionable conception of measurement uses clocks not rulers, because of the impossibility of Born-rigid objects. The Landau–Lifshitz radar metric [25, §79] achieves this using null rays to probe nearby space, defining spatial distance using the time of their return journey. This leads to a 3-metric $^{(3)}ds^2 = \gamma_{ij}dx^i dx^j$, where

$$\gamma_{ij} := g_{ij} - \frac{g_{0i} g_{0j}}{g_{00}} \tag{19}$$

in a given coordinate system. Since all null rays move at c this might seem a preferred/absolute measure of spacetime, however, it also depends on the motion and proper time of the radar device. Equation 19 presumes the device is comoving with the coordinate system: $x^i = $ const, for $i = 1, 2, 3$. In Schwarzschild coordinates the radar metric gives the $dt = 0$ slice, and since it is the static observers which are comoving, the slice $dt = 0$ including Eq. 17 is the measurement of static observers. The generalised Gullstrand–Painlevé coordinates also have

---

[3] We write ds for both, but these should not be equated, as they are restrictions of the full spacetime metric along different 4-vectors.

static observers as comoving, so yield the same radar metric. For the falling
frames we require instead the Lemaître coordinates, for which the radar metric
is the line element Eq. 9 with $dT = 0$. This must equal the Gullstrand–Painlevé
line element with $dT = 0$, as both coordinates express the same 4-metric
tensor $\mathbf{g}$. In particular, falling observers measure the radial distance Eq. 18 as
claimed.

Another approach uses the spatial projector tensor

$$P_{\mu\nu} := g_{\mu\nu} + u_\mu u_\nu \tag{20}$$

for a given observer $\mathbf{u}$, to derive its spatial metric $P_{\mu\nu}dx^\mu dx^\nu$ [7, §6.1]. The radial
distance is given by contracting over the radial vector, which naively would be the
coordinate basis vector ∂_r, which picks out the P_{rr}-component. For static observers
and Schwarzschild coordinates, P_{rr} leads to Eq. 17, whereas for falling observers in
Gullstrand–Painlevé coordinates, P_{rr} leads to Eq. 18 as before.

In fact the radar metric is simply a special case of the spatial projector, or vice
versa the spatial projector is the fully covariant generalisation of the radar metric.
To see this, compute $P_{\mu\nu}$ for the comoving observer $u^\mu = \pm((-g_{00})^{-1/2}, 0, 0, 0)$.
This matches the radar metric if Eq. 19 is reinterpreted as 4-dimensional, which
amounts to padding the matrix of components with zeroes. Since the components
agree within one coordinate system (any comoving one), they must agree in all
coordinate systems, if the radar metric is to transform as a tensor.

A major advantage of the new line element is the faller's radial distance is
clear from inspection (Eq. 18). This provides much-needed contrast with the static
measurement which is clear from inspection of Schwarzschild coordinates. Some
of the most esteemed coordinates are null, so not insightful in this way. For
other coordinate systems with x^0 timelike, including Kerr-Schild and Novikov
coordinates, either the corresponding observers or their measurement are not
clear from inspection. One might protest computations can be performed in any
coordinates, which of course is true, however, history shows this has not been
sufficiently achieved in practice in this context. In fact, Schwarzschild coordinates
could easily be misapplied to give

$$ds = |e|\left(1 - \frac{2M}{r}\right)^{-1}dr \tag{21}$$

for the faller's measurement. To arrive at this, express the falling observer's spatial
metric in Schwarzschild coordinates, either by evaluating the projector directly, or
transforming the radar metric from Lemaître coordinates. Now the correct radial
vector to contract over is Eq. 23, as this lies in the observer's local 3-space; it
leads to $dr/|e|$ as before. The mistaken choice is the Schwarzschild coordinate basis
vector ∂_r, which is not orthogonal to $\mathbf{u}$. Contracting over this vector anyway picks
out the component $P_{rr} = e^2(1 - 2M/r)^{-2}$, hence Eq. 21. In fact this quantity
is salvageable under a different physical interpretation: it relates a falling ruler to

the coordinate gradient dr, but as determined within the *static frame*.[4] By contrast Eq. 18 relates a falling ruler to dr within its own *falling frame*. Equation 17 relates a static ruler to dr as determined within *any frame*, because static particles remain at fixed r-values so foliation cannot alter this association. We will illustrate this in future work. Note the three measurements are consistent with the usual length-contraction formula, where the local Lorentz factor between the static and falling frames is:

$$\gamma = -\mathbf{u} \cdot \mathbf{u}_{\text{static}} = |e|(1 - 2M/r)^{-1/2} \tag{22}$$

However, there are pitfalls and conceptual challenges when passing from the textbook $\int ds$ proper distance or $1/\gamma$ length-contraction formula.

The results are 1-volume elements, giving the ratio of proper distance to coordinate gradient. Much pedagogy sets up a false dichotomy that dr is not *the* distance but $(1 - 2M/r)^{-1/2}dr$ *is*. For example Eq. 17 is termed "radial ruler distance" [43, §11.2] or "actual radial distance between two radial coordinates r_A and r_B" [36, §9]. These lack qualification as to which local frames do the measuring, or the intuition behind the foliation choice $t = $ const. Conversely, others claim "r is not the radial distance" [22, §9.7], and similarly that the isotropic coordinate r_{iso} is not radial distance. However, we can find frames within which r or r_{iso} are exactly the radial distance. Better descriptions include the $dt = 0$ slice as "the correct spatial distance in the three-dimensional space defined by the static Killing vector" [48], and that Schwarzschild coordinates "are adapted to observers at rest" [20, §7.14]. Also, while there are entire books on relativistic measurement—in theoretical physics [8, §9] [7], astrometry [23, 50], and philosophy [4]—our results fill an independent niche.

In forthcoming work, we use orthonormal frames to extend the formulae to angular momentum. What these various approaches to distance have in common is they measure within the local 3-space orthogonal to the observer. Radar is not without its own limitations [3], in fact locally the approaches here give identical results. Some suggest the coordinate dependency on r be removed, but should then the usual quantity $(1 - 2M/r)^{-1/2}dr$ be excised? There are certainly quantities defined independently of coordinates, such as the expansion tensor, or strain as predicted by gravitational waves. These are intrinsic properties, but often an external standard of reference is useful, and coordinates are literally such a "map" (chart/atlas) of spacetime. Finally, the lesson from introductory special relativity remains valid, that distance is relative to the observer's motion.

---

[4]By "ruler" we mean technically a vector orthogonal to $\mathbf{u}$ in the local tangent space, but intended as an approximation to an extended object on the manifold. This could be a hypothetical construction based on radar results, or a "resilient" physical rod [42, §2.5] whenever the *rod hypothesis* is justified.

6   Space and Time Coordinates

As is well known Schwarzschild t becomes spacelike inside the horizon, and r timelike. Hence one might question the use of a timelike coordinate to describe spatial distance in Eq. 18. This section clarifies various properties of coordinate vectors and coordinate hypersurfaces for non-diagonal line elements, and ties up some remaining questions about distance measurement.

It is common to state space and time swap roles at the horizon, but this is also rightly criticised as a Schwarzschild coordinate property: "Space and time themselves do not interchange roles: Coordinates do." [53, §3.7] Recall the nature of a coordinate Φ is based on its level sets $\Phi = $ const, a definition implied in various sources but rarely stated explicitly. We say Φ is timelike/null/spacelike when its orthogonal hypersurfaces are spacelike/null/timelike, meaning they have timelike/null/spacelike normals respectively. The gradient $d\Phi$ is one such normal 1-form, with dual $(d\Phi)^\sharp$ a normal vector. Its nature is given by the sign of the squared-norm $(d\Phi)^\sharp \cdot (d\Phi)^\sharp = d\Phi \cdot d\Phi = g^{\Phi\Phi}$, which is simply a component of the inverse metric. Table 3 summarises this for the present coordinates.

However, the nature of *coordinate vectors* is distinct, in general. These are elements of the coordinate basis and have components $(\partial_\Phi)^\mu = \delta^\mu_\Phi$ for given Φ. Hence each vector has squared norm $\partial_\Phi \cdot \partial_\Phi = g_{\Phi\Phi}$, so its nature is given by the sign of this metric component. If the metric is diagonal in a given coordinate system, then $g_{\Phi\Phi} = (g^{\Phi\Phi})^{-1}$ and so each coordinate vector has the same nature as the coordinate hypersurfaces. But in non-diagonal line elements there is room for misconception. See Table 4. If sticking to purely geometric properties, the most one can state is all Killing vector fields are spacelike inside the horizon, and in particular the Killing vector field which is timelike at infinity becomes spacelike inside the horizon.

A surprising feature of Table 4 is that ∂_r for Gullstrand–Painlevé coordinates is a distinct vector from ∂_r in Schwarzschild coordinates. This is despite the r-coordinate being identical in both cases—in the sense of a scalar field which

Table 3 The nature of coordinate hypersurfaces

Coordinate hypersurface	Squared-norm of normal	Interpretation for $r > 2M$ / $r = 2M$ / $r < 2M$
Schwarzschild t	$g^{tt} = -(1 - 2M/r)^{-1}$	Timelike/undefined/spacelike
Gullstrand–Painlevé T	$g^{TT} = -1$	Timelike everywhere
r	$g^{rr} = 1 - \frac{2M}{r}$	Spacelike/null/timelike
Lemaître ρ	$g^{\rho\rho} = \frac{e^2}{e^2 - 1 + \frac{2M}{r}}$	Spacelike everywhere
θ	$g^{\theta\theta} = \frac{1}{r^2}$	Spacelike everywhere
ϕ	$g^{\phi\phi} = \frac{1}{r^2 \sin^2\theta}$	Spacelike everywhere

The Gullstrand–Painlevé coordinate system has two timelike coordinates inside the horizon. The Lemaître coordinate system has consistent nature everywhere

Table 4 The nature of coordinate basis vectors

Coordinate vector	Squared norm	Interpretation for $r > 2M$ / $r = 2M$ / $r < 2M$
Schwarzschild ∂_t	$g_{tt} = -(1 - 2M/r)$	Timelike/null/spacelike
Gullstrand–Painlevé ∂_T	$g_{TT} = -\frac{1}{e^2}(1 - 2M/r)$	As above
Lemaître ∂_T	$g_{TT} = -1$	Timelike everywhere
Schwarzschild ∂_r	$g_{rr} = (1 - 2M/r)^{-1}$	Spacelike/null/timelike
Gullstrand–Painlevé ∂_r	$g_{rr} = \frac{1}{e^2}$	Spacelike everywhere
Lemaître ∂_ρ	$g_{\rho\rho} = \frac{1}{e^2}\left(e^2 - 1 + \frac{2M}{r}\right)$	Spacelike everywhere

For Gullstrand–Painlevé coordinates all basis vectors are spacelike inside the horizon. Omitted are ∂_θ and ∂_ϕ, which have the same nature as the gradients $d\theta$ and $d\phi$ above, because all our line elements are diagonal in θ and ϕ.

matches on chart overlaps, and both having the same components $(0, 1, 0, 0)$ in their respective systems. The standard vector transformation law yields:

$$\partial_r^{(GP)} = \pm\frac{1}{e}\left(1 - \frac{2M}{r}\right)^{-1}\sqrt{e^2 - 1 + \frac{2M}{r}}\,\partial_t^{(Schw)} + \partial_r^{(Schw)} \tag{23}$$

There are few mentions of this potential error-causing subtlety in the literature [13, §1] [2], although there are related comments about each basis vector depending on *every* element of the dual basis [47, §3.3], or ambiguities with partial derivative notation. Recall a basis and its dual are related by

$$dx^\mu(\partial_\nu) = \delta^\mu_\nu. \tag{24}$$

A coordinate dual basis element $d\Phi$ depends only on the coordinate Φ, since it is related to the level sets $\Phi = \mathrm{const}$. This is not the case for the vector ∂_Φ, as without specifying the accompanying coordinates all one can say is $d\Phi(\partial_\Phi) = 1$. Note duality of bases is distinct from duality of individual vectors: $(d\Phi)^\sharp \neq \partial_\Phi$ in general. We could instead define a unique coordinate vector depending only on Φ:

$$\partial_\Phi^{(unique)} := \frac{(d\Phi)^\sharp}{d\Phi \cdot d\Phi} \tag{25}$$

which is orthogonal to the level sets $\Phi = \mathrm{const}$ and satisfies $d\Phi(\partial_\Phi^{(unique)}) = 1$. For diagonal line elements it coincides with the default coordinate vector. However, in our context the flexibility of ∂_r is actually helpful! Both spatial measurement and Einstein simultaneity require vectors orthogonal to an observer. From Eq. 24, $\partial_r^{(GP)}$ is orthogonal to dT, so this vector lies in the 3-space of a falling observer. Similarly $\partial_r^{(Schw)}$ is orthogonal to dt, so it lies in the 3-space of a static observer. This justifies the naive contractions in Sect. 5. Our well-suited coordinates glossed over this requirement automatically. This also partly answers the earlier question:

even though r is timelike inside the horizon, the measurement direction $\partial_r^{(GP)}$ is spacelike.

So why does the proper distance $dr/|e|$ diverge as $e \to 0$? This is because the r-coordinate, which is timelike for $r < 2M$, is *purely* timelike to the zero Killing energy observers, meaning $(dr)^\sharp$ is parallel to the 4-velocity **u**. In an observer's frame, distances are measured along vectors **v** orthogonal to **u**, but for $e = 0$ observers the gradient of r is zero in all these directions: $dr(\mathbf{v}) = 0$, so cannot describe any spatial measurement. This situation is much less exotic than it appears. The same thing occurs for static observers outside the horizon, for whom the t-coordinate is purely timelike so cannot describe space. In fact this situation occurs for spacelike coordinates also, if they are orthogonal to the ruler direction: none of our observers can describe *radial* distance in terms of the θ or ϕ-coordinates! Returning to the $e = 0$ observers, one can set $dr = 0$ in Eq. 1 to obtain the radial proper distance $\sqrt{2M/r - 1}\, dt$ in terms of the t-coordinate gradient.

Our proper distance expressions are only defined along the 1-dimensional ruler direction. If $dr/|e|$ were instead interpreted as a 1-form on an entire 4-dimensional tangent space, indeed this would be a timelike 1-form inside the horizon. This would mean the vector $(dr/|e|)^\sharp$, the direction of steepest gradient, is timelike. But when restricted to the 1-dimensional subspace of the tangent space parallel to the ruler, $dr/|e|$ and its dual are spatial.

7   Simultaneity, and Time at Infinity

Schwarzschild t is called the "time at infinity". At infinity this is unambiguously true, as the proper time for a static observer is $d\tau = \sqrt{1 - 2M/r}\, dt \to dt$ as $r \to \infty$. However, at events with $r < \infty$, the interpretation of their t-coordinate as the time at infinity *right now* makes an implicit simultaneity assumption. This is justified as one of the most natural choices, based on the static Killing vector field. But the falling observers lead to a different convention based on the hypersurfaces orthogonal to the congruence. Under this alternative, the time passing at infinity is only *finite* for a falling test particle to cross $r = 2M$. This goes beyond the usual statement that the falling particle's proper time is finite, to a simultaneity convention extended across spacetime. This alternate convention is conceptually motivated, helps avoid common misconceptions, and may have practical results for calculations which depend on foliation.

Recall in general relativity, any spatial hypersurface can be considered a "simultaneous" instant of time. Some authors allow the hypersurface to be null or even go timelike in places. In symmetric and physically relevant spacetimes, one would expect to pick out natural choices from amongst the infinitude of options. For example in static spacetimes, the hypersurfaces orthogonal to the timelike Killing vector make a natural choice of simultaneity. In spacetimes with a vorticity-

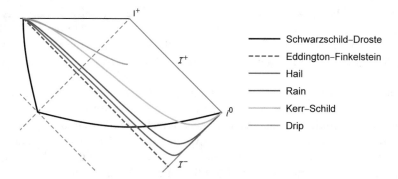

Fig. 4 Simultaneity conventions defined by $x^0 = $ const in various coordinate systems, in the physical regions I and II. Each line shows the events considered simultaneous with a given event near $r = 0$. The rain, hail, and drip hypersurfaces are orthogonal to the worldlines in Fig. 1, and for $r > 2M$ the Schwarzschild choice is orthogonal to static observers. Those simultaneity choices reach to spatial infinity, but with different angles of approach. The drip hypersurface extends only to r_{max}, and the Eddington–Finkelstein coordinate is null

free congruence, the hypersurfaces orthogonal to it make a natural choice, as is the approach in cosmology. (Orthogonality to the observer is a consequence of Einstein–Poincaré simultaneity.)

Figure 4 shows some possible conventions for Schwarzschild. Note the drip, rain, and hail simultaneity curves appear from top to bottom in that order. This means the events considered simultaneous to the given $r \approx 0$ event are earlier for hail than for rain, for example. Conversely, for a fixed location r_0 say, simultaneous events at $r < r_0$ are later for hail than rain, so events at $r < r_0$ occur earlier under hail simultaneity; vice versa for $r > r_0$. This is consistent with simultaneity in special relativity, as applied to each local frame. Recall that for a Lorentz-boosted frame, events ahead of them (in the 3-direction of the boost) occur earlier than they do for the un-boosted frame, and vice versa. These statements of causal order are not absolute, even though we seek natural choices. The only absolute standard of causality is based on the light cone.

In Schwarzschild, consider a radially falling test particle. Its t-coordinate diverges as it passes the horizon, then in the black hole interior t *decreases* towards the future, at least for $e > 0$ (Eq. 3). Historically this led to much confusion, but modern pedagogy stresses the proper time is finite and the divergence at $r = 2M$ is merely a Schwarzschild coordinate issue. However, it is still said t is the time at infinity, rarely with any qualifiers. But how would this view deal with the old misconception that time runs backwards in the black hole interior? One might then limit the interpretation of t as time at infinity to region I. But the pathology at $r \rightarrow 2M$ remains: would a distant observer say the particle freezes at the horizon, or that black holes never form? It seems observers at infinity still hold to pre-1960s views. Note we do not mean the visual appearance of the observer, which fades exponentially as is well known [35, §32.3], as simultaneity is not determined only by when null signals reach an observer, but also accounts for travel time of the signals [46]. Note also t does not coincide with the arrival time of photons at distant r, which is described by the outgoing Eddington–Finkelstein null coordinate which has only qualitatively similar behaviour to t for the falling particle.

There is a *local* physical interpretation for t. For a static observer, it is their proper time divided by the redshift factor $\sqrt{\boldsymbol{\xi} \cdot \boldsymbol{\xi}}$ to compensate for gravitational time-dilation. Relying on the time-symmetry of their worldlines through spacetime, static observers can determine their relative time-dilation factors by signalling one another. We can even find a physical interpretation of t for the falling particle. Consider a line of static observers along the faller's worldline, where each records the time interval for the faller to pass them. The particle crosses an interval dr in proper time dr/u^r, but according to the local static frame the faller is time-dilated so the actual time passed is increased by the Lorentz factor between the frames (Eq. 22). Now if the static observer were to further increase this quantity to compensate for their gravitational time-dilation as above, the resulting "time at infinity" would be $dr \cdot u^t/u^r$ which is dt. Note other frames give different results, and this interpretation relies on static frames which are impossible for $r \leq 2M$ and physically unreasonable at $2M + \epsilon$ due to extreme acceleration. Compare the discussion of Eq. 21.

Simultaneity defined from the faller congruence is more physically realistic, and extends inside the horizon. Locally, a falling test particle has coordinate time matching its proper time: $dT = d\tau$. Under the new simultaneity convention, the same coordinate interval dT passes everywhere. At infinity (which requires $e \geq 1$), this corresponds to a proper time dT/e for a static observer. Hence we conclude the time at infinity is T/e. In particular, the time at infinity for the particle to cross the horizon is finite! This avoids misconceptions such as the particle freezing at the horizon, black holes never forming, having unlimited time to fly down and rescue a falling astronaut, and so on. The $e = 1$ convention is especially natural.

The $e = 0$ congruence leads to qualitatively different simultaneity. The orthogonal hypersurfaces $r = $ const are the infinite 3-cylinders mentioned in Sect. 4, except for $r = 2M$ as Fig. 5 shows. Incidentally, one might wonder what the "radial" direction is for these observers. The remaining spatial direction is orthogonal to ∂_θ and ∂_ϕ, hence must be ∂_t. However, in the 3-dimensional space, ∂_t points along the axis of the cylinder $\mathbb{R} \times S^2$, so it is better conceived as a translation vector not radial. Likewise in 4 dimensions one thinks of ∂_t as translational, at least in regions I and III where it is translation in time.

Typical pedagogy can leave the impression of a single global time slicing, in certain aspects. An exception is Frolov & Novikov [15, §3.2.2], who make clear the existence of different choices. They advocate the usual t-slices, rejecting the "Lemaître frame" ($e = 1$) because it is not a rigid congruence. However, one would not expect rigidity for a freefalling congruence, as tidal forces are a sign of gravity.

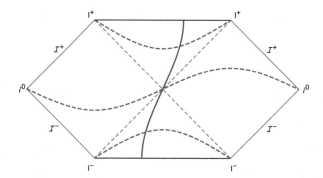

Fig. 5 Simultaneity for the zero Killing energy congruence. A representative worldline $t = $ const is shown (blue), and orthogonal hypersurfaces (red) at three selected events. The upper and lower curves are $r = $ const. At the bifurcate horizon, the orthogonal hypersurface leaves the congruence but can be extended geodesically into the exterior regions, where it concurs with static simultaneity

8 Conclusion

The coordinates and foliation induced by timelike radial geodesic observers provide an insightful contrast to Schwarzschild coordinates and the static foliation. We examined this contrast for the curvature of 3-dimensional space and its embedding diagram, measurement of radial proper distance, simultaneity and the time at infinity, as well as other areas. A special case of these radial geodesics and their accompanying foliation has already proved useful for Hawking radiation and analogue gravity, and we anticipate further applications in these areas in the general case.

References

1. F. Belgiorno, S. Cacciatori, D. Faccio, *Hawking Radiation: from Astrophysical Black Holes to Analogous Systems in Lab* (World Scientific, Singapore, 2018)
2. D. Bini, A. Geralico, R.T. Jantzen, Gen. Relativ. Gravit. **44**, 603 (2012)
3. D. Bini, L. Lusanna, B. Mashhoon, Int. J. Mod. Phys. D **14**, 1413 (2005)
4. H.R. Brown, Physical relativity. Space-time structure from a dynamical perspective (Oxford University Press, Oxford, 2005)
5. B. Carter, in *Black Holes (Les Astres Occlus)*, ed. by C. DeWitt, B.S. DeWitt (Gordon & Breach Science, New York, 1973), pp. 57–214
6. S. Chandrasekhar, *The Mathematical Theory of Black Holes* (Springer, Dordrecht, 1983)
7. F. de Felice, D. Bini, *Classical Measurements in Curved Space-Times* (Cambridge University Press, Cambridge, 2010)
8. F. de Felice, C.J.S. Clarke, *Relativity on Curved Manifolds*. (Cambridge University Press, Cambridge, 1990)
9. J. Ehlers, Gen. Relativ. Gravit. **25**, 1225 (1993)
10. J. Eisenstaedt, *Einstein and the History of General Relativity*, ed. by D. Howard, J. Stachel (1989), pp. 277–292

11. G.F.R. Ellis, R. Maartens, M.A.H. MacCallum *Relativistic Cosmology* (Cambridge University Press, Cambridge, 2012)
12. F. Estabrook, H. Wahlquist, S. Christensen, B. Dewitt, L. Smarr, E. Tsiang, Phys. Rev. D **7**, 2814 (1973)
13. T.K. Finch, Gen. Relativ. Gravit. **47**, 56 (2015)
14. L. Flamm, Gen. Relativ. Gravit. **47**, 72 (2015)
15. V.P. Frolov, I.D. Novikov, *Black Hole Physics : Basic Concepts and New Developments* (Springer, Berlin, 1998)
16. R. Gautreau, B. Hoffmann, Phys. Rev. D **17**, 2552 (1978)
17. E. Gourgoulhon, *3+1 Formalism in General Relativity* (Springer, Berlin, 2012)
18. A. Gullstrand, *Allgemeine lösung des statischen einkörperproblems in der Einsteinschen gravitationstheorie* (Almqvist & Wiksell, Stockholm, 1922)
19. Y. Hagihara, *Celestial Mechanics. Vol. 1: Dynamical Principles and Transformation Theory* (MIT Press, London, 1970)
20. A.J.S. Hamilton, *General Relativity, Black Holes, and Cosmology* (Oxford University Press, Oxford, 2015)
21. S.W. Hawking, Commun. Math. Phys. **43**, 199 (1975)
22. M.P. Hobson, G.P. Efstathiou, A.N. Lasenby, *General Relativity* (Cambridge University Press, Cambridge, 2006)
23. S. Kopeikin, M. Efroimsky, G. Kaplan, *Relativistic Celestial Mechanics of the Solar System* (Wiley, Hoboken, 2011)
24. P. Kraus, F. Wilczek, Some applications of a simple stationary line element for the Schwarzschild geometry. Mod. Phys. Lett. A **9**(40), 3713–3719 (1994)
25. L. Landau, E. Lifshitz, *Field Theory* (GITTL, Moscow, 1941)
26. G. Lemaître, Publication du Laboratoire d'Astronomie et de Géodésie de l'Université de Louvain **9**, 171–205 (1932)
27. S. Liberati, G. Tricella, M. Visser, Class. Quantum Grav. **35**, 155004 (2018)
28. H.-C. Lin, C. Soo, Gen. Relativ. Gravit. **45**, 79 (2013)
29. D.A. Lowe, J. Polchinski, L. Susskind, L. Thorlacius, J. Uglum, Phys. Rev. D **52**, 6997 (1995)
30. C. MacLaurin (2018), Mimicking a black hole in flat spacetime. ColinsCosmos.com
31. D.B. Malament, *Philosophy of Physics. Part A.* Handbook of the Philosophy of Science (Elsevier, Amsterdam, 2006), pp. 229
32. D. Marolf, Gen. Relativ. Gravit. **31**, 919 (1999)
33. K. Martel, E. Poisson, Am. J. Phys. **69**, 476 (2001)
34. S.D. Mathur, What exactly is the information paradox?, in *Physics of Black Holes*, vol. 3 (Springer, Berlin, 2009)
35. C.W. Misner, K.S. Thorne, J.A. Wheeler, Gravitation (W.H. Freeman and Co., New York, 1973)
36. T. Moore, *A General Relativity Workbook* (University Science Books, Sausalito, 2012)
37. T. Mueller, F. Grave, Catalogue of spacetimes (2010). arXiv: 0904.4184
38. B. O'Neill, *The Geometry of Kerr Black Holes* (Courier Corporation, North Chelmsford, 1995)
39. P. Painlevé, CR Acad. Sci. Paris (serie non specifiee) **173**, 677 (1921)
40. M.K. Parikh, F. Wilczek, Phys. Rev. Lett. **85**, 5042 (2000)
41. E. Poisson, A relativist's toolkit : the mathematics of black-hole mechanics (Cambridge University Press, Cambridge, 2004)
42. W. Rindler, *Essential Relativity. Special, General, and Cosmological.* (Springer, Berlin, 1977)
43. W. Rindler, *Relativity: Special, General and Cosmological*, 2nd edn. (Springer, New York, 2006)
44. K. Rosquist, Gen. Relativ. Gravit. **41**, 2619 (2009)
45. R.K. Sachs, H.-H. Wu, *General Relativity for Mathematicians* (Springer, New York, 1977), pp. 52–59
46. R.E. Scherr, P.S. Shaffer, S. Vokos, Am. J. Phys. **70**, 1238 (2002)
47. B. Schutz, *A First Course in General Relativity* (Cambridge University Press, Cambridge, 2009)

48. J.M.M. Senovilla, Gen. Relativ. Gravit. **39**, 685 (2007)
49. L. Smarr, J.W. York, Phys. Rev. D **17**, 2529 (1978)
50. M.H. Soffel, *Relativity in Astrometry, Celestial Mechanics and Geodesy* (Springer, Berlin, 1989)
51. J.L. Synge, Proc. Lond. Math. Soc. 2 **43**, 376 (1937)
52. J.L. Synge, *Relativity: The General Theory* (North-Holland, Amsterdam, 1960)
53. E.F. Taylor, J.A. Wheeler, *Exploring Black Holes: Introduction to General Relativity* (Pearson, London, 2000)
54. C. Vilain, *Studies in the History of General Relativity* **3**, 419 (1992)
55. R.M. Wald, *General Relativity* (University of Chicago Press, Chicago, 1984)
56. R.M. Wald, *Quantum Field Theory in Curved Spacetime and Black Hole Thermodynamics* (University of Chicago Press, Chicago, 1994)

Crystal Spacetimes with Discrete Translational Symmetry

Jiří Ryzner and Martin Žofka

Abstract The aim of this work is to construct exact solutions of Einstein–Maxwell(-dilaton) equations possessing a discrete translational symmetry. We present two approaches to the problem. The first one is to solve Einstein–Maxwell equations in 4D, the second one relies on dimensional reduction from 5D. We examine the geometry of the solutions, their horizons and singularities and compare them.

Keywords Majumdar–Papapetrou · Extreme black hole · Einstein–Maxwell solutions · Scalar field · Discrete symmetry

Mathematics Subject Classification (2000) 83C15, 83C22, 83C57

1 Introduction

The multiple extremal[1] black hole solutions first appeared in the 1940s due to the works of Majumdar [1] and Papapetrou [2], yet their interpretation had to wait until the 1970s and the classical paper by Hartle with Hawking [3]. These solutions are a manifestation of the curious fact that the balance between electrostatic and gravitational forces in classical physics is preserved even in general relativity. This extends even to the linearity of the field equations and thus the principle

---

Grants GAUK 80918, GACR 14-37086G.

[1]The term extremal refers to the fact that the charges of the black holes are equal to their masses, rendering their horizons degenerate.

J. Ryzner (✉) · M. Žofka
Institute of Theoretical Physics, Charles University, Prague, Czech Republic
e-mail: zofka@mbox.troja.mff.cuni.cz

© Springer Nature Switzerland AG 2019
S. Cacciatori et al. (eds.), *Einstein Equations: Physical and Mathematical Aspects of General Relativity*, Tutorials, Schools, and Workshops in the Mathematical Sciences, https://doi.org/10.1007/978-3-030-18061-4_10

of superposition holds in GR as well. In our previous works on the subject [4] and [5], we were interested in solutions exhibiting axial and cylindrical symmetry, respectively. The cylindrically symmetric solution was the field of an infinitely extended extremally charged string located along the axis of symmetry. One of the questions arising from the paper was whether it is possible to produce the same field asymptotically far from the axis of axial symmetry while the sources would be an infinite number of isolated, equidistantly distributed, identical point sources of a mass equal to their charge. The interesting feature of the resulting spacetime would be that it has a discrete translational symmetry everywhere while asymptotically it would be fully cylindrically symmetric. One could then study, for instance, various averaging techniques and whether they reproduce the correct asymptotic symmetry. We study here several possible approaches to constructing such a solution and study its properties.

The paper is organized as follows: In the first chapter we review the M-P solution in an arbitrary dimension. In Chap. "Lectures on Linear Stability of Rotating Black Holes" we look at the first example which is an infinite "crystal" constructed of alternating positive and negative charges. Since the solution is in the form of an infinite series of functions we investigate its convergence, asymptotics, and derivatives to be able to infer the symmetries of the spacetime and its interpretation. In Chap. "The Bianchi Classification of the Three-Dimensional Lie Algebras and Homogeneous Cosmologies and the Mixmaster Universe" we study the case of a uniform crystal composed of identical charges. Unlike in the previous case, the solution is more subtle as it diverges far from the axis of symmetry. For both the alternating and uniform crystal we plot the equipotential and singular surfaces of the two spacetimes. Chapter "The Physics of LIGO–Virgo" presents a way of constructing the 4D infinite crystal out of a closed-form 5D solution. In the final Chap. "Generation of Initial Data for General-Relativistic Simulations of Charged Black Holes" we apply this procedure to a uniform crystal and study the properties of the resulting spacetime. We conclude with some final remarks and open questions.

1.1 Majumdar–Papapetrou Solution

The metric $^D g$ of the Majumdar–Papapetrou solution in arbitrary dimension $D = n + 1, n \geq 3$,[2] reads [6]

$$^D g = -U^{-2}\mathrm{d}t^2 + {}^n h_{ij}\mathrm{d}x^i\mathrm{d}x^j, \tag{1.1}$$

[2]The number of space-like dimensions is denoted as n.

where t is a time-like Killing coordinate, so that the metric is static with the function $U = U(x^i)$ only depending on Cartesian-like spatial coordinates x^i.[3] The spatial metric $^n h_{ij}$ is conformally flat

$$^n h = U^{\frac{2}{n-2}} \cdot {}^n \delta_{ij} dx^i dx^j. \tag{1.2}$$

These coordinates describe well the region above the horizons. The electromagnetic potential A and the electromagnetic field tensor F read

$$A = c_n \frac{dt}{U}, \quad F = dA = -c_n \sum_{i=1}^{n} \frac{U_{,i}}{U^2} dx^i \wedge dt \tag{1.3}$$

with $c_n = \sqrt{\frac{n-1}{2(n-2)}}$. The corresponding stress-energy tensor, T, is

$$T^{\mu\nu} = \frac{1}{4\pi} \left(F^\mu_{\ \beta} F^{\nu\beta} - \frac{\mathcal{F}}{4} g_{\mu\nu} \right), \tag{1.4}$$

where

$$\mathcal{F} = F_{\mu\nu} F^{\mu\nu} = c_n^2 (n-2)^2 \sum_{i=1}^{n} \left[\frac{\partial \left(U^{\frac{-1}{(n-2)}} \right)}{\partial x^i} \right]^2 \tag{1.5}$$

is the Maxwell scalar. The non-vanishing components of the stress-energy tensor are

$$16\pi T_0^{\ 0} = \mathcal{F}, \quad 4\pi T_i^{\ j} = -c_n^2 U^{\frac{2-2n}{n-2}} U_{,i} U_{,j} - \frac{\mathcal{F}}{4} \delta_i^{\ j}. \tag{1.6}$$

Einstein and Maxwell equations then have the form

$$\text{Ric}_{\mu\nu} - \frac{R}{2} g_{\mu\nu} = 8\pi T_{\mu\nu}, \quad \nabla_\nu F^{\mu\nu} = 4\pi J^\mu, \quad \nabla_\mu J^\mu = 0. \tag{1.7}$$

The 4-current J^μ due to the charge density $\rho(x)$ leads to a single Einstein–Maxwell equation

$$J^\mu = -c_n \frac{\rho(x)}{\sqrt{-g}} \delta_0^\mu \Rightarrow \Delta_\delta U = \sum_{i=1}^{n} U_{,ii} = 4\pi \rho(x). \tag{1.8}$$

---

[3]Latin indices range over $1, \ldots, n$ and label only spatial components, Greek indices are $0, \ldots, n$.

Here, $\mathfrak{g}$ is the determinant of the metric $g_{\mu\nu}$ and Δ_δ denotes the flat-space Laplacian.[4] In case of Majumdar–Papapetrou, $\rho(x)$ is assumed to be a distribution of point charges which means that $J^\mu = 0$ away from the sources. One particular solution, in which we are interested, is a multi-black hole spacetime of the form

$$U(x) = 1 + \sum_{i=1}^{N} \frac{M_i}{r_i^{n-2}}, \quad r_i^2 = \sum_{a=1}^{n} (x^a - x_i^a)^2, \tag{1.9}$$

with the corresponding charge current [6]

$$\sqrt{-\mathfrak{g}} J^0 = -\frac{c_n}{4\pi} \Delta_\delta U = \frac{c_n \pi^{\frac{n}{2}-1}}{\Gamma\left(\frac{n}{2}-1\right)} \sum_{i=1}^{N} M_i \cdot {}^n\delta(x - x_i). \tag{1.10}$$

Here M_i are constants, Γ is the gamma function and ${}^n\delta$ is the n-dimensional Dirac delta function. It can be shown that M_i is the mass and also charge of each black hole and for $M_i > 0$ the puncture located at $r_i = 0$ looks like a point, but in fact it represents the surface of a sphere $\mathbb{S}^{n-1}$ of dimension $n - 1$ (and for $M_i < 0$ the surface $r_i = 0$ corresponds to the location of a naked singularity). In $D = 4$ there exists a coordinate transformation, which regularizes the metric at a (arbitrarily chosen) horizon $r_i = 0$ and the horizon is smooth [3]. However, in $D > 4$ this holds only for a single black hole ($N = 1$). For $N = 2, 3$ it was shown that the horizon is not smooth [7] while for a higher number of black holes the situation is still unclear.

2 Alternating Crystal

Working in the usual four-dimensional setting, we now investigate the construction of a solution with an infinite number of punctures distributed equidistantly along an axis with a separation constant k. Due to the alignment of the punctures we call these solutions a "crystal", as it resembles an infinite one-dimensional crystallographic structure. We first introduce an alternating crystal, where neighbouring punctures have opposite signs. We use cylindrical coordinates in which the metric and potential read

$$g = -U^{-2}dt^2 + U^2\left(d\rho^2 + \rho^2 d\phi^2 + dz^2\right), \; A = \frac{dt}{U}, \, U = 1 + \frac{Q}{k}\chi, \tag{2.1}$$

where χ satisfies Laplace's equation (1.8) away from the sources:

$$\chi_{,\rho\rho} + \frac{\chi_{,\rho}}{\rho} + \chi_{,zz} = 0. \tag{2.2}$$

---

[4]The Laplacian for the spatial metric h is defined as $\Delta_h f \equiv h^{ij}\nabla_i\nabla_j f = \frac{1}{\sqrt{\mathfrak{h}}}\left(\sqrt{\mathfrak{h}}h^{ij}f_{,i}\right)_{,j}$.

We assume χ to be of the following form:

$$\chi = \sum_{n=-\infty}^{\infty} (-1)^n \hat{\chi}_n, \ \hat{\chi}_n = \frac{1}{r_n}, \ r_n = \sqrt{\rho^2 + (z-n)^2}, \tag{2.3}$$

where r_n is the coordinate distance from the puncture located at $z = n$.

2.1 Convergence of the Potential

We shall prove that the sum and its second derivatives converge uniformly for

$$\mathcal{R} = \{0 \le \rho, 0 \le z \le 1/2, 0 \le \phi \le 2\pi\}. \tag{2.4}$$

Uniform convergence is crucial since derivatives then commute with the sums and it enables us to prove that the resulting potential satisfies Laplace's equation. Using the symmetries of the potential, we shall extend the definition of the potential for any ρ and z, obtaining thus a full solution of the Einstein–Maxwell equations.

Inspecting the Potential

It is convenient to rewrite the potential χ as

$$\chi \equiv \chi_0 + \chi_{\neg 0}, \ \chi_{\neg 0} \equiv \sum_{n=1}^{\infty} (-1)^n \chi_n, \ \chi_n \equiv \hat{\chi}_n + \hat{\chi}_{-n}, \ \chi_0 \equiv \hat{\chi}_0. \tag{2.5}$$

We now need to find "good" bounds for χ. Luckily, it is easy to check that the terms are decreasing with n as follows:

$$|\chi_n| \le \frac{4n-1}{n(2n-1)}, \ \frac{\partial \chi_n}{\partial n} = \frac{z-n}{r_{-n}^3} - \frac{z+n}{r_n^3} < 0 \text{ for } \rho \in \mathcal{R}. \tag{2.6}$$

Then it simply follows that[5] $\chi_n \rightrightarrows 0$ and thanks to the Leibniz theorem,[6] the sum $\sum_{n=1}^{\infty} \chi_n$ converges uniformly for $\rho \in \mathcal{R}$.

---

[5] Let $f_k(x) : I \to \mathbb{C}$ be functions and assume that there exist a_k such that $|f_k(x) - f(x)| \le a_k$ for $\forall x \in I$ and $a_k \to 0$ when $k \to \infty$. Then $f_k(x) \rightrightarrows f(x)$ in I.
[6] Leibniz theorem: Let $f_k(x) : I \to \mathbb{C}$ such that $f_k \rightrightarrows 0$ in I and let $f_k(x)$ be monotonous for $\forall x \in I$. Then $\sum(-1)^k f_k(x)$ converges uniformly in I.

Inspecting the First Derivatives

The ρ derivative of χ_n reads

$$\frac{\partial \chi_n}{\partial \rho} = -\frac{\rho}{r_n^3} - \frac{\rho}{r_{-n}^3}. \tag{2.7}$$

The bounds on the respective terms are

$$0 \le \frac{\rho}{r_n^3} \le \frac{8}{3\sqrt{3}(1-2n)^2}, \, 0 \le \frac{\rho}{r_{-n}^3} \le \frac{2}{3\sqrt{3}n^2}, \, \left|\frac{\partial \chi_n}{\partial \rho}\right| \le \frac{10}{3\sqrt{3}n^2}. \tag{2.8}$$

So according to Weierstrass M-test,[7] $\sum \chi_{n,\rho}$ converges absolutely uniformly for $\rho \in \mathcal{R}$. We also get that

$$\frac{\partial \chi_{\neg 0}}{\partial \rho} = \frac{\partial}{\partial \rho} \sum_{n=1}^{\infty} (-1)^n \chi_n = \sum_{n=1}^{\infty} (-1)^n \frac{\partial \chi_n}{\partial \rho} \text{ for } \rho \in \mathcal{R}. \tag{2.9}$$

In Laplace's equation, we have the term $\chi_{n,\rho}/\rho$. But it behaves also well, because

$$\left|\frac{1}{\rho} \frac{\partial \chi_n}{\partial \rho}\right| = \left|-\frac{1}{r_n^3} - \frac{1}{r_{-n}^3}\right| \sim \frac{1}{n^3}. \tag{2.10}$$

The z derivative of χ_n reads

$$\frac{\partial \chi_n}{\partial z} = \frac{n-z}{r_n^3} - \frac{n+z}{r_{-n}^3}. \tag{2.11}$$

The derivative is bounded as follows:

$$0 \le \frac{n+z}{r_{-n}^3} \le \frac{2n+1}{2n^3}, \, 0 \le \frac{n-z}{r_n^3} \le \frac{n+3}{n^3}, \, \left|\frac{\partial \chi_n}{\partial z}\right| \le \frac{4n+7}{n^3}. \tag{2.12}$$

Again, we found that $\sum \chi_{n,z}$ converges absolutely uniformly for $\rho \in \mathcal{R}$ and

$$\frac{\partial \chi_{\neg 0}}{\partial z} = \frac{\partial}{\partial z} \sum_{n=1}^{\infty} (-1)^n \chi_n = \sum_{n=1}^{\infty} (-1)^n \frac{\partial \chi_n}{\partial z} \text{ for } \rho \in \mathcal{R}. \tag{2.13}$$

---

[7] Weierstrass M-test: Let $f_k(x) : I \to \mathbb{C}$ be functions and assume that there exist a_k such that $|f_k(x)| \le a_k$ and $\left|\sum a_k\right| < \infty$. Then $\sum f_k(x)$ converges absolutely uniformly in I.

Inspecting the Second Derivatives

The second ρ derivative of χ_n reads

$$\frac{\partial^2 \chi_n}{\partial \rho^2} = \frac{3\rho^2}{r_n^5} + \frac{3\rho^2}{r_{-n}^5} - \frac{1}{r_n^3} - \frac{1}{r_{-n}^3}. \tag{2.14}$$

The bounds are

$$0 \le \frac{3\rho^2}{r_n^5} \le \frac{5}{n^3}, 0 \le \frac{3\rho^2}{r_{-n}^5} \le \frac{1}{n^3}, 0 \le \frac{1}{r_n^3} \le \frac{8}{n^3}, 0 \le \frac{1}{r_{-n}^3} \le \frac{1}{n^3}. \tag{2.15}$$

Combining the terms, we get that $\chi_{n,\rho\rho} \rightrightarrows 0$ and thus its sum converges uniformly:

$$\left| \frac{\partial^2 \chi_n}{\partial \rho^2} \right| \le \frac{15}{n^3} \Rightarrow \sum \chi_{n,\rho\rho} \rightrightarrows \chi_{\neg 0,\rho\rho} \text{ for } \rho \in \mathcal{R}. \tag{2.16}$$

The second z derivative of χ_n reads

$$\frac{\partial^2 \chi_n}{\partial z^2} = \frac{3(n-z)^2}{r_n^5} - \frac{1}{r_n^3} - \frac{1}{r_{-n}^3} + \frac{3(n+z)^2}{r_{-n}^5}. \tag{2.17}$$

Combining the terms, we get that $\chi_{n,zz} \rightrightarrows 0$ and its sum converges uniformly:

$$\left| \frac{\partial^2 \chi_n}{\partial z^2} \right| \le \frac{36}{n^3} \Rightarrow \sum \chi_{n,zz} \rightrightarrows \chi_{\neg 0,zz} \text{ for } \rho \in \mathcal{R}. \tag{2.18}$$

Putting It All Together

We have proved that the infinite sum and its first and second derivatives converge for $\rho \in \mathcal{R}$. Now it is time to extend the definition of the potential for all values of z. First we notice the mirror symmetry:

$$\chi_n(\rho, z) = \chi_n(\rho, -z) \Rightarrow \chi(\rho, z) = \chi(\rho, -z). \tag{2.19}$$

Using this we can calculate the potential for $-1/2 \le z \le 1/2$. The final step is to show (anti-)periodicity in z.

$$\chi(\rho, z+1) = \hat{\chi}_{-1}(\rho, z) + \sum_{n=1}^{\infty} (-1)^n \left(\hat{\chi}_{n-1}(\rho, z) + \hat{\chi}_{-n-1}(\rho, z) \right) = \tag{2.20}$$

$$= \hat{\chi}_{-1}(\rho, z) - \hat{\chi}_0(\rho, z) + \hat{\chi}_1(\rho, z) - \sum_{l=2}^{\infty} (-1)^l \chi_l(\rho, z) = -\chi(\rho, z).$$

We see that the potential has an anti-period 1 and a period 2 in z. Based on the periodicity and mirror symmetry, we define the potential for all values of z. Using the results so far, we can exchange the Laplacian with infinite sum:

$$\Delta \chi = \Delta \chi_0 + \sum_{n=1}^{\infty} (-1)^n \Delta \chi_n. \tag{2.21}$$

We thus proved that χ is the solution of Laplace's equation with the charge density

$$\rho(x, y, z) = \sum_{n=-\infty}^{\infty} (-1)^n \cdot {}^3\delta(x, y, z - n). \tag{2.22}$$

Plots of χ are shown in Fig. 1a, b.

Bounds for $\rho \to \infty$

Since the sums of $\chi_{n,\rho}$ converge absolutely and uniformly, we can split the derivative of $\chi_{\neg 0}$ into two sums:

$$\frac{\partial \chi_{\neg 0}}{\partial \rho} = \sum_{n=1}^{\infty} \frac{\partial \chi_{2n}}{\partial \rho} - \sum_{n=1}^{\infty} \frac{\partial \chi_{2n-1}}{\partial \rho}. \tag{2.23}$$

The derivatives are negative and increasing:

$$\frac{\partial \chi_n}{\partial \rho} < 0, \quad \frac{\partial^2 \chi_n}{\partial \rho \partial n} > 0 \text{ for } \rho \in \mathcal{R}. \tag{2.24}$$

Using integral estimates, we get

$$\frac{\partial \chi_1}{\partial \rho} + \int_1^{\infty} \frac{\partial \chi_{2n-1}}{\partial \rho} dn \leq \sum_{n=1}^{\infty} \frac{\partial \chi_{2n-1}}{\partial \rho} \leq \int_1^{\infty} \frac{\partial \chi_{2n-1}}{\partial \rho} dn, \tag{2.25}$$

$$\frac{\partial \chi_2}{\partial \rho} + \int_1^{\infty} \frac{\partial \chi_{2n}}{\partial \rho} dn \leq \sum_{n=1}^{\infty} \frac{\partial \chi_{2n}}{\partial \rho} \leq \int_1^{\infty} \frac{\partial \chi_{2n}}{\partial \rho} dn, \tag{2.26}$$

and the following bounds:

$$-\frac{2}{\rho^2} + \frac{6z^2 + 17}{2\rho^4} + O\left(\frac{1}{\rho^5}\right) \leq \frac{\partial \chi}{\partial \rho} \leq \frac{2}{\rho^2} - \frac{6z^2 + 13}{2\rho^4} + O\left(\frac{1}{\rho^5}\right). \tag{2.27}$$

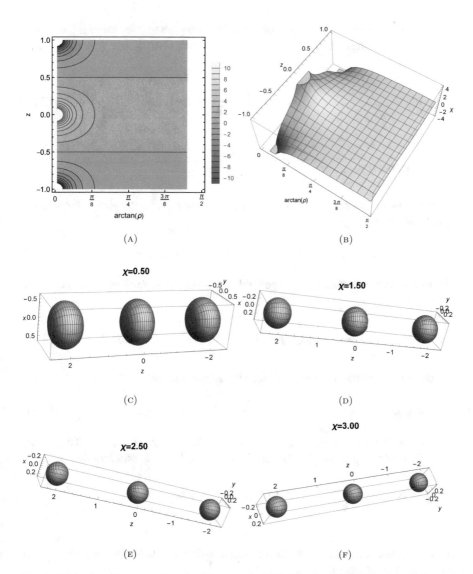

Fig. 1 (**a**) Conformal contour plot of χ; (**b**) conformal 3D plot of χ; (**c**)–(**f**) 3D plots of singular surfaces $\chi = -k/Q$ for various values of Q/k. The surface is always disconnected and always encloses naked singularity

We thus conclude that $\chi_{,\rho} \sim \rho^{-2}$ or faster. We also see that for large ρ the dependence on z vanishes. In fact, numerical calculation reveals that $\chi \sim ce^{-\pi\rho}/\sqrt{\rho}$, which implies

$$U = 1 + c\frac{Q}{k}\frac{e^{-\pi\rho}}{\sqrt{\rho}}, \rho \gg 1. \tag{2.28}$$

2.2 Geometry

We can now investigate the geometry of the solution. We have shown the translational and mirroring properties of χ:

$$\chi(\rho, z) = \chi(\rho, -z) = -\chi(\rho, z+1) = -\chi(\rho, z-1). \tag{2.29}$$

The sum has no closed formula in general, but we can evaluate it on the axis, $\rho = 0$. We get

$$\chi(0, z) = \frac{1}{z} + \chi_{\neg 0}(0, z), \chi(0, 1/2) = 0, \chi_{\neg 0}(0, 0) = -2\ln 2, \tag{2.30}$$

and obtain a formula involving the Hurwitz–Lerch transcendent[8]

$$\chi_{\neg 0}(0, z) = -\Phi(-1, 1, 1-z) - \Phi(1, 1, 1+z), 0 \le z \le 1/2. \tag{2.31}$$

The first derivative with respect to ρ is zero on the axis: $\chi_{,\rho}(0, z) = 0$. The second derivative with respect to ρ on the axis involves the Hurwitz zeta function[9]:

$$8\chi_{,\rho\rho}(0, z) = \zeta_3\left(\frac{z+1}{2}\right) - \zeta_3\left(\frac{z}{2}+1\right) + \zeta_3\left(\frac{1-z}{2}\right) - \zeta_3\left(1-\frac{z}{2}\right) - \frac{8}{z^3}. \tag{2.32}$$

If we use spherical coordinates in the neighbourhood of the origin and expand the potential in r, we obtain the following series:[10]

$$\chi(r, \theta) = \frac{1}{r} - 2\ln 2 - \frac{\bar{\zeta}(3)}{8}[1 - 3\cos(2\theta)]r^2 + O(r^4), r \ll 1. \tag{2.33}$$

---

[8]The Hurwitz–Lerch transcendent is defined as $\Phi(z, s, a) = \sum_{k=0}^{\infty} z^k(a+k)^{-s}$.

[9]The Hurwitz zeta function, $\zeta_s(a) = \sum_{k=0}^{\infty}(a+k)^{-s}$, has singularities at $a = -n$ for non-negative integers n.

[10]Here $\bar{\zeta}(s) = \sum_{k=1}^{\infty} k^{-s}$ is the Riemann zeta function.

On the other hand, far away from the axis, $\rho \gg 1$, we find the following limits based on the Moore–Osgood theorem[11]:

$$\lim_{\rho \to \infty} \chi = \lim_{\rho \to \infty} \chi_{,z} = \lim_{\rho \to \infty} \chi_{,zz} = \lim_{\rho \to \infty} \chi_{,\rho} = \lim_{\rho \to \infty} \chi_{,\rho\rho} = 0. \tag{2.34}$$

Thus the spacetime is asymptotically flat. Since the range of χ is $\mathcal{R}$ then for any given Q there always exists a surface $\chi = -k/Q$ that is singular. This surface is always disconnected and represents a naked singularity.

2.3 Physical Interpretation

The charge enclosed in a sphere centred at $r = 0$ is

$$Q_{sph} = -\frac{1}{8\pi} \int_0^\pi \int_0^{2\pi} r^2 U_{,r} \sin\theta d\theta d\phi = \frac{Q}{k} + O(r^2). \tag{2.35}$$

We see that Q corresponds to the charge of each individual black hole and the naked singularities have an opposite charge due to the anti-periodicity of U. From the Taylor expansion of χ near the origin it follows that the geometry approaches the extremal Reissner–Nordström solution. Interpretation of charge and mass can also be seen from circular electrogeodesics[12] in the weak-field limit, i.e., $Q/k \to 0$ and $\rho \to \infty$ within the mirror plane, $z = 0$. The equations of motion for a time-like particle with a charge-to-mass ratio q then read

$$\rho^2 \omega^2 U^2 - \frac{\gamma^2}{U^2} = -1, \ddot{t} = \ddot{\phi} = 0 \Rightarrow t = \gamma\tau, \phi = \omega\tau, \tag{2.36}$$

$$U_{,z} = 0, U_{,\rho}\left(-\gamma q U + \rho^2 \omega^2 U^4 + \gamma^2\right) + \rho\omega^2 U^5 = 0. \tag{2.37}$$

The solution for γ and ω reads

$$\omega^2 = \frac{\gamma^2 - U^2}{\rho^2 U^4}, \gamma = \frac{U\left(q\rho U_{,\rho} - \sqrt{(q^2+8)\,\rho^2 U_{,\rho}^2 + 12\rho U_{,\rho}U + 4U^2}\right)}{2\left(2\rho U_{,\rho} + U\right)}. \tag{2.38}$$

---

[11] Moore–Osgood: Let f, f_n be defined in a punctured neighbourhood of x_0. Let $f_n \rightrightarrows f$ in I and $c_n = \lim_{x \to x_0} f_n(x)$ be finite. Then $\lim_{x \to x_0} \lim_{n \to \infty} f_n(x) = \lim_{n \to \infty} \lim_{x \to x_0} f_n(x)$.

[12] Equations of motion for electrogeodesics are derived from the Lagrangian $\mathcal{L} = \frac{1}{2}g_{\mu\nu}\dot{x}^\mu\dot{x}^\nu + q A_\mu \dot{x}^\mu$, the dot denotes derivative with respect to affine parameter τ.

We plug in the asymptotic expansion (2.28) and see that γ and ω falls off exponentially. The alternating crystal thus approaches Minkowski faster than any isolated system.

3 Uniform Crystal

We now turn to the case of an infinite number of identical punctures distributed again equidistantly along an axis with a separation constant k. This has been already studied in [8] for $D > 4$, however, as we shall show, in $D = 4$ (and more generally in even dimensions) there are some difficulties. We use cylindrical coordinates and since the punctures are equidistant we scale all coordinates by k. The metric and electromagnetic potential are of the same form as for the alternating crystal (2.1) while the potential appearing here is now denoted φ to distinguish it from the alternating crystal.

3.1 Convergence of the Potential

Inspecting the Potential

Contribution of each hole goes as $1/n$ and we thus need to subtract such term which will improve the convergence. Without loss of generality, we choose

$$\varphi(\rho, z) = \sum_{n=-\infty}^{\infty} \hat{\varphi}_n, \hat{\varphi}_0 = \frac{1}{r_0}, \hat{\varphi}_n = \frac{1}{r_n} - \frac{1}{\sqrt{n^2}} \ \forall n \neq 0. \tag{3.1}$$

Function r_n is the Cartesian distance from a specific puncture, defined in (2.3). It is convenient to rewrite the sum for $n \geq 1$ and take out the φ_0 term, since it is divergent at the origin. We thus obtain

$$\varphi(\rho, z) \equiv \varphi_0 + \varphi_{\neg 0}, \varphi_{\neg 0} = \sum_{n=1}^{\infty} \varphi_n, \varphi_n = \hat{\varphi}_n + \hat{\varphi}_{-n}, \varphi_0 = \hat{\varphi}_0. \tag{3.2}$$

We get bounds

$$-\frac{1}{n} \leq \hat{\varphi}_n \leq \frac{1}{n(2n-1)}, -\frac{1}{n} \leq \hat{\varphi}_{-n} \leq 0, |\varphi_n| \leq \frac{2}{n}. \tag{3.3}$$

However, this is not enough, as there are insufficiently strong bounds for φ_n. If we restrict $\rho \leq R$, then we have

$$\left|\hat{\varphi}_{-n}\right| \leq \frac{R^2 + \frac{1}{4}}{2n^3} + \frac{2}{n^2}, \left|\hat{\varphi}_n\right| \leq 3\frac{R^2 + \frac{1}{4}}{2n^3} + \frac{2}{n^2}. \tag{3.4}$$

We see that for $\rho \leq R$ the series converges uniformly. To make it convergent in $\mathcal{R}$, we redefine the potential as

$$\varphi_{\neg 0} \equiv f(\rho) \sum_{n=1}^{\infty} \frac{\varphi_n}{f(\rho)}. \tag{3.5}$$

Here f is a regulator function. If f and its derivative $f_{,\rho}$ are well behaved, then we extend the uniform-convergence region and the derivatives will act[13] only on φ_n. Our choice is $f = e^{2\rho}$. Then the estimates of φ_n read

$$\left|e^{-2\rho}\hat{\varphi}_n\right| \leq \frac{1}{8n^3} + \frac{2}{n^2}, \left|e^{-2\rho}\hat{\varphi}_{-n}\right| \leq \frac{3}{8n^3} + \frac{2}{n^2}. \tag{3.6}$$

Inspecting the Derivatives

We can use results from the alternating crystal case since

$$\varphi_n = \chi_n - \frac{2}{n} \Rightarrow \nabla\varphi_n = \nabla\chi_n. \tag{3.7}$$

Because we chose an exponential as our regulator function, we have

$$\nabla\varphi_{\neg 0} = e^{2\rho} \sum_{n=1}^{\infty} e^{-2\rho}\nabla\varphi_n. \tag{3.8}$$

We recall now the bounds from the alternating crystal (2.9) and (2.12):

$$\left|\frac{\partial\varphi_n}{\partial z}\right| \sim \left|\frac{\partial\varphi_n}{\partial\rho}\right| \sim \frac{1}{n^2} \text{ for } \rho \in \mathcal{R}. \tag{3.9}$$

---

[13]Formally $\nabla\varphi_{\neg 0} = (\nabla f)\sum f^{-1}\varphi_n - f\sum\varphi_n f^{-2}\nabla f + f\sum f^{-1}\nabla\varphi_n$. Under suitable assumptions the first two terms cancel each other.

These bounds are strong in $\mathcal{R}$—they hold for any n and are independent of ρ and z. This allows us to take out the factor $e^{-2\rho}$ from the sum, so we can write

$$\nabla\varphi_{\neg0} = e^{2\rho}\sum_{n=1}^{\infty}e^{-2\rho}\nabla\varphi_n = \sum_{n=1}^{\infty}\nabla\varphi_n. \tag{3.10}$$

This also holds for the second derivatives, as (2.16) and (2.18) tell us that

$$\left|\frac{\partial^2\varphi_n}{\partial z^2}\right| \sim \left|\frac{\partial^2\varphi_n}{\partial\rho^2}\right| \sim \frac{1}{n^3} \text{ in } \mathcal{R}. \tag{3.11}$$

So we conclude

$$\nabla_\mu\nabla_\nu\varphi_{\neg0} = e^{2\rho}\sum_{n=1}^{\infty}e^{-2\rho}\nabla_\mu\nabla_\nu\varphi_n = \sum_{n=1}^{\infty}\nabla_\mu\nabla_\nu\varphi_n. \tag{3.12}$$

Putting It All Together

We have proved that the infinite sum and its first and second derivatives converge for $\rho \in \mathcal{R}$. Now it is time to extend the definition of potential for all values of z. First we notice the mirror symmetry:

$$\varphi_n(\rho, z) = \varphi_n(\rho, -z) \Rightarrow \varphi(\rho, z) = \varphi(\rho, -z). \tag{3.13}$$

Using this, we can calculate the potential for $-1/2 \leq z \leq 1/2$. The final step is to show periodicity in z.

$$\varphi(\rho, z+1) = r_{-1}(\rho, z) + e^{2\rho}\sum_{n=1}^{\infty}e^{-2\rho}\left(\frac{1}{r_{n-1}(\rho, z)} + \frac{1}{r_{-n-1}(\rho, z)} - \frac{2}{n}\right). \tag{3.14}$$

We take some terms out of the sum, rewrite it using a new summation index, and insert and subtract $2/k$. This yields

$$\hat{\varphi}_0 - 1 + \frac{1}{r_1} - \frac{1}{2} + e^{2\rho}\sum_{k=2}^{\infty}e^{-2\rho}\varphi_k + \sum_{k=2}^{\infty}\left(\frac{2}{k} - \frac{1}{k+1} - \frac{1}{k-1}\right). \tag{3.15}$$

The last sum cancels with $-1/2$ and we get

$$\varphi(\rho, z + 1) = \ldots \tag{3.16}$$

$$\ldots = \hat{\varphi}_{-1}(\rho, z) + \hat{\varphi}_0(\rho, z) + \hat{\varphi}_1(\rho, z) + e^{2\rho} \sum_{k=2}^{\infty} e^{-2\rho} \varphi_k(\rho, z) = \varphi(\rho, z).$$

We see that the potential has a period 1 in z. Using the periodicity and mirror symmetry we define the potential for all values of z. Plots of φ are shown in Fig. 2a, b.

Bounds for $\rho \rightarrow \infty$

We notice that the functions $\varphi_{n,\rho}$ are negative and increasing:

$$\frac{\partial \varphi_n}{\partial \rho} < 0, \quad \frac{\partial^2 \varphi_n}{\partial \rho \partial n} > 0 \text{ for } \rho \in \mathcal{R}. \tag{3.17}$$

Using integral estimates, we find

$$\frac{\partial \varphi_0}{\partial \rho} + \frac{\partial \varphi_1}{\partial \rho} + \int_1^\infty \frac{\partial \varphi_n}{\partial \rho} dn \leq \frac{\partial \varphi}{\partial \rho} \leq \frac{\partial \varphi_0}{\partial \rho} + \int_1^\infty \frac{\partial \varphi_n}{\partial \rho} dn. \tag{3.18}$$

We now evaluate the integral to obtain

$$-\frac{2}{\rho} - \frac{1}{\rho^2} + \frac{3z^2 + 4}{2\rho^4} + O\left(\frac{1}{\rho^5}\right) \leq \frac{\partial \varphi}{\partial \rho} \leq -\frac{2}{\rho} + \frac{1}{\rho^2} - \frac{3z^2 + 2}{2\rho^4} + O\left(\frac{1}{\rho^5}\right). \tag{3.19}$$

It then follows that $\varphi_{,\rho} \rightarrow 0$ for $\rho \rightarrow \infty$ and that $\varphi_{,\rho} \sim -2/\rho$. We can also see that for large ρ the dependence on z vanishes.

3.2   Geometry

Let us now investigate the geometry of the solution. We have shown the translational properties of φ:

$$\varphi(\rho, z) = \varphi(\rho, z + 1) = \varphi(\rho, z - 1) = \varphi(\rho, -z). \tag{3.20}$$

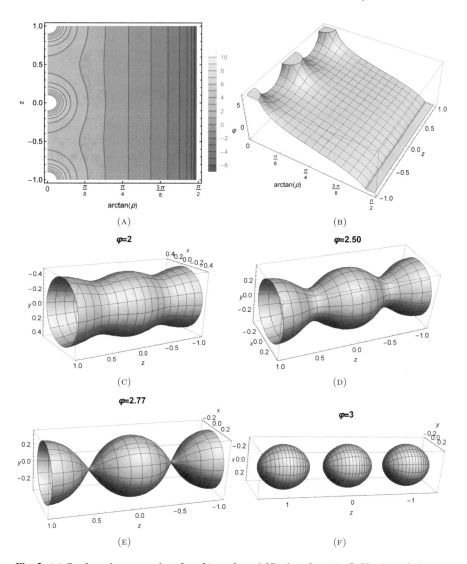

Fig. 2 (**a**) Conformal contour plot of φ; (**b**) conformal 3D plot of φ; (**c**)–(**f**) 3D plots of singular surfaces $\varphi = -k/Q$ for various values of Q/k. For $k/Q \lesssim -2.77$ the surface is disconnected

As with the alternating crystal, we have closed formulae for special cases only. On the axis we have an expression involving the harmonic number function[14]

$$\varphi(0, z) = \frac{1}{z} - H(z) - H(-z), 0 < z \leq 1/2. \tag{3.21}$$

---

[14]Harmonic number $H(z) = \gamma_e + \frac{d(\ln \Gamma(z+1))}{dz}$, where γ_e is the Euler gamma constant and Γ is the Euler gamma function.

The first derivative with respect to ρ is zero on the axis, $\chi_{,\rho}(0, z) = 0$. In the vicinity of the origin of (spherical) coordinates, $r \ll 1$, we have

$$\varphi(r, \theta) = \frac{1}{r} + \frac{\bar{\zeta}(3)}{2} [1 - 3 \cos(2\theta)] r^2 + O(r^4), r \ll 1. \tag{3.22}$$

Again, φ ranges throughout $\mathcal{R}$, so for non-zero Q the surface $\varphi = -k/Q$ is singular. For $k/Q \lesssim -2.77$ the surface is disconnected, otherwise it is connected.

The spacetime has obvious Killing vectors $\xi^{(t)} = \frac{\partial}{\partial t}$ and $\xi^{(\phi)} = \frac{\partial}{\partial \phi}$. The vector field $\xi^{(z)} = \frac{\partial}{\partial z}$ is an asymptotic Killing vector for $\rho \rightarrow \infty$, since then U only depends on ρ.

3.3 Physical Interpretation

The charge enclosed in a sphere around the origin of (spherical) coordinates is

$$Q_{sph} = -\frac{1}{8\pi} \int_0^\pi \int_0^{2\pi} r^2 U_{,r} \sin\theta d\theta d\phi = \frac{Q}{k} + O(r^2). \tag{3.23}$$

We see that Q corresponds to the charge of each individual black hole. If we study the asymptotic form of circular electrogeodesics far away from the axis as we did for the alternating crystal, we find an asymptotic expression (3.19) for the function $U = 1 - 2\frac{Q}{k} \ln \rho + O(\rho^{-1})$, while for the Lorentz factor and angular velocity (2.38) we get

$$\gamma = 1 + \frac{Q(1 - q - 2 \ln \rho)}{k} + O\left(\frac{Q^2}{k^2}\right), \omega^2 = \frac{2(1 - q)Q}{k\rho^2} + O\left(\frac{Q^2}{k^2}\right). \tag{3.24}$$

Compared to electrogeodesics in the field of an extremally charged infinite string, the ECS spacetime [5], we see that the source has a linear mass and charge density Q/k. Indeed, the two solutions share the same asymptotics for $\rho \gg 1$.

4 Dimensional Reduction

There are various approaches enabling us to take a solution of Einstein's equations of a certain dimension and transforming it to a different solution of (sometimes modified) Einstein's equations in a different dimension, such as direct products of metrics, warp products or the Kaluza–Klein theory. Here we follow the same procedure as in [9] and apply a dimensional reduction, removing a space-like, compact dimension in $D + 1$ dimensions and producing thus a D-dimensional spacetime. The initial assumption of the procedure is that the $D + 1$ metric, ^{D+1}g, can be decomposed[15] as

$$^{D+1}g_{AB}dx^A dx^B = {}^D\bar{g}_{\mu\nu}(x^\alpha)dx^\mu dx^\nu + \Phi(x^\alpha)^2 d\xi^2, \qquad (4.1)$$

where ξ is a compact Killing coordinate and $\bar{g}_{\mu\nu} = g_{\mu\nu}$. The metric $^{D+1}g_{AB}$ solves $D + 1$ Einstein's equations with action

$$^{D+1}S = \frac{1}{16\pi} \int d^{D+1}x \sqrt{|^{D+1}g|} \left[L_{EH}(^{D+1}g_{AB}) + 16\pi L_M(F_{AB}) \right], \qquad (4.2)$$

where L_{EH} is Einstein–Hilbert Lagrangian of gravity and $L_M(F_{AB})$ is the Lagrangian of the electromagnetic field:

$$L_{EH}(g) = R(g) - 2\Lambda, \ 16\pi L_M(F) = -F_{AB}F^{AB}. \qquad (4.3)$$

Now we treat Φ as an independent scalar field and factor it from the metric determinant as $\sqrt{|^{D+1}g|} = \Phi\sqrt{|^D\bar{g}|}$. Since ξ is a Killing coordinate, it does not appear in the action and we can integrate the action ^{D+1}S with respect to ξ (which is in fact equivalent to multiplication by a constant, as ξ is compact). To define the new action, we also need to decompose all quantities in the action in terms of D-dimensional quantities only.

4.1 Decomposition of Electromagnetic Field

The electromagnetic tensor is decomposed as

$$F_{AB}dx^A \wedge dx^B = F_{\mu\nu}dx^\mu \wedge dx^\nu + \frac{1}{2}\psi_{,\nu}dx^\nu \wedge d\xi, \qquad (4.4)$$

where we put $2F_{\nu\xi} \equiv \psi_{,\nu}$. We thus have

$$F_{AB}F^{AB} = F_{\mu\nu}F^{\mu\nu} + \frac{1}{2\Phi^2}\bar{g}^{\mu\nu}\psi_{,\mu}\psi_{,\nu}, \qquad (4.5)$$

where we interpret $F_{\mu\nu}$ as an electromagnetic field in D dimensions and ψ as an additional scalar field.

---

[15]Capital Latin indices range over $0, \ldots, D + 1$, Greek indices are $0, \ldots, D$.

4.2 Decomposition of Ricci Scalar

Using the decomposition of Ricci scalar [10] we obtain

$$R(^{D+1}g) = \bar{R} - \frac{2}{\Phi}\Box_{\bar{g}}\Phi, \tag{4.6}$$

where $\bar{R} = R(^D\bar{g})$ is the shorthand for Ricci scalar of metric $^D\bar{g}$. Now we can rewrite the term with $R(^{D+1}g)$ in the action and get

$$\int d^{D+1}x\sqrt{|^{D+1}g|}R(g^{D+1}) = \int d\xi \int d^D x\sqrt{|^D\bar{g}|}\Phi\bar{R}, \tag{4.7}$$

since the integral of the d'Alembertian[16] $\Box\Phi$ vanishes thanks to Gauss's theorem.

4.3 Equations of Motion

Using results from the previous sections, we define the new action in terms of the fields Φ, Ψ, $F_{\mu\nu}$, $^D\bar{g}_{\mu\nu}$ in the following way:

$$^D S = \frac{1}{16\pi}\int d^D x\sqrt{|^D\bar{g}|}\Phi\left[\bar{R} - 2\Lambda - F^{\mu\nu}F_{\mu\nu} - \frac{1}{2\Phi^2}\Psi^{,\mu}\Psi_{,\mu}\right]. \tag{4.8}$$

From now on, we only work with D-dimensional quantities, drop the bars ($\bar{g} \to g$), and do not mention the dimension explicitly. The action can be rewritten in terms of Lagrangians of the respective fields as

$$S = \int d^D x\sqrt{|g|}\left(\Phi L_{EH} + \Phi L_M + \frac{1}{\Phi}L_S\right), \tag{4.9}$$

where $32\pi L_S = -\psi_{,\mu}\psi^{,\mu}$ is the Lagrangian of a massless scalar field ψ. We thus obtain a set of equations of motion:

$$R_{\mu\nu} = \frac{R}{2}g_{\mu\nu} - \Lambda g_{\mu\nu} + 8\pi(T^M_{\mu\nu} + T^\Phi_{\mu\nu} + \frac{1}{\Phi}T^\psi_{\mu\nu}), \tag{4.10}$$

$$8\pi T^\Phi_{\mu\nu} = \frac{1}{\Phi}\left(\Phi_{;\mu\nu} - g_{\mu\nu}\Box\Phi\right), \tag{4.11}$$

---

[16]$\Box_g$ denotes the d'Alembertian of a Lorentzian metric g, which is defined as $\Box_g f \equiv g^{\mu\nu}\nabla_\mu\nabla_\nu f = \frac{1}{\sqrt{|g|}}\left(\sqrt{|g|}g^{\mu\nu}f_\mu\right)_{,\nu}$.

$$8\pi T^{\psi}_{\mu\nu} = \psi_{,\mu}\psi_{,\nu} - \frac{1}{2}g_{\mu\nu}\psi_{,\alpha}\psi^{,\alpha}, \tag{4.12}$$

$$\frac{D}{\Phi}\Box\Phi = \frac{D-1}{2}R - (D+1)\Lambda + 8\pi T^M + \frac{8\pi}{\Phi}T^{\psi}, \tag{4.13}$$

where the last equation for Φ is in this case not independent and follows from the trace of the first equation. Equations of motion for the fields $F_{\mu\nu}$ and ψ read

$$\nabla_{\beta}\left(\Phi F^{\alpha\beta}\right) = 0, \ \psi^{,\mu}(\Phi\psi)_{,\mu} = 0. \tag{4.14}$$

Equations of motion for the electromagnetic field yield the conservation of charge:

$$4\pi Q = \oint_S \Phi \star F. \tag{4.15}$$

5   Uniform Reduced Crystal

We take the 5D Majumdar–Papapetrou solution in cylindrical coordinates

$$^5g = -U^{-2}dt^2 + U\left(d\rho^2 + \rho^2 d\phi^2 + \rho^2 \sin^2\phi \ d\xi^2 + dz^2\right), \ ^5A = \frac{\sqrt{3}}{2}\frac{dt}{U}, \tag{5.1}$$

where the function U for a uniform crystal satisfying the 4D Laplace's equation [8] reads

$$U = 1 + \frac{M}{L^2}\eta, \eta = \sum_{n=-\infty}^{\infty} \frac{1}{\rho^2 + (z-n)^2} = \frac{\pi}{\rho}\frac{\sinh(2\pi\rho)}{\cosh(2\pi\rho) - \cos(2\pi z)}. \tag{5.2}$$

We take advantage of the fact that in the 5D case U has a closed form. We apply the dimensional reduction from the previous chapter and obtain the 4D fields

$$^4\bar{g} = -U^{-2}dt^2 + {}^3\bar{h}, {}^3\bar{h} = U\left(d\rho^2 + \rho^2 d\phi^2 + dz^2\right), \tag{5.3}$$

$$\psi \equiv 0, \ \Phi = \rho\sqrt{U}\sin^2\phi, \tag{5.4}$$

$$^4A = \frac{\sqrt{3}}{2}\frac{dt}{U}, \ ^4F = \frac{\sqrt{3}}{2}\frac{dt}{U^2} \wedge \left(U_{,\rho}d\rho + U_{,z}dz\right). \tag{5.5}$$

The metric $^4\bar{g}$ satisfies (4.10) and the functions U and Φ now satisfy (4.13)

$$\Delta_{3\delta} U = -\frac{U_{,\rho}}{\rho}, \quad 3\Box_{\bar{g}}\Phi = \Phi\bar{R}. \tag{5.6}$$

5.1 Geometry

Function η has the following obvious symmetries:

$$\eta(\rho, z) = \eta(\rho, -z) = \eta(\rho, z+1) = \eta(\rho, z-1). \tag{5.7}$$

On the axis and in the mirror plane we have

$$\eta(0, z) = \frac{\pi^2}{\sin^2(\pi z)}, \quad \eta(\rho, 0) = \frac{\pi}{\rho}\frac{\sinh(2\pi\rho)}{\cosh(2\pi\rho) - 1}. \tag{5.8}$$

For $r \ll 1$ the series expansion of η reads

$$\eta(r, \theta) = \frac{1}{r^2} + \frac{\pi^2}{3} - \frac{\pi^4}{45}[2\cos(2\theta) - 1]r^2 + O(r^4), r \ll 1. \tag{5.9}$$

From the series expansion for large ρ we see that the spacetime is cylindrically asymptotically flat:

$$\eta = \frac{\pi}{\rho} + O(\rho^{-2}), \rho \gg 1. \tag{5.10}$$

For $M > 0$ the function η is always positive, so there is no singularity. Plots of η are shown in Fig. 3a, b.

5.2 Physical Interpretation

With (4.15), the charge enclosed in a sphere around $r = 0$ reads

$$Q = -\int_0^\pi \int_0^{2\pi} \frac{\sqrt{3}r^3 \sin^2\theta \, |\sin\phi| \, U_{,r}}{8\pi} d\theta d\phi = \frac{\sqrt{3}}{2}\frac{M}{L^2} + O(r). \tag{5.11}$$

This suggests that the charge of each black hole in the crystal is $Q = \frac{\sqrt{3}}{2}\frac{M}{L^2}$. Again, we calculate circular electrogeodesics for large ρ within the mirror plane, $z = 0$.

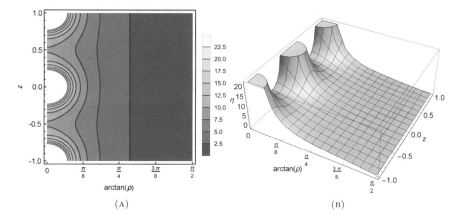

Fig. 3 (**a**) Conformal contour plot of η; (**b**) conformal 3D plot of η

This time the geometry is different, so the equations of motion read

$$\rho^2 \omega^2 U - \frac{\gamma^2}{U^2} = -1, \ddot{t} = \ddot{\phi} = 0 \Rightarrow t = \gamma\tau, \phi = \omega\tau, \qquad (5.12)$$

$$U_{,z} = 0, U_{,\rho}\left(-\sqrt{3}\gamma q U + \rho^2\omega^2 U^3 + 2\gamma^2\right) + 2\rho\omega^2 U^4 = 0. \qquad (5.13)$$

The solution for γ and ω reads

$$\omega^2 = \frac{\gamma^2 - U^2}{\rho^2 U^3}, \gamma = U\frac{\sqrt{3U_{,\rho}^2\left(q^2 + 4\right)\rho^2 + 32U_{,\rho}\rho U + 16U^2} + \sqrt{3}U_{,\rho}q\rho}{6\rho U_{,\rho} + 4U}.$$

$$(5.14)$$

We insert $U = 1 + \pi M/(L^2\rho) + O(\rho^{-2})$ following from (5.10) and get

$$\gamma = 1 + \frac{M\pi\left(6 - \sqrt{3}q\right)}{4L^2\rho} + O\left(\frac{M^2}{L^4}\right), \omega^2 = \frac{\pi M}{L^2\rho^3}\left(1 - \frac{\sqrt{3}}{2}q\right) + O\left(\frac{M^2}{L^4}\right).$$

$$(5.15)$$

We see that far away from the axis the source behaves as a point source of mass $\pi M/L^2$ and charge $\sqrt{3}\pi M/(2L^2)$. The leading order in ω^2 is ρ^{-3} because of the asymptotic flatness and presence of the scalar field Φ.

6 Conclusions and Summary

In this proceedings contribution, we have studied the properties of spacetimes exhibiting a discrete translational symmetry. We constructed them based on the Majumdar–Papapetrou solution of Einstein–Maxwell equations, consisting of an infinite number of identical black holes located on a line and separated by constant distances. Each black hole has an electric charge equal to its mass so that locally it is the extreme Reissner–Nordström charged black hole. One needs to deal with issues arising from the (non)convergence of the series.

The asymptotic properties of the spacetime depend on the average charge contained in a large enough volume containing the source located along the axis of symmetry. We first investigated the case of alternating positive and negative charges, forming an infinite spacetime "crystal". Far away from the axis, this solution is flat and the sums of point-particle potentials converge uniformly along with their first and second derivatives. Therefore, the sum represents a solution of Einstein–Maxwell equations everywhere, although there is no closed-form expression for the sum. We further proved that, as expected, the solution indeed exhibits two discrete translational symmetries along the axis and that, at radial infinity, there is an additional Killing vector along the axis. There is a closed-form expression for the sum on the axis where it diverges at grid points. The spacetime features disconnected singular surfaces enveloping the individual grid points on the axis. The area of the singular surfaces is zero so, in fact, they represent point singularities rather than 2D regions.

We then explored the crystal consisting of identical point charges, distributed equidistantly along a line, forming thus the axis of the spacetime. To achieve the convergence of the infinite sum, we employed a regulator function in the form of an exponential, redefining the original terms but identical for any finite sum. It allowed us to prove uniform convergence for the sum and its first and second derivatives, showing thus that it is a solution of Einstein–Maxwell equations. The solution is again periodic along the axis and features an additional Killing vector asymptotically far away from the axis. On the axis, the solution can be written in a closed form using special functions. Similarly to the alternating crystal, there are singular surfaces where the potential vanishes but their surface area is zero again and thus they are actually point-like and, for $k/Q \lesssim -2.77$, they form an infinite line. In fact, both for the alternating and uniform crystals, the grid points along the axis are not point singularities but rather extended horizons. Therefore, one may view the solution from the opposite point of view: it represents a string of point-like singularities (represented in the original coordinates as surfaces) hidden behind a horizon (located at any grid point on the axis) and seen from the outside by an asymptotic observer (living in a region of negative radial coordinates).

Finally, we took advantage of the fact that in even dimensions there exists a closed-form expression for the corresponding sum of point-charge potentials. We took the 5D uniform solution and reduced it to four dimensions along with Einstein–Maxwell equations. This entails an additional scalar field but retains the closed form of the solution. Therefore, we were able to study its properties analytically. Far away from the axis, the spacetime is asymptotically flat despite the fact that all charges are of the same sign. This is due to the scalar field, which does not vanish asymptotically. It is also of interest that the scalar does not share the axial symmetry of the gravitational field.

In our future work, we plan to try another approach to this problem and construct directly a solution of the Laplace equation, expanding it in the fundamental functions of the separated form of the equation. One imposes the discrete translational symmetry, which selects only some of the basis functions. Since we obtain the Bessel equation for the radial sector, we have solutions diverging at infinity or along the axis. To construct a regular solution, we can continue an interior solution regular along the axis through a tube-like surface to an exterior solution regular at infinity. In this manner, the source in the form of charged matter would be located on the separating interface. We also plan to investigate infinite superposition of Yukawa potential, which decays exponentially in infinity, so we expect regular behaviour there.

Acknowledgements J.R. was supported by grant No. 80918 of Charles University Grant Agency. M.Ž. Acknowledges support by GACR 17-13525S.

References

1. S.D. Majumdar, A class of exact solutions of Einstein's field equations. Phys. Rev. **72**, 390–398 (1947)
2. A. Papaetrou, A static solution of the equations of the gravitational field for an arbitrary charge distribution. Proc. R. Irish Acad. (Sect. A) **A51**, 191–204 (1947)
3. J.B. Hartle, S.W. Hawking, Solutions of the Einstein-Maxwell equations with many black holes. Commun. Math. Phys. **26**(2), 87–101 (1972)
4. J. Ryzner, M. Žofka, Electrogeodesics in the di-hole Majumdar-Papapetrou spacetime. Class. Quant. Gravit. **32**(20), 205010 (2015)
5. J. Ryzner, M. Žofka, Extremally charged line. Class. Quant. Gravit. **33**(24), 245005 (2016)
6. J.P.S. Lemos, V.T. Zanchin, Class of exact solutions of Einstein's field equations in higher dimensional spacetimes, $d \geq 4$: Majumdar-Papapetrou solutions. Phys. Rev. D **71**, 124021 (2005)
7. D.L. Welch, On the smoothness of the horizons of multi-black-hole solutions. Phys. Rev. D **52**, 985–991 (1995)
8. V.P. Frolov, A. Zelnikov, Scalar and electromagnetic fields of static sources in higher dimensional Majumdar-Papapetrou spacetimes. Phys. Rev. D **85**, 064032 (2012)
9. J.P.S. Lemos, V.T. Zanchin, Rotating charged black strings and three-dimensional black holes. Phys. Rev. D **54**, 3840–3853 (1996)
10. R. Deszcz, M. Głogowska, J. Jełowicki, G. Zafindratafa, Curvature properties of some class of warped product manifolds. Int. J. Geom. Methods Mod. Phys. **13**(1), 1550135 (2016)

Electrogeodesics and Extremal Horizons in Kerr–Newman–(anti-)de Sitter

Jiří Veselý and Martin Žofka

Abstract We review some properties of the Kerr–Newman–(anti-)de Sitter solution. We present admissible extremal configurations, but the main focus of this work is charged test particle motion in the equatorial plane and along the spacetime's axis of rotation, with emphasis on static positions and effective potentials.

Keywords Kerr–Newman–(anti-)de Sitter · Electrogeodesics · Extreme horizons

Mathematics Subject Classification (2000) Primary 83-06; Secondary 83C10, 83C15

1 Introduction

Black holes were one of the most revolutionary predictions of general relativity, forever changing the way we look at the Universe. The first exact black hole spacetime, the Schwarzschild solution, was discovered mere months after Einstein's publication of the theory in 1915. The Schwarzschild solution describes a single static black hole in vacuum. However, to model astrophysical black holes, one has to take into account their rotation. The corresponding model was not discovered until 1963 by Kerr and was subsequently expanded to contain electric charge (usually considered negligible in astrophysics) in the Kerr–Newman solution of 1965. Eventually a non-zero cosmological constant has been added to the solution, yielding the most general black hole spacetimes: the Kerr–Newman–de Sitter solution for a positive cosmological constant and the Kerr–Newman–anti-de Sitter spacetime for a negative one. This addition has far-reaching consequences for the

J. Veselý (✉) · M. Žofka
Institute of Theoretical Physics, Faculty of Mathematics and Physics, Charles University, Prague, Czech Republic
e-mail: jiri.vesely@utf.mff.cuni.cz; zofka@mbox.troja.mff.cuni.cz

S. Cacciatori et al. (eds.), *Einstein Equations: Physical and Mathematical Aspects of General Relativity*, Tutorials, Schools, and Workshops in the Mathematical Sciences, https://doi.org/10.1007/978-3-030-18061-4_11

spacetime and should not be ignored, as the cosmological constant of our Universe indeed seems to be non-zero according to current observations.

One possible way to examine a spacetime is to follow the motion of test particles. By doing so we are, for example, able to make sure that the horizon of the Schwarzschild solution is only an apparent singularity due to an inappropriate choice of coordinates. Particles, however, can move past the said surface. While the motion of uncharged test particles in the Kerr–(anti-)de Sitter spacetime and of charged test particles in the Kerr–Newman spacetime has already been thoroughly analyzed at the University of Bremen [1, 2], the last logical step remains to be done—to consider both charge and the cosmological constant, which is the aim of this work. We do not attempt to provide so complete an analysis, however, as we focus only on a few selected problems.

In Sect. 2 we provide an overview of some of the general properties of the Kerr–Newman–(anti-)de Sitter solution. Unlike the simpler black hole models without the cosmological constant, the studied spacetime has multiple extremal configurations, which are the focus of Sect. 3. Finally, in Sect. 4 we proceed to study charged test particle motion. After discussing the constants of motion we examine possible static positions in the equatorial plane and along the spacetime's axis of rotation, before introducing an effective potential to analyze radial motion as well.

2   The Spacetime

The spacetime we shall be investigating is the Kerr–Newman–(anti-)de Sitter solution (KN(a)dS) with the standard Boyer–Lindquist-type coordinates [3]. The line element reads

$$ds^2 = -\frac{\Delta_r}{\Xi^2 \rho^2} \left(dt - a \sin^2 \theta \, d\phi \right)^2 + \frac{\rho^2}{\Delta_r} dr^2 + \frac{\rho^2}{\Delta_\theta} d\theta^2 + \frac{\Delta_\theta \sin^2 \theta}{\Xi^2 \rho^2} \left(a dt - (r^2 + a^2) d\phi \right)^2$$

$$(2.1)$$

with

$$\rho^2 = r^2 + a^2 \cos^2 \theta \, , \tag{2.2a}$$

$$\Delta_r = (r^2 + a^2) \left(1 - \tfrac{1}{3} \Lambda r^2 \right) - 2mr + q^2 \, , \tag{2.2b}$$

$$\Delta_\theta = 1 + \tfrac{1}{3} \Lambda a^2 \cos^2 \theta \, , \tag{2.2c}$$

$$\Xi = 1 + \tfrac{1}{3} \Lambda a^2 \, . \tag{2.2d}$$

This spacetime describes a rotating electrically charged black hole with mass m, angular momentum per unit energy[1] a, and charge q in a background universe with

---

[1] The orientation of the ϕ coordinate is chosen in such a way that angular momentum a is positive.

a non-zero cosmological constant Λ. The spacetime has a ring singularity located at $r = 0$, $\theta = \pi/2$. It is a stationary and axially-symmetrical electrovacuum solution of the Einstein–Maxwell equations with the four-potential

$$A = -\frac{qr}{\Xi\rho^2}\left(\mathrm{d}t - a\sin^2\theta\,\mathrm{d}\phi\right).\tag{2.3}$$

In order to retain the Lorentzian signature of the metric for all $\theta \in [0, \pi]$, we require

$$\tfrac{1}{3}\Lambda a^2 > -1 \Leftrightarrow \Xi > 0.\tag{2.4}$$

Then, $\Delta_\theta(\theta)$ is always positive. If $\Lambda > 0$, this condition is irrelevant as it is fulfilled for any a.

These coordinates make it very easy to see that setting a particular set of parameters equal to zero yields a simpler "household-name" spacetime: e.g., one gets the Reissner–Nordström solution by considering $\Lambda = a = 0$.

The solution can contain up to four distinct horizons, which can be found as the roots of $\Delta_r(r)$. Aside from the inner and outer black hole horizons R_I and R_O, which can be found in spacetimes of the Kerr and Reissner–Nordström families even for a vanishing Λ, for a positive Λ the two so-called cosmological horizons R_{C-} and R_{C+} appear. If all four horizons are present, their radii satisfy

$$R_{C-} < 0 < R_I < R_O < R_{C+}.\tag{2.5}$$

However, for a particular combination of the spacetime's parameters, certain horizons may merge, leading to extremal scenarios, or disappear altogether.

The addition of the cosmological constant to the Kerr–Newman solution is a much more dramatic change than the transition from the Kerr to the Kerr–Newman spacetime by adding charge, not only because of the addition of the two horizons for $\Lambda > 0$. Due to the presence of the cosmological term, this solution is not asymptotically flat. Current observations [4] suggest that we live in a universe with a positive cosmological parameter

$$\Omega_\Lambda \equiv \frac{\Lambda c^2}{3H_0^2} = 0.6889 \pm 0.0056,\tag{2.6}$$

where $H_0 = (67.66 \pm 0.42)$ km s^{-1} Mpc^{-1} is the present-day Hubble parameter. From there, we get

$$\Lambda = (1.11 \pm 0.02)\,10^{-52}\,\mathrm{m}^{-2}.\tag{2.7}$$

However, the most massive astrophysical black holes ever observed have masses of the order of [5]

$$M_{\max}^{\mathrm{BH}} = 10^{10}M_\odot \approx 1.5 \times 10^{13}\,\mathrm{m}.\tag{2.8}$$

As astrophysical black holes satisfy $a \leq m$, then

$$\Lambda a^2 \leq \Lambda m^2 \leq \Lambda \left(M_{\text{max}}^{\text{BH}}\right)^2 \approx 2.5 \times 10^{-26} \ll 1 \qquad (2.9)$$

and Λ can therefore be treated locally as a perturbation of the Kerr–Newman metric. Nonetheless, its addition has far-reaching consequences for the spacetime. For it to have a measurable effect, we need $\Lambda r^2 > 1$, which occurs for sufficiently large radii.

Unlike the standard Euclidean spherical coordinates, the Boyer–Lindquist radial coordinate r is extended into negative values. These are hidden behind the black hole horizons (if there are any) within the inner region of the black hole. It is natural for us to assume that we live in the outer region.[2] We shall, therefore, place greater emphasis on positive values of r.

For illustration, a slice of constant t and ϕ of the region with positive r of one possible spacetime configuration can be found in Fig. 1, where the vertical axis represents the axis of rotation and the horizontal one represents the equatorial plane. It is the most general configuration with all four horizons (only three of which are located in the area with $r > 0$), represented by black circles. They separate the dark gray non-stationary areas of the spacetime (with $\Delta_r < 0$) and the lighter stationary areas (with $\Delta_r > 0$). Light gray is the ergosphere (with $g_{tt} > 0$), which can be interconnected (like in the figure) or disjoint, depending on the spacetime's parameters. As there can be no static observers in the ergosphere (their four-velocity would be spacelike), the dotted curve separating the ergosphere from the rest of the stationary area is called the static limit. The black area near the singularity is where closed timelike curves can be found. It can be shown that it must lie below the innermost part of the ergosphere lying in the area with positive r, if there is an ergosphere in the particular spacetime configuration to begin with.

3   Extremal Configurations

In contrast to the Schwarzschild solution with only one horizon, black holes with multiple horizons can become extremal when two originally distinct horizons merge. This happens only for specific combinations of spacetime's parameters, which means that the number of independent physical parameters describing the solution is reduced, simplifying its description.

For simpler black hole models, the conditions under which they become extremal are obvious from the formulae for their horizon radii. For example, in the Kerr solution described by the Boyer–Lindquist coordinates, the horizons are located at $r_{\pm} = m \pm \sqrt{m^2 - a^2}$, which makes it clear that the spacetime is extremal for

---

[2]A static observer in the area with negative r would be, among other things, always subjected to a naked singularity.

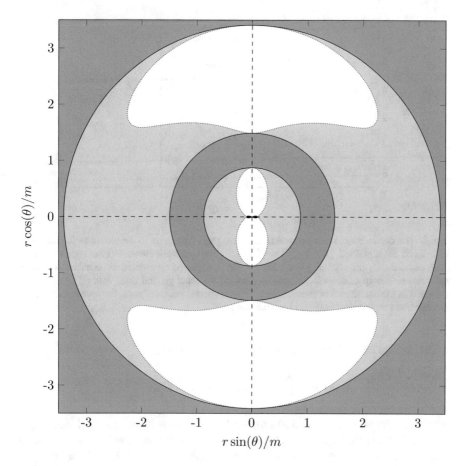

Fig. 1 A slice of constant t and ϕ of the region with $r > 0$ of the most general spacetime configuration containing all four horizons. A detailed description can be found in the last paragraph of Sect. 2. $\{\Lambda\, m^2, a/m, q/m\} = \{0.121, 10/11, 5/11\}$

$m = a$. However, the horizons of the studied spacetime are the roots of a fourth-degree polynomial $\Delta_r(r) = 0$ (2.2b), which are known to be rather unwieldy when expressed analytically. That, of course, does not mean that the spacetime cannot become extremal, see Fig. 2 illustrating a dependence of the horizon radii on one parameter in the metric, while the others are kept fixed.

We can avoid working with the horizon radii completely by comparing a more convenient form of (2.2b),

$$\Delta_r(r) = -\frac{\Lambda}{3}\left(r^4 + \left(a^2 - \frac{3}{\Lambda}\right)r^2 + \frac{6m}{\Lambda}r - \frac{3(a^2 + q^2)}{\Lambda}\right) \qquad (3.1)$$

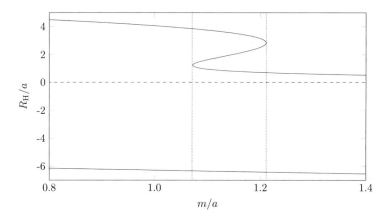

Fig. 2 The dependence of the horizon radii on the mass parameter in the metric as the remaining parameters are kept fixed, $\{\Lambda\, a^2, q/a\} = \{0.1, 0.5\}$. The spacetimes in the left and right parts of the chart contain only the cosmological horizons and can thus be considered to be naked singularity spacetimes. The spacetimes in the middle area are the most general ones with all four horizons present. On the boundaries of these regions, for two special values of m/a, the spacetimes become extremal, as two originally distinct horizons merge

to the desired factorization, the most general of which is

$$\Delta_r(r) = -\frac{\Lambda}{3}(r - A)(r - B)(r - C)(r - D),\qquad(3.2)$$

where constants A, B, C, and D represent the horizons if they are real. Should we take an interest in, say, a spacetime with a double horizon, we can set $C = D$, expand the above form in powers of r and compare it to (3.1). In this way we get a set of 4 conditions on the physical parameters of the spacetime that ensure the given structure.

The higher the multiplicity of the horizons, the less independent parameters; for example, with a given positive cosmological constant[3] and one triple horizon, there is just one more physical parameter, which may be taken to be the mass of the black hole. However, even these parameters may need to satisfy certain inequalities to represent physically acceptable spacetimes.

We describe the causal structure of the spacetime as, e.g., $(-1 \oplus 2 + 1-)$. The numbers in the employed notation represent horizons (in order of increasing r) and their multiplicity, the brackets represent the negative "(" and positive ")" radial infinity and the circle denotes $r = 0$. The signs tell us whether the area between two horizons or a horizon and a radial infinity is stationary "+" or not "−."

---

[3]We generally prefer to keep Λ as a free parameter because, from the astrophysical point of view, its value is known and fixed for all black holes.

It turns out that it is possible to construct four types of extremal KN(a)dS black holes. We list them below, using a set of free parameters that yields the simplest expressions.

$(-1 \oplus 2 + 1-)$ An extreme black hole immersed in de Sitter spacetime with horizons located at

$$R_{1-} = -\alpha_+ - \beta_-,$$
$$R_2 = \beta_-, \tag{3.3}$$
$$R_{1+} = +\alpha_+ - \beta_-,$$

with

$$\alpha_\pm = \sqrt{\frac{1}{3\Lambda}\left(6 - 2a^2\Lambda \pm \sqrt{9 - 42a^2\Lambda + a^4\Lambda^2 - 36\Lambda q^2}\right)},$$
$$\tag{3.4}$$
$$\beta_\pm = \sqrt{\frac{1}{6\Lambda}\left(3 - a^2\Lambda \pm \sqrt{9 - 42a^2\Lambda + a^4\Lambda^2 - 36\Lambda q^2}\right)}.$$

This is the most realistic extremal case allowing observers outside of the black hole but well below the cosmological horizon located generally at a large distance from the black hole. It requires

$$\Lambda > 0,$$
$$a^2 \in \left[0, \frac{21 - 12\sqrt{3}}{\Lambda}\right],$$
$$q^2 \in \left[0, \frac{9 - 42a^2\Lambda + a^4\Lambda^2}{36\Lambda}\right], \tag{3.5}$$
$$m = \frac{\Lambda\alpha_+^2\beta_-}{3}.$$

If we set $a = q = 0$ we obtain the non-extremal de Sitter spacetime.

$(-1 \oplus 1 - 2-)$ Here, we get a degenerate cosmological horizon resulting from a merger of the cosmological and outer black hole horizons.

$$R_{1-} = -\alpha_- - \beta_+,$$
$$R_{1+} = +\alpha_- - \beta_+, \tag{3.6}$$
$$R_2 = \beta_+.$$

The corresponding physical parameters are

$$\Lambda > 0 \,,$$

$$a^2 \in \left[0, \frac{21 - 12\sqrt{3}}{\Lambda} \right] \,,$$

$$q^2 \in \left[0, \frac{9 - 42a^2\Lambda + a^4\Lambda^2}{36\Lambda} \right] \,,$$

$$m = \frac{\Lambda \alpha_-^2 \beta_+}{3} \,.$$

$(\oplus 2+)$ This can only occur for a negative cosmological constant. The horizon is located at

$$R_2 = \beta_- \tag{3.8}$$

and the parameters are

$$\Lambda < 0 \,,$$

$$a^2 \in \left[0, \frac{3}{|\Lambda|} \right) \,,$$

$$q^2 \in [0, \infty) \,,$$

$$m = \frac{\Lambda \alpha_+^2 \beta_-}{3} \,.$$

Setting both $a = q = 0$ leads to the anti-de Sitter spacetime and the horizon disappears.

$(-\mathbf{1} \oplus \mathbf{3}-)$ This is the only triply degenerate horizon allowed in the present family of spacetimes. If we choose Λ and m as our parameters, we can write

$$R_1 = -3R_3 \,,$$

$$R_3 = \sqrt[3]{\frac{3}{4} \frac{m}{\Lambda}} \,, \tag{3.10}$$

while

$$\Lambda > 0 \,,$$

$$m^2 \in \left[\frac{16(26\sqrt{3} - 45)}{3\Lambda}, \frac{2}{9\Lambda} \right] \,, \tag{3.11}$$

$$a^2 = \frac{3}{\Lambda} - 6R_3^2,$$

$$q^2 = \Lambda R_3^4 - a^2.$$

If, instead, we parameterize the spacetime using m and a, we find

$$R_1 = -3R_3,$$

$$R_3 = \frac{m}{2}\left(x + 1 + x^{-1}\right) \tag{3.12}$$

with

$$x = \frac{m^{2/3}}{\left(m^2 + a^2 + a\sqrt{2m^2 + a^2}\right)^{1/3}} \tag{3.13}$$

and the parameters are constrained as follows:

$$m > 0,$$

$$a^2 \in \left[0, \frac{3}{16}\left(3 + 2\sqrt{3}\right)m^2\right],$$

$$\Lambda = \frac{3}{4}\frac{m}{R_3^3}, \tag{3.14}$$

$$q^2 = \frac{3}{4}mR_3 - a^2.$$

One generally differentiates between naked and non-naked singularities based on whether an asymptotic observer can see the singularity. However, this notion is traditionally defined for asymptotically flat spacetimes that do not admit the cosmological constant. Here, we thus employ the most distant static observer available. We then conclude that naked singularities occur in scenarios $(-1 \oplus 1 - 2-)$ and $(-1 \oplus 3-)$.

4 Electrogeodesics

In order to bring a different perspective to the physics of the examined spacetime, we shall make a few remarks on selected problems in charged test particle motion using the Lagrangian formalism. In it, a particle with charge-to-mass ratio κ can be described by Lagrangian density [6]

$$\mathcal{L} = \frac{1}{2} g_{\mu\nu} \dot{x}^\mu \dot{x}^\nu + \kappa \dot{x}^\mu A_\mu , \tag{4.1}$$

where $\dot{x}^\mu$ is the particle's four-velocity, i.e., a total derivative of its four-position x^μ with respect to the variable used to parameterize motion.

Using the Lagrange equations of the second kind, one obtains the equations of motion

$$\ddot{x}^\mu + \Gamma^\mu_{\ \rho\sigma} \dot{x}^\rho \dot{x}^\sigma = \kappa F^\mu_{\ \nu} \dot{x}^\nu , \tag{4.2}$$

which shall be referred to as the electrogeodesic equations in the following. Take note that these equations do not take into account the radiation of energy of charged particles during their acceleration, that is, no self-force (neither electromagnetic, nor gravitational) is included.

4.1 Constants of Motion

Seeing that the metric $g_{\mu\nu}$ (2.1) and the four-potential A_μ (2.3) do not depend on coordinate time t and angle ϕ, neither is $\mathcal{L}$ a function of these variables. These two coordinates are therefore cyclic and are associated with conservation of certain quantities,

$$\begin{aligned} -E &\equiv \frac{\partial \mathcal{L}}{\partial \dot{t}} = g_{tt} \dot{t} + g_{t\phi} \dot{\phi} + \kappa A_t , \\ L &\equiv \frac{\partial \mathcal{L}}{\partial \dot{\phi}} = g_{t\phi} \dot{t} + g_{\phi\phi} \dot{\phi} + \kappa A_\phi . \end{aligned} \tag{4.3}$$

We shall call E the particle's energy and L its angular momentum parallel to the axis of rotation of the black hole purely out of analogy with asymptotically flat spacetimes. However, the studied spacetime is not asymptotically flat and the interpretation of the constants is not clear.

Due to the used signature of the metric, the normalization of the four-velocity for massive particles is

$$g_{\mu\nu} \dot{x}^\mu \dot{x}^\nu = -1 . \tag{4.4}$$

For the sake of completeness, two further constants of motion can be found by venturing beyond the boundaries of the Lagrangian formalism for a bit, even though we shall not make use of them in any of the following. First, as $\mathcal{L}$ does not depend explicitly on the affine parameter, neither does the Hamiltonian $\mathcal{H}$. Hence, its value is constant and it can be shown to be proportional to the used normalization of four-velocity δ, $\mathcal{H} = \delta/2$. Moreover, it turns out that the terms containing r and θ in the

Hamilton–Jacobi equation can be separated and the equation rewritten as

$$
- \delta a^2 \cos^2 \theta + \Delta_\theta(\theta) \big(\Theta'(\theta) \big)^2 + \frac{\Xi^2}{\Delta_\theta(\theta) \sin^2 \theta} \Big(L - Ea \sin^2 \theta \Big)^2 =
$$

$$
= \delta r^2 - \Delta_r(r) \big(R'(r) \big)^2 + \frac{\Xi^2}{\Delta_r(r)} \Big(La - E(a^2 + r^2) + \frac{r}{\Xi} q\kappa \Big)^2 \equiv K , \quad (4.5)
$$

with both sides equal to the Carter constant K. $R(r)$ is the radial and $\Theta(\theta)$ the polar part of the separated action.

4.2 Static Particles

First, we shall investigate particles static in our coordinate system. In general relativity we have considerable freedom in choosing coordinates, and as such it is generally pointless to look for particles static in a particular set of coordinates. The used Boyer–Lindquist coordinates, however, can be considered an extension of the classical spherical coordinate system (at least in the stationary areas of the spacetime) and it is, therefore, not entirely unreasonable to say that this is the system in which a "truly" static particle ought to have constant spatial coordinates, that is, its world line should be $x^\mu = (t, r_0, \theta_0, \phi_0)$ with constant r_0, θ_0, and ϕ_0. The spatial components of the particle's four-velocity vanish, $\dot{x}^i = 0$. From the normalization of the four-velocity (4.4), we get $\dot{t} = 1/\sqrt{-g_{tt}(r_0, \theta_0)}$, yielding

$$
\dot{x}^\mu = \left(\frac{1}{\sqrt{-g_{tt}(r_0, \theta_0)}}, 0, 0, 0 \right). \quad (4.6)
$$

We are looking for points where (4.2) yields $\ddot{x}^\mu = 0$ with the above ansatz, which is applicable wherever $g_{tt} < 0$, elsewhere it would not describe a timelike particle. Such locations must clearly exhibit some symmetry and we thus consider the equatorial plane and the axis where the only remaining nontrivial electrogeodesic equation is the one for $\ddot{r}$, imposing conditions on the position of static particles. The subscript "0" shall be omitted in the following.

The Equatorial Plane

For $\theta = \pi/2$, static particles have to fulfill

$$
\frac{\Delta_r(r)}{\Delta_r(r) - a^2} \big[\Lambda s r^5 - q\kappa \Lambda r^4 - \Big(3ms + q\kappa(a^2 \Lambda - 3) \Big) r^2
$$

$$
+ 3q(qs - 2\kappa m)r + 3q^3\kappa \big] \big[r^4 s \big]^{-1} = 0 \quad (4.7)
$$

with

$$s = \sqrt{\frac{\Delta_r(r) - a^2}{\rho^2(r, \theta = \pi/2)}} \, . \tag{4.8}$$

The left-hand side of the equation is well-defined wherever

$$\Delta_r > a^2 \Leftrightarrow g_{tt}\big|_{\theta=\pi/2} < 0 \, , \tag{4.9}$$

i.e., in the stationary areas of the spacetime but not inside the ergosphere, which corresponds to our ansatz.

Instead of looking for r solving the equation for a given κ, it is considerably easier to look for a "properly-charged" particle that would remain still at a given r in a given spacetime, as the equation is linear in κ and easy to solve:

$$\kappa = -\frac{\Lambda r^4 - 3mr + 3q^2}{3qrs} \, . \tag{4.10}$$

Divergences for $r = 0$ (i.e., the singularity) and $q = 0$ are to be expected, as for an uncharged black hole the charge of the particle is irrelevant and fine-tuning of κ cannot help in any way to make the particle static.

For $\theta = \pi/2$ and $\dot{\phi} = 0$, the integrals of motion (4.3) are

$$E = \frac{(\Delta_r(r) - a^2)\dot{t} + \Xi qr\kappa}{\Xi^2 \rho^2(r, \theta = \pi/2)} \, ,$$
$$L = \frac{a\big((\Delta_r(r) - r^2 - a^2)\dot{t} + \Xi qr\kappa\big)}{\Xi^2 \rho^2(r, \theta = \pi/2)} \, , \tag{4.11}$$

with $\dot{t}$ given by (4.6) and κ by (4.10) for static particles. Unless $a \neq 0$, angular momentum L is generally non-vanishing and, therefore, not all static particles in the used coordinates are zero angular momentum observers, or ZAMO's, which is a much more covariant particle characteristic than staticity in one set of coordinates. By expressing κ from $L = 0$ and comparing it to (4.10), one can obtain a condition on the radial coordinate of ZAMO's,

$$2\Lambda r^3 + \Lambda a^2 r + 3m = 0 \, . \tag{4.12}$$

For $\Lambda > 0$ (and thus also for our universe) this condition manifestly forbids the existence of such particles in the area with $r > 0$. On the other hand, for $\Lambda < 0$ the polynomial always has exactly one positive root, as can be shown using either Descartes' rule of signs or the more intricate theorem of Sturm. However, one must bear in mind that condition (4.12) does not mean that the positive root satisfies $\Delta_r(r) - a^2 > 0$ as required by s. This inequality is to be viewed as an additional constraint on the spacetime if it were to contain static electrogeodesic ZAMO's in the equatorial plane.

The Axis

For $\theta \in \{0, \pi\}$, we have

$$\left[\Lambda s r^7 - \kappa q \Lambda r^6 + 3\Lambda a^2 s r^5 - 3(ms - \kappa q)r^4 + 3\left((a^4\Lambda + q^2)s - 2\kappa q m\right)r^3 \right.$$
$$+ \kappa q(a^4\Lambda + 3q^2)r^2 + a^2\left((a^4\Lambda + 3q^2)s + 6\kappa q m\right)r$$
$$\left. + 3a^2\left(a^2 m s - \kappa q(a^2 + q^2)\right) \right]\left[(a^2 + r^2)^3 s\right]^{-1} = 0$$

$$(4.13)$$

with redefined

$$s = \sqrt{\frac{\Delta_r(r)}{\rho^2(r, \theta = 0)}}, \qquad (4.14)$$

well-defined only in the stationary areas of the spacetime where $\Delta_r > 0$.

Once again, one can find the required static particle's κ,

$$\kappa = -\frac{\Lambda r^5 + 2a^2\Lambda r^3 - 3mr^2 + (a^4\Lambda + 3q^2)r + 3a^2 m}{3q(r + a)(r - a)s}. \qquad (4.15)$$

If $q \neq 0$, this expression is well-defined in the stationary areas of the spacetime with the possible exception of $r = \pm a$, which may or may not be in the stationary area.[4]

For $r = \pm a$, the terms with κ in (4.13) vanish, and the rest of the equation is remarkably reduced into

$$\frac{4a^4\Lambda + 3q^2}{a^3} = 0. \qquad (4.16)$$

If the spacetime's parameters fulfill this condition, any particle can remain at rest at $r = \pm a$ regardless of its charge.[5] The problem with $a = 0$ in the condition is of no relevance, as that would mean our desired static particles are in the singularity, which for $a = 0$ is pointlike at $r = 0$ and all θ. Assuming thus $a \neq 0$, we may recast

---

[4]In order to avoid confusion, let us point out that positive values of r are not "above" and negative "below" the black hole on its axis—that is determined by θ. Negative r represent the analytically extended part of the spacetime.

[5]Take note that the particle may even be uncharged.

requirement (4.16) as

$$\Lambda = -\frac{3}{4}\frac{q^2}{a^4},\tag{4.17}$$

whence it is obvious that Λ has to be non-positive. That unfortunately prohibits astrophysical black holes from having these particles at rest at $r = \pm a$ as the universe is thought to have a positive Λ, cf. (2.7). Comparing with (2.4), we obtain

$$q^2 < 4a^2.\tag{4.18}$$

Whether $r = \pm a$ is in the stationary area of the spacetime, as required by our ansatz, is given by the sign of

$$\Delta_r(r = \pm a)\Big|_{\Lambda = -\frac{3q^2}{4a^4}} = 2a^2 + \frac{3}{2}q^2 \mp 2ma.\tag{4.19}$$

The negative solution is thus always in the stationary area, while the positive one further requires

$$m < \frac{4a^2 + 3q^2}{4a}.\tag{4.20}$$

Take note that while these static positions may exist in the Kerr and the KNadS solutions, not only are they nowhere to be found in the KNdS spacetime, as we have already commented, but they also do not exist in the asymptotically flat Kerr–Newman solution, as can be seen from (4.17): non-zero a and q manifestly require a non-zero Λ as well. Likewise, the same can also be said of the Kerr–(anti-)de Sitter solutions, because having a non-zero Λ and vanishing q violates (4.17) as well.

Now that we have established the possibility of the existence of static uncharged particles, we are left with the question what holds them in place, because it is neither electromagnetism (as the particles stay in place regardless of their charge) nor a repulsive cosmological constant (as the required negative Λ is actually attractive, to make the situation even more puzzling) in a kind of an equilibrium with gravity. It turns out [7, 8] that repulsive gravity is a common occurrence in naked singularity spacetimes. If the above conditions on the parameters result in a naked singularity, it is natural to assume that we have just discovered yet another manifestation of this phenomenon.

For $r < 0$ the only stationary area invariantly contains the singularity, meaning static particles at $r = -a$ always experience a naked singularity. For particles at

$r = +a$, after inserting (4.17) and inequality (4.20) into Δ_r we obtain

$$\Delta_r(r)\Big|_{\text{repulsion}} > \frac{(r-a)^2(q^2r^2 + 2aq^2r + 4a^4 + 4a^2q^2)}{4a^4}. \tag{4.21}$$

The polynomial acting as a lower bound is positive almost everywhere with the sole exception of $r = a$, where it is equal to zero.[6] Due to the strict inequality, Δ_r is positive everywhere in the spacetime. As it has no real roots, there are indeed no horizons to separate particles at $r = a$ (or any r, for that matter) and the singularity.[7]

For $\theta \in \{0, \pi\}$, the integrals of motion (4.3) become

$$E = \frac{\Delta_r(r)\dot{t} + \Xi qr\kappa}{\Xi^2\rho^2(r, \theta = 0)}, \tag{4.22}$$

$$L = 0$$

regardless of the particle's $\dot{\phi}$, as the ϕ coordinate is degenerate on the axis, $g_{\phi\mu} = 0 \ \forall\mu$. To evaluate E for static particles, $\dot{t}$ is to be substituted from (4.6) and for $r \neq \pm a$, κ is given by (4.15). All particles static on the spacetime's axis in our coordinate system are also ZAMO's.

4.3 Effective Potential for Radial Motion

In order to analyze radial motion, we can establish an effective potential V by means of the integrals of motion. One can thus obtain a formula for the radial component of the particle's four-velocity without actually solving the equations of motion, which generally can prove to be a system of differential equations impossible to be solved analytically. We define the effective potential as

$$\frac{1}{2}(\dot{r})^2 = -V. \tag{4.23}$$

A given particle can exist at a given r if and only if its $\dot{r}$ is real, or, equivalently, the corresponding effective potential is non-positive. It can be used to find turning points (with disappearing $\dot{r}$) at $V = 0$. Stationary circular orbits further require $dV/dr = 0$. They are stable if the potential has a local minimum at the given radius (i.e., $d^2V/dr^2 > 0$) and unstable if there is a local maximum ($d^2V/dr^2 < 0$).

---

[6]The quadratic polynomial $q^2r^2 + 2aq^2r + 4a^4 + 4a^2q^2$ has a negative discriminant $-4(4a^4q^2 + 3a^2q^4)$, barring real roots.

[7]For the Kerr solution this observation can be done directly from (4.20), which for $q = 0$ reduces into $m < a$, the known condition for the singularity's nakedness.

The Equatorial Plane

After inserting $\theta = \pi/2$ into (4.3), one can easily find

$$\dot{t} = \frac{\Xi}{r^2 \Delta_r}\left[-\Xi\left(a^2\Delta_r - \left(a^2 + r^2\right)^2\right)E + a\Xi\left(\Delta_r - \left(a^2 + r^2\right)\right)L - \left(a^2 + r^2\right)q\kappa r\right]$$

$$(4.24)$$

and

$$\dot{\phi} = \frac{\Xi}{r^2 \Delta_r}\left[-a\Xi\left(\Delta_r - \left(a^2 + r^2\right)\right)E + \Xi\left(\Delta_r - a^2\right)L - aq\kappa r\right]. \quad (4.25)$$

To obtain $\dot{r}$, we consider a more general form of the normalization equation

$$g_{\mu\nu}\dot{x}^\mu\dot{x}^\nu = \delta, \quad (4.26)$$

keeping in mind that we are interested in particles in the equatorial plane and with $\dot{\theta} = 0$, which leads to

$$\frac{1}{2}(\dot{r})^2 = -V(r; E, L, \kappa), \quad (4.27)$$

where V is the sought effective potential,

$$V(r; E, L, \kappa) = \frac{\Xi^2}{2r^4}\left((\Delta_r(r) - (a^2 + r^2))(aE - L)^2 - (a^2 + r^2)r^2E^2 + r^2L^2\right)$$

$$+ \frac{q\kappa}{2r^3}\left(2\Xi(a^2 + r^2)E - 2\Xi aL - q\kappa r\right) - \delta\frac{\Delta_r(r)}{2r^2}. \quad (4.28)$$

The potential can be used both for massive particles by setting $\delta = -1$ and for photons with $\delta = \kappa = 0$. The latter case has already been thoroughly analyzed in previous literature [9, 10].

After inserting the four-velocity of a static massive particle into E and L, the resulting potential expectedly vanishes. By expressing κ from the first derivative of V and inserting E and L we obtain the familiar result (4.10). Unfortunately, determining the static particle's stability from the second derivative of V requires us to deal with a polynomial of order eight in r, necessitating numerical computations.

Be warned that having $V \leq 0$ does not guarantee that the corresponding $\dot{t}$ (4.24) is positive, which needs to be checked numerically every time the potential is used, see Fig. 3 for an illustration of both a stable circular orbit and this phenomenon.

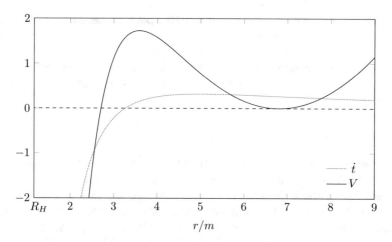

Fig. 3 Stable circular orbit at $r/m \approx 6.8$ for spacetime $\{\Lambda m^2, a/m, q/m\} = \{-0.729, 17/27, 1/9\}$ and particle with $\kappa = 200$ and $\dot{\phi} = 4 \times 10^{-4}$ corresponding to $E \approx 7.469$ and $L \approx 138.1$. R_H is the position of the outer black hole horizon. The dotted curve of $\dot{t}$ shows that a non-positive V is not enough to guarantee the possibility of the existence of a particle with given $\{E, L, \kappa\}$ at a particular r

The Axis

The situation here is simpler than in the equatorial plane, as L vanishes for particles bound to the axis. Further, because $g_{t\phi}(\theta \in \{0, \pi\}) = 0$, one can use (4.3) to express $\dot{t}$ directly from E,

$$\dot{t} = \frac{\Xi}{\Delta_r} \left(\Xi \rho_0^2(r) E - q\kappa r \right), \tag{4.29}$$

where

$$\rho_0^2(r) \equiv \rho^2(r, \theta \in \{0, \pi\}) = r^2 + a^2 . \tag{4.30}$$

Substituting into the normalization equation, we obtain, similarly as before,

$$\frac{1}{2} (\dot{r})^2 = -V(r; E, \kappa) \tag{4.31}$$

with

$$V(r; E, \kappa) = -\frac{1}{2\rho_0^4(r)} \left(\Xi \rho_0^2(r) E - q\kappa r \right)^2 - \frac{\Delta_r(r)}{2\rho_0^2(r)} \delta . \tag{4.32}$$

Take note that for photons (with $\delta = \kappa = 0$) the potential becomes $-\Xi^2 E^2/2$. As it has no roots,[8] there are unsurprisingly no turning points for photons bound to the axis, as that would violate the normalization of photon four-velocity.

Unlike in the equatorial plane, with a little trick we can separate the dependence on E from the potential here to obtain a separated effective potential W which has particles satisfy

$$E \geq W(r; \kappa). \tag{4.33}$$

For particles bound to the axis (which must satisfy $\dot{\theta} = 0$ in order to stay on the axis), $\dot{r}$ is the only relevant non-zero spatial component of the particle's four-velocity, as the ϕ coordinate is degenerate on the axis and $\dot{\phi}$ disappears from both the normalization and the equations of motion. Therefore, massive particles (considering $\delta = -1$ from now on) reaching turning points must become momentarily static, while in the equatorial plane they still can have arbitrary $\dot{\phi}$. To obtain W, we can insert the static value of $\dot{t} = 1/\sqrt{-g_{tt}}$ into (4.29) and express $E(= W)$, resulting in

$$W(r; \kappa) = \frac{1}{\Xi \rho_0^2(r)} \left(q\kappa r + \sqrt{\Delta_r(r)\rho_0^2(r)} \right). \tag{4.34}$$

By computing the derivatives of $V(r, W(r)) = 0$ it can be shown that one can use the derivatives of W instead of V for finding static points and their stability, as $\mathrm{sgn}(V'') = \mathrm{sgn}(W'')$ at static points. Furthermore, due to our derivation, W automatically selects particles with $\dot{t} > 0$, which for V must be checked numerically.

Returning to particles at $r = \pm a$, we can use both potentials to reproduce the results from the previous section. Moreover, from the second derivatives of the potentials,

$$\left. \frac{\mathrm{d}^2 V}{\mathrm{d}r^2} \right|_{r=\pm a} = \frac{1}{4a^4} \left(\pm 2am + 2q^2 \mp q\kappa\sqrt{4a^2 \mp 4am + 3q^2} \right), \tag{4.35a}$$

$$\left. \frac{\mathrm{d}^2 W}{\mathrm{d}r^2} \right|_{r=\pm a} = \frac{2}{a} \frac{\pm 2am + 2q^2 \mp q\kappa\sqrt{4a^2 \mp 4am + 3q^2}}{(4a^2 - q^2)\sqrt{4a^2 \mp 4am + 3q^2}}, \tag{4.35b}$$

we can see that while the value of κ was irrelevant in the question of staticity of particles at $r = \pm a$, it plays a key role for the stability of said positions.[9] There is a

---

[8]Note that according to (4.29) photons on the axis must have $E > 0$, otherwise they would move backwards in time or not move in time at all.

[9]Recall that $4a^2 - q^2 > 0$, (4.18). Then indeed $\mathrm{sgn}(V'') = \mathrm{sgn}(W'')$.

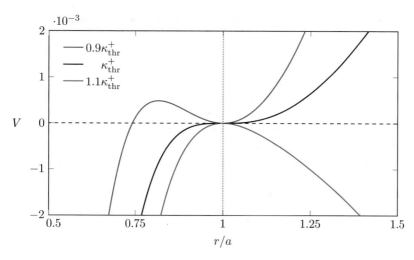

Fig. 4 The effective potentials V in the vicinity of $r = a$ in spacetime $\{\Lambda a^2, m/a, q/a\} = \{-12/49, 20/21, 4/7\}$ allowing for static particles at $r = \pm a$ for three particles differing by their charge and E. $\kappa_{thr}^+ = 47/\sqrt{129} \approx 4.14$, given by (4.36), is the charge-to-mass ratio of the black particle with $E \approx 1.88$ and it represents a marginally stable static position. The blue particle with lower $q\kappa$ and $E \approx 1.75$ is in a stable static position and the red particle with higher $q\kappa$ and $E \approx 2.01$ is in an unstable static position

threshold κ for which the particle stability changes,

$$\kappa_{thr}^{\pm} = \frac{2\left(am \pm q^2\right)}{q\sqrt{4a^2 \mp 4am + 3q^2}}.\tag{4.36}$$

For $r = +a$ stable positions (in the minima of the potentials) require $q\kappa < q\kappa_{thr}^+$ and unstable $q\kappa > q\kappa_{thr}^+$. Curiously, for $r = -a$ the inequalities are swapped and stable orbits require $q\kappa > q\kappa_{thr}^-$, unstable $q\kappa < q\kappa_{thr}^-$. The situation for $r = a$ is illustrated in terms of V in Fig. 4.

5 Conclusion

In this work we provided a review of some of the properties of the Kerr–Newman–(anti-)de Sitter solution. In Sect. 3 we listed the four plausible extremal configurations of the KN(a)dS family of spacetimes along with the parameter combinations that result in each scenario. In Sect. 4 we examined charged test particle motion using the Lagrangian formalism, first determining the constants of motion. We then focused on the equatorial plane and the axis of rotation, where we investigated static particles and established effective potentials for radial motion, managing to separate it for the axis. Furthermore, as expected, in the KNadS spacetime involving a naked singularity we discovered a manifestation of repulsive gravity for particles on the axis at $r = \pm a$, a case we studied in greater detail.

In the near future we aspire to publish a paper expanding the present work substantially. In it, we aim to compare the extremal scenarios to the extremal Kerr spacetime, examine the effects of parametric perturbations and illustrate the structure of the spacetime with conformal diagrams. We shall also prove our claim that CTC's must lie below the innermost part of the ergosphere. Regarding test particle motion, we shall also examine closed circular orbits in the equatorial plane, provide some insight about photons, and, above all, present a more useful method of separating the effective potential that would work for the equatorial plane as well.

Acknowledgements J.V. was supported by Charles University, project GA UK No. 80918. M.Ž. acknowledges support by GACR 17-13525S.

References

1. E. Hackmann, C. Lämmerzahl, V. Kagramanova, J. Kunz, Analytical solution of the geodesic equation in Kerr–(anti-) de Sitter space-times. Phys. Rev. D **81**, 044020 (2010)
2. E. Hackmann, H. Xu, Charged particle motion in Kerr–Newmann space-times. Phys. Rev. D **87**, 124030 (2013)
3. J.B. Griffths, J. Podolský, *Exact Space-Times in Einstein's General Relativity* (Cambridge University Press, Cambridge, 2009)
4. Planck Collaboration, Planck 2018 results. VI. Cosmological parameters. arXiv:1807.06209 [astro-ph.CO] (2018)
5. T.R. Lauer et al., The masses of nuclear black holes in luminous elliptical galaxies and implications for the space density of the most massive black holes. Astrophys. J. **662**, 808 (2007)
6. H. Goldstein, C. Poole, J. Safko, *Classical Mechanics*, 3rd edn. (Addison-Wesley, Boston, 2000)
7. O. Luongo, H. Quevedo, Characterizing repulsive gravity with curvature eigenvalues. Phys. Rev. D **90**, 084032 (2014)
8. K. Boshkayev, E. Gasperín, A.C. Gutiérrez-Piñeres, H. Quevedo, S. Toktarbay, Motion of test particles in the field of a naked singularity. Phys. Rev. D **93**, 024024 (2016)
9. Z. Stuchlík, G. Bao, E. Østgaard, S. Hledík, Kerr–Newman–de Sitter black holes with a restricted repulsive barrier of equatorial photon motion. Phys. Rev. D **58**, 084003 (1998)
10. Z. Stuchlík, S. Hledík, Equatorial photon motion in the Kerr–Newman spacetimes with a non-zero cosmological constant. Class. Quant. Gravit. **17**, 4541 (2000)

Hearing the Nature of Compact Objects

Sebastian H. Völkel and Kostas D. Kokkotas

Abstract In this article we present results on the quasi-normal mode spectra of compact relativistic objects and discuss under what conditions they can be used to solve the inverse spectrum problem. We start from the one-dimensional wave equation for axial gravitational perturbations of spherically symmetric and non-rotating compact objects in general relativity. It is then shown how WKB methods can be used to approximate the spectrum and reconstruct the underlying perturbation potential by using different Bohr–Sommerfeld rules. The uniqueness and properties of the reconstructed relations are discussed as well.

Keywords Gravitational waves · Inverse spectrum problem · WKB

1 Introduction

The emerging field of gravitational wave physics has been manifested with the repeated detections of binary black hole and binary neutron star mergers by the current generation of gravitational wave detectors LIGO and Virgo [1–6]. These first direct detections of gravitational waves confirm Einstein's general theory of relativity 100 years after its birth in an unprecedented way. At the same time it strongly supports the existence of black holes and neutron stars as compact objects. However, due to the challenging technical aspects of the measurements, deviations from general relativity or the nature of the compact objects might still be buried in the experimental inaccuracies. These will be decreased in future observations. Besides the precise modeling of systems like black holes and neutron stars, it is also of interest to study how signals of alternative compact objects, so-called exotic compact objects (ECOs), would differ [7, 8]. This question triggered much attention recently, due to claims that such deviations have been found in gravitational wave

S. H. Völkel (✉) · K. D. Kokkotas
University of Tübingen, Theoretical Astrophysics, Tübingen, Germany
e-mail: sebastian.voelkel@uni-tuebingen.de; kostas.kokkotas@uni-tuebingen.de

© Springer Nature Switzerland AG 2019   333
S. Cacciatori et al. (eds.), *Einstein Equations: Physical and Mathematical Aspects of General Relativity*, Tutorials, Schools, and Workshops in the Mathematical Sciences, https://doi.org/10.1007/978-3-030-18061-4_12

detections [9, 10]. However, these claims are highly disputed [11, 12] but will be answered with future observations. Since serious realistic models for ECOs do not exist and contain many extrapolations from well understood standard physics, the only way to describe deviations from black holes and neutron stars is to study simplified toy models of such systems. These models should be seen as an effective way to capture new and qualitatively different features, for which one then has to search for in gravitational wave observations. Understanding the potential deviations from the standard picture is important, because one "knows" what should be found in the observed data. It will also be of paramount importance if one wants to understand the nature of such compact objects.

In this work we present a semi-analytical study of the so-called quasi-normal mode (QNM) spectra of different types of compact objects. By restricting ourselves to linear perturbations of spherically symmetric and non-rotating objects, it is possible to describe the QNM spectra with the one-dimensional wave equation. We use different Bohr–Sommerfeld (BS) rules, which are results of WKB theory, to approximate them in the direct problem, but also address the inverse spectrum problem.

The article is organized as follows. In Sect. 2 we outline the physical problem of linear perturbations of compact objects relevant to our study. One possible tool to solve the resulting direct and inverse problems is presented in Sects. 3 and 4. We show and discuss our results for different objects in Sect. 5. Our conclusions can be found in Sect. 6. Throughout the paper we set $\hbar = G = c = 1$.

2 Gravitational Perturbations

The study of linear perturbations of compact objects traces back to the seminal work of Regge and Wheeler in 1957 [13] and Zerilli in 1970 [14], who studied linear perturbations (first axial and then polar perturbations) of the Schwarzschild black hole. This approach has first been extended to the perturbations of relativistic stars by Thorne and Campolattaro in 1967 [15]. The two different types of gravitational perturbations are the so-called axial and polar ones, which are defined by their parity. Axial perturbations do not couple to fluid oscillations, which makes them simpler for analytical studies. This is not the case for polar perturbations and the resulting equations are more involved. It can be shown that the wave equation describing axial perturbations of spherically symmetric and non-rotating stars is given by

$$\frac{d^2}{dr^{*2}}\Psi(r) + \left[\omega_n^2 - V(r)\right]\Psi(r) = 0, \tag{2.1}$$

where $\Psi(r)$ is related to the linear perturbation $h_{\mu\nu}$ of the background metric, ω_n^2 is the QNM spectrum, $V(r)$ acts as effective potential and $r^*(r)$ is the so-called tortoise coordinate

$$r^*(r) \equiv \int^r \sqrt{\frac{g_{11}(r')}{g_{00}(r')}} \, dr'.$$ (2.2)

This equation, with small changes in the potential $V(r)$, represents also the perturbations (axial and polar) of non-rotating black holes. A full discussion of the derivation of Eq. (2.1) is beyond the scope of this proceeding, but can be found in much detail in these reviews [16–18]. From now on we will use Eq. (2.1) as basis for our further considerations, which are twofold. First, we want to show how the QNM spectrum can be approximated for different types of objects, assuming that the underlying potential is known. Second, we are interested in the inverse problem, where we assume that the QNM spectrum is provided, but not the potential. The second case is much more involved and in general not unique. In astrophysics it is actually more important, since any observation would measure the QNM spectrum and not the potential itself. In Sects. 3 and 4 we show how the BS rules can be applied to study the spectra of different types of compact objects in order to address both kind of problems.

3   Bohr–Sommerfeld Rules

The effective potential appearing in the wave equation, together with the non-trivial tortoise coordinate transformation, make exact analytic studies very complicated, if one is interested in explicit results for the spectrum. Therefore, it is common practice to find numerical solutions, once a model and its parameters are specified, or to use approximate methods, which might allow for semi-analytic results. Since we are also interested in the inverse problem, we are following the second approach. A well-known technique for finding approximate solutions of the wave equation is WKB theory. Here one assumes that the solution can be written as a series for the phase, whose individual contributions are then related to integral equations. For an overview of WKB theory we refer to [19, 20]. The validity of the WKB solutions depends on the type of potential and does usually not hold throughout the whole domain. Special care has to be taken at classical turning points, where the WKB solutions have to be matched with the exact solution of a local approximation of the potential. It turns out that for different types of potentials, it is possible to derive rather simple integral equations to approximate the eigenvalues ω_n^2 directly. These so-called BS rules or quantization conditions[1] can be used to approximate the eigenvalues, without doing explicit calculations for the wave function $\Psi(r)$ itself.

---

[1]The approach was known before quantum mechanics, but it became popular there. We want to note that although the terminology might relate it to quantum mechanics, there is no such aspect assumed here. It is an approximate method for finding solutions of the wave equation.

In the following we discuss some of the existing BS rules that are of relevance for different types of compact objects. Defining $\omega_n^2 \equiv E_n$ brings the problem closer to the standard WKB literature. Furthermore we set x as the one-dimensional variable of the wave equation. The classical BS rule is given by

$$\int_{x_0(E_n)}^{x_1(E_n)} \sqrt{E_n - V(x)}\,dx = \pi\left(n + \frac{1}{2}\right),$$ (3.1)

where $(x_0(E_n), x_1(E_n))$ are the classical turning points and $n = (0, 1, 2, \dots)$. The BS rule is valid for pure potential wells, which we describe as functions admitting two classical turning points and one minimum. If the integration can be carried out analytically for a given potential $V(x)$, one has to solve the resulting equation for E_n to obtain the spectrum. In this case one finds real eigenvalues. However, there is no compact object that admits such a potential. Actually, the potential corresponding to perturbations of the Schwarzschild spacetime is a potential barrier and it is known how the bound states of the inverted potential can be related to the eigenvalues of the barrier themselves [21–23]. The QNM spectrum of the Schwarzschild black hole has been studied extensively in the literature and we only refer to the aforementioned reviews [16–18]. The more interesting cases for this work are neutron stars and ultra-compact objects. In the first case one deals with a one turning point problem and it is possible to derive a generalized BS rule, which gives the right properties of the spectrum if compared with numerical results. It is given by

$$\int_{x_0(E_n)}^{R} \sqrt{E_n - V(x)}\,dx = \pi\left(n + \frac{3}{4}\right) + i\,\mathrm{arctanh}\left(\frac{\sqrt{E_n - V_-}}{\sqrt{E_n - V_+}}\right),$$ (3.2)

and is valid for eigenvalues that are above the potential values (V_-, V_+) at the surface discontinuity R. Here we assumed that the potential has a discontinuity at the surface, which appears in stars with a solid crust or most equations of state. The density, which contributes to the potential, is not smoothly going to zero at the surface. The above shown BS rule also works for approximating the real part of E_n for continuous potentials. This can be the case for polytropes. The derivation and application of Eq. (3.2) will be presented soon in a future publication [24].

The second case, i.e., the ultra-compact objects, is related to ECOs. In Fig. 1 we show two such potentials. The three turning point potential in the left panel is relevant for ultra-compact stars and ECOs. The four turning point potential in the right panel appears for some wormhole models, for which we studied the inverse problem in [25]. In both type of potentials, a new family of modes that does not exist for normal neutron stars emerges [26, 27]. These so-called trapped modes appear for objects with $R \lesssim 3M$. These modes can be described as quasi-stationary states related to the eigenvalues of the potential well. In a dynamical picture, where incoming waves scatter at the potential, a new phenomenon emerges. Some of the waves are trapped between the center of the object (or second barrier) and the external potential barrier. It has first been studied in these works [28–30], but

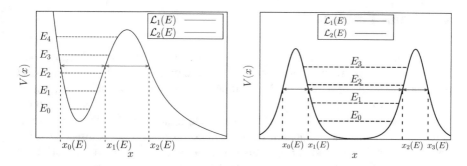

Fig. 1 Left: Qualitative example for the three turning point potentials studied in this work. Figure taken from S. H Völkel and K. D. Kokkotas 2017 Class. Quantum Grav. 34 175015, DOI: https://doi.org/10.1088/1361-6382/aa82de. **Right:** Qualitative example for the four turning point potentials studied for the inverse problem in this work. Figure taken from S. H. Völkel and K. D Kokkotas 2018 Class. Quantum Grav. 35 105018, DOI: https://doi.org/10.1088/1361-6382/aabce6

receives renewed interest in the context of ECOs, see [31] for a recent review. It turns out that the first reflected waves can mimic the Schwarzschild black hole, but a repeated signal of "echoes" would predict an alternative ultra-compact object. A generalized BS rule for the trapped modes in a three turning point potential is given by

$$\int_{x_0}^{x_1} \sqrt{E_n - V(x)}dx = \pi \left(n + \frac{1}{2} \right) - \frac{i}{4} \exp\left(2i \int_{x_1}^{x_2} \sqrt{E_n - V(x)}dx \right), \quad (3.3)$$

which has first been derived for quantum mechanical systems in [32]. The trapped modes are described by a real part that is similar to pure bound states, and an imaginary part that is exponentially small and corresponds to a slow damping of the bound states. We have applied this BS rule to ultra-compact constant density stars and thin shell gravastars in [33] to approximate the spectrum in different ways. A confirmation using precise numerical solutions is presented as well. Using analytical approximations for the potential we were furthermore able to provide explicit analytical approximations for the trapped QNM spectrum of different objects.

So far one could have argued that alternative methods are available as well and a full numerical solution of the wave equation could be achieved nowadays with moderate effort. Although this is true, we demonstrate the analytical power of the method in the next section, where we present an explicit way for addressing the inverse spectrum problem. Note that most direct methods being used to calculate the spectrum cannot be applied to this more complicated type of problem.

4 Inverse Problem

Inverse problems arise in various fields in physics and beyond. Probably, the most famous example in mathematical physics has been coined by Kac for asking "Can one hear the shape of a drum?" [34]. Although the problem is much older, it took a long time to construct explicit examples that demonstrate the non-uniqueness of the question [35]. More technically expressed, it asks whether the spectrum of eigenvalues of the wave equation being defined on a surface enclosed by some boundary with Dirichlet conditions, can be used to uniquely reconstruct the shape of the boundary. In our problem there is only the one-dimensional wave equation, but we include an additional term, that is, the axial perturbation potential of the object.

From now on we will assume that the spectrum E_n for some system is known, but not the corresponding potential $V(x)$. Such a situation is closer to any observation of a system, since eigenvalues usually correspond to the observables, but not the potential itself. In the literature it is well known that the classical BS rule can be "inverted" to reconstruct the width of the potential well, but not the potential itself [36, 37]. The result is given by

$$\mathcal{L}_1(E) = x_1 - x_0 = 2\frac{\partial}{\partial E} \int_{E_{\min}}^{E} \frac{n(E') + 1/2}{\sqrt{E - E'}} dE', \tag{4.1}$$

where $E_{\min}$ is the potential minimum that has to extrapolated from $n(E_{\min}) = -1/2$. A similar result is known for pure potential barriers, where the so-called transmission, but not the spectrum is being provided [38, 39]. The width of the potential barrier is found to be

$$\mathcal{L}_2(E) = x_2 - x_1 = -\frac{1}{\pi} \int_{E}^{E_{\max}} \frac{\left(dT(E')/dE'\right)}{T(E')\sqrt{E' - E}} dE', \tag{4.2}$$

where $T(E)$ is the semi-classical notion of the transmission. Here it is worth making two comments. First, the reconstruction of the width only yields a unique solution for the potential, if one of the two functions describing the turning points is provided. This might be the case in physical applications if some general aspects of the system are known, but not in general. Second, it is obvious that there are in general infinitely many so-called WKB spectrum equivalent potentials [40–42]. These can be constructed by assuming one of the two turning point functions. In the special case that the width is a constant, describing the box potential, there is only one reasonable unique solution. This case is like the one-dimensional drum, a finite and uniform string that is fixed at its ends. In this case there is only a unique shape for the trivial drum.

More interesting for our type of spectra is the solution of the inverse problem of neutron star like potentials and the ones for ultra-compact stars. In both cases a BS rule is known, but the exact inversion is more involved. We have shown how the situation can be simplified by expanding the generalized BS rule in the small imaginary part of the spectra in [43]. The main result is that it is approximately possible to first reconstruct the width of the potential well $\mathcal{L}_1$ by studying the real part of the complex QNM spectrum. In a second step, one can use this width together with the imaginary part of the spectrum to find the unknown transmission $T(E)$ from the spectrum and then reconstruct the width of the barrier $\mathcal{L}_2$. Assuming the validity of Birkhoff's theorem in a third step, it is possible to find "approximate" but "unique" solutions for the inverse problem. The reconstruction of the potential can in this case be done from the minimum up to the value of the potential barrier.

In a more recent work [25] we have extended a similar idea to double barrier potentials, which appear for some symmetric wormhole models. In this case one deals with a four turning point potential. There are two main results in this case. First, for symmetric potentials it is possible to reconstruct both, the width of the well and the two barriers. Second, we have found that one could in principle, for the same spectrum, also reconstruct a three turning point potential with slightly different widths. For the comprehensive discussion we refer to [25, 43]. For a class of wormholes with a two turning point potential, an alternative way to solve the inverse problem for some parts of the metric has been proposed recently [44].

5   Results and Discussion

In this section we summarize our main results that can be found in more detail along with extended discussions in [25, 33, 43, 45].

The generalized BS rule for quasi-stationary states described by Eq. (3.3) can be further simplified since the imaginary part for most states is exponentially small. Doing so one finds two equations, one for the real and one for the imaginary part of E_n. The equations can be solved numerically as root finding problem in E_n or be solved analytically for a simple choice of the potential. Both approaches have been studied along with numerically obtained precise eigenvalues in [33]. It was found that the BS rule along with its WKB next correction term give good results. For a quantitative discussion we refer to the details presented there.

The inverse spectrum problem has been studied for constant density stars and thin shell gravastars in [43]. The extension to symmetric double barrier potentials, which appear for some toy models of wormhole space-times can be found in [25]. For a more comprehensive investigation of the inverse method for quasi-stationary states, without the application to compact objects but analytical examples for different spectra, we refer to [45]. In the left and right panel of Fig. 2 we show one example for the reconstructed potentials of each type (three and four turning points) together with the model parameters. In the first case the reconstruction of the potential, using the trapped modes spectra, recovers the internal region of the potential between the local minimum and the maximum of the barrier. In the second case, the same is possible. But, since the barrier maximum corresponds also to the maximum of the potential, the whole potential can be reconstructed. In both cases one can see that the reconstruction (red dashed lines) agrees very well with the exact (black solid

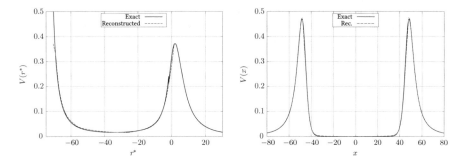

Fig. 2 Left: Reconstructed axial perturbation potential of a constant density star with $R = 2.26M$ and $l = 3$, using the trapped modes spectrum. Figure taken from S. H. Völkel and K. D. Kokkotas 2017 Class. Quantum Grav. 34 175015, DOI: https://doi.org/10.1088/1361-6382/aa82de. **Right:** Reconstructed scalar perturbation potential for a Damour–Solodukhin wormhole with $\lambda = 10^{-5}$ and $l = 3$. Figure taken from S. H. Völkel and K. D. Kokkotas 2018 Class. Quantum Grav. 35 105018, DOI: https://doi.org/10.1088/1361-6382/aabce6

lines) potential, but not perfectly. This is due to the intrinsic non-exactness of WKB theory and the fact that for numerical data of the spectrum, the continuous spectrum $n(E)$, which is needed for integration, has to be interpolated. For more details we refer to the discussions provided in the aforementioned works [25, 43, 45].

The analysis of the one turning point potential, which appears for neutron stars, will be presented soon in a future publication [24]. In this case we are also able to put constraints on the equation of state and approximate the g_{00} component of the metric inside the star.

6 Conclusions

In this article we have summarized possible ways to address the direct and inverse spectrum problem for different types of compact objects. Most of the work has been published recently [25, 33, 43, 45] or will be submitted soon. In the regime of linear perturbations on a known background metric, it was shown that WKB theory along with its generalized BS rules offer a promising tool. Depending on the type of compact objects (black holes, ECOs or neutron stars), qualitatively different potentials for the axial perturbations arise. By using known results in the literature and extending them to different potentials (3 and 4 turning points), we were able to approximate the spectrum of ECOs in a semi-analytic way and address the conceptually more complicated inverse problem. It was discussed how an inversion of the BS rules can be used to reconstruct the widths of different potentials, but not in general the potential itself. The related problem for neutron stars, a one turning point problem, will be analyzed in a future publication. As it is known from Kac's famous question: "Can one hear the shape of a drum?", such a result should not be surprising. However, by providing additional physically motivated information to the problem, an approximate but unique reconstruction can be possible. Approximations arise due to the non-exactness of the underlying WKB theory, but also due to interpolation that is necessary if only numerical data for the spectrum is provided.

Acknowledgements SV wants to thank the organizers and all participants of Domoschool 2018 for creating an unforgettable week of learning and cultural exchange in the beautiful town of Domodossola. The authors also want to thank Daniela D. Doneva for useful discussions. SV is indebted to the Baden-Württemberg Stiftung for the financial support of this research project by the Eliteprogramme for Postdocs and receives the PhD scholarship Landesgraduiertenförderung.

References

1. B.P. Abbott et al. (LIGO Scientific Collaboration and Virgo Collaboration), Observation of gravitational waves from a binary black hole merger. Phys. Rev. Lett. **116**, 061102 (2016)
2. B.P. Abbott et al. (LIGO Scientific and Virgo Collaborations), Tests of general relativity with GW150914. Phys. Rev. Lett. **116**, 221101 (2016)
3. B.P. Abbott et al. (LIGO Scientific Collaboration and Virgo Collaboration), Gw151226: observation of gravitational waves from a 22-solar-mass binary black hole coalescence. Phys. Rev. Lett. **116**, 241103 (2016)
4. B.P. Abbott et al. (LIGO Scientific and Virgo Collaboration), Gw170104: observation of a 50-solar-mass binary black hole coalescence at redshift 0.2. Phys. Rev. Lett. **118**, 221101 (2017)
5. B.P. Abbott et al. (LIGO Scientific Collaboration and Virgo Collaboration), Gw170814: a three-detector observation of gravitational waves from a binary black hole coalescence. Phys. Rev. Lett. **119**, 141101 (2017)
6. B.P. Abbott et al. (LIGO Scientific Collaboration and Virgo Collaboration), Gw170817: observation of gravitational waves from a binary neutron star inspiral. Phys. Rev. Lett. **119**, 161101 (2017)
7. V. Cardoso, E. Franzin, P. Pani, Is the gravitational-wave ringdown a probe of the event horizon? Phys. Rev. Lett. **116**, 171101 (2016)
8. V. Cardoso, S. Hopper, C.F.B. Macedo, C. Palenzuela, P. Pani, Gravitational-wave signatures of exotic compact objects and of quantum corrections at the horizon scale. Phys. Rev. D **94**, 084031 (2016)
9. J. Abedi, H. Dykaar, N. Afshordi, Echoes from the abyss: tentative evidence for Planck-scale structure at black hole horizons. Phys. Rev. D **96**, 082004 (2017)
10. R.S. Conklin, B. Holdom, J. Ren, Gravitational wave echoes through new windows. Phys. Rev. D **98**, 044021 (2018)
11. G. Ashton, O. Birnholtz, M. Cabero, C. Capano, T. Dent, B. Krishnan, G.D. Meadors, A.B. Nielsen, A. Nitz, J. Wester- weck, Comments on: "Echoes from the abyss: Evidence for Planck-scale structure at black hole horizons" (2016). arXiv:1612.05625
12. J. Westerweck, A. Nielsen, O. Fischer-Birnholtz, M. Cabero, C. Capano, T. Dent, B. Krishnan, G. Meadors, A.H. Nitz, Low significance of evidence for black hole echoes in gravitational wave data (2017). arXiv:1712.09966
13. T. Regge, J.A. Wheeler, Stability of a Schwarzschild singularity. Phys. Rev. **108**, 1063–1069 (1957)
14. F.J. Zerilli, Effective potential for even-parity regge-wheeler gravitational perturbation equations. Phys. Rev. Lett. **24**, 737–738 (1970)

15. K.S. Thorne, A. Campolattaro, Non-radial pulsation of general-relativistic stellar models. I. Analytic analysis for L ≥ 2. ApJ **149**, 591 (1967)
16. K.D. Kokkotas, B.G. Schmidt, Quasi-normal modes of stars and black holes. Living Rev. Relativ. **2**, 2 (1999)
17. H.-P. Nollert, Topical review: quasinormal modes: the characteristic 'sound' of black holes and neutron stars. Class. Quantum Grav. **16**, R159–R216 (1999)
18. E. Berti, V. Cardoso, A.O. Starinets, Topical Review: quasinormal modes of black holes and black branes. Class. Quantum Grav. **26**, 163001 (2009)
19. C.M. Bender, S.A. Orszag, *Advanced Mathematical Methods for Scientists and Engineers* (McGraw-Hill, New York, 1978)
20. B.M. Karnakov, V.P. Krainov, *WKB Approximation in Atomic Physics* (Springer, Berlin, 2013)
21. H.-J. Blome, B. Mashhoon, Quasi-normal oscillations of a schwarzschild black hole. Phys. Lett. A **100**, 231–234 (1984)
22. V. Ferrari, B. Mashhoon, Oscillations of a black hole. Phys. Rev. Lett. **52**, 1361–1364 (1984)
23. V. Ferrari, B. Mashhoon, New approach to the quasinormal modes of a black hole. Phys. Rev. D **30**, 295–304 (1984)
24. S.H. Völkel, K.D. Kokkotas, On the inverse spectrum problem of neutron stars. Class. Quantum Grav. **36**(11), 115002 (2019)
25. S.H. Völkel, K.D. Kokkotas, Wormhole potentials and throats from quasi-normal modes. Class. Quantum Grav. **35**, 105018 (2018)
26. S. Chandrasekhar, V. Ferrari, On the non-radial oscillations of a star. III - a reconsideration of the axial modes. Proc. Royal Soc. Lond. Ser. A **434**, 449–457 (1991)
27. K.D. Kokkotas, Axial modes for relativistic stars. Mon. Not. R. Astron. Soc. **268**, 1015 (1994)
28. K.D. Kokkotas, Pulsating relativistic stars, Relativistic gravitation and gravitational radiation, in *Proceedings, School of Physics, Les Houches, France* (1995), pp. 89–102
29. K. Tominaga, M. Saijo, K. Maeda, Gravitational waves from a test particle scattered by a neutron star: axial mode case. Phys. Rev. D **60**, 024004 (1999)
30. V. Ferrari, K.D. Kokkotas, Scattering of particles by neutron stars: time evolutions for axial perturbations. Phys. Rev. D **62**, 107504 (2000)
31. V. Cardoso, P. Pani, Tests for the existence of horizons through gravitational wave echoes. Nat. Astron. **1**, 586–591 (2017)
32. V.S. Popov, V.D. Mur, A.V. Sergeev, Quantization rules for quasistationary states. Phys. Lett. A **157**, 185–191 (1991)
33. S.H. Völkel, K.D. Kokkotas, A semi-analytic study of axial perturbations of ultra compact stars. Class. Quantum Grav. **34**, 125006 (2017)
34. M. Kac, Can one hear the shape of a drum? Am. Math. Mon. **73**, 1–23 (1966)
35. C. Gordon, D. Webb, S. Wolpert, Isospectral plane domains and surfaces via Riemannian orbifolds. Invent. Math. **110**, 11–22 (1992)
36. J.A. Wheeler, *Studies in Mathematical Physics: Essays in Honor of Valentine Bargmann.* Princeton Series in Physics (Princeton University Press, Princeton, 2015), pp. 351–422
37. K. Chadan, P.C. Sabatier, *Inverse Problems in Quantum Scattering Theory*, 2nd ed. Texts and Monographs in Physics (Springer, New York, 1989)
38. J.C. Lazenby, D.J. Griffiths, Classical inverse scattering in one dimension. Am. J. Phys. **48**, 432–436 (1980)
39. S.C. Gandhi, C.J. Efthimiou, Inversion of Gamow's formula and inverse scattering. Am. J. Phys. **74**, 638–643 (2006)
40. D. Bonatsos, C. Daskaloyannis, K. Kokkotas, WKB equivalent potentials for the q-deformed harmonic oscillator. J. Phys. A **24**, 795–801 (1991)
41. D. Bonatsos, C. Daskaloyannis, K. Kokkotas, WKB equivalent potentials for q-deformed harmonic and anharmonic oscillators. J. Math. Phys. **33**, 2958–2965 (1992)
42. D. Bonatsos, C. Daskaloyannis, K. Kokkotas, WKB equivalent potentials for q-deformed anharmonic oscillators. Chem. Phys. Lett. **193**, 191–196 (1992)
43. S.H. Völkel, K.D. Kokkotas, Ultra compact stars: reconstructing the perturbation potential. Class. Quantum Grav. **34**, 175015 (2017)

44. R.A. Konoplya, How to tell the shape of a wormhole by its quasinormal modes. Phys. Lett. B **784**, 43–49 (2018)
45. S.H. Völkel, Inverse spectrum problem for quasi-stationary states. J. Phys. Commun. **2**, 025029 (2018)

Minisuperspace Quantisation via Conditional Symmetries

Adamantia Zampeli

Abstract We review the canonical quantisation of minisuperspace models by promoting to operators the constraints as well as the additional symmetries of the metric of the configuration space of variables. We describe the classical and quantum formulation of the theory and give an application of this approach to the FLRW spacetime coupled to a massless scalar field.

Keywords Minisuperspace · FLRW universe · Bohmian interpretation

1 Introduction

The construction of simplified models due to the appearance of symmetries has always been considered in physics. Despite the simplifications, they can still describe important features of the full models and allow for their study. Minisuperspace models are simplified versions of the full gravity which usually represent cosmological models, without of course restricting to them. One usually starts with the full gravitational action, which might also contain matter fields

$$S_{tot} = \int d^4x \sqrt{-g} \left(R - \frac{1}{2} g^{\mu\nu} \partial_\mu \phi \partial_\nu \phi \right) \tag{1.1}$$

Prepared for the proceedings of the 1st Domoschool, Domodossola.

A. Zampeli (✉)
Institute of Theoretical Physics, Faculty of Mathematics and Physics, Charles University, Prague, Czech Republic
e-mail: azampeli@phys.uoa.gr

© Springer Nature Switzerland AG 2019
S. Cacciatori et al. (eds.), *Einstein Equations: Physical and Mathematical Aspects of General Relativity*, Tutorials, Schools, and Workshops in the Mathematical Sciences, https://doi.org/10.1007/978-3-030-18061-4_13

and inserts an ansatz for the geometry of the spacetime. In the following consideration, we will assume that the line element is of the form[1]

$$d s^2 = \pm N(x)^2 \, d x^2 + g_{ab}(x) \, \sigma_i^a(x^i) \sigma_j^b(x^i) \, dx^i \, dx^j, \quad i, j = 1, 2, 3. \quad (1.2)$$

where x is the independent variable which can be either the time t or radial r
coordinate, so that the line element corresponds to either spatially homogeneous
and/or point-like geometries, g_{ab} is the 3-metric. After inserting (1.2) into (1.1), the
action reduces to the form

$$S = \int dx \left(\frac{1}{2N} G_{\alpha\beta}(q) \dot{q}^\alpha \dot{q}^\beta - N V(q) \right) \quad (1.3)$$

The degrees of freedom of the theory, collectively denoted as $q^\alpha(x)$, are usually the
components of the spatial 3-metric g_{ab} together with matter degrees of freedom.[2]
The rest of the variables denote the lapse function $N(x)$ and the invariant one-
forms $\sigma_i^a(x)$. Finally, $G_{\alpha\beta}$ is the metric tensor on the configuration space of the
dependent variables q^α, known as supermetric and $V(q)$ is a function of only
q^α's called superpotential. It corresponds to the Ricci scalar and other possible
contributions in the action from e.g. fields potentials. The Lagrangian (1.3) has
the form kinetic plus potential part and the presence of the lapse function clearly
indicates the reparametrisation invariance of the theory which means that it is form
invariant under a reparametrisation

$$x = f(\tilde{x}) \quad (1.4)$$

together with the following transformations of the dependent variables $q^\alpha(x)$, $N(x)$:

$$N(x) \rightarrow \tilde{N}(\tilde{x}) = N(f(\tilde{x})) \, f'(\tilde{x}), \quad q^\alpha(x) \rightarrow \tilde{q}^\alpha(\tilde{x}) = q^\alpha(f(\tilde{x})). \quad (1.5)$$

i.e. the lapse function transforms as a density, while the scale factors as scalars under
these transformations. The above changes of $N(x)$ and $q^\alpha(x)$ imply that one of these
variables can be prescribed at will by suitably choosing the freedom encoded in the
arbitrariness of $f(x)$.

It is important to note that one should always perform the check of validity of the
action (1.3), i.e. that the corresponding Euler–Lagrange equations are equivalent to
the Einstein field equations $G_{\mu\nu} = T_{\mu\nu}$. A well-known case that this equivalence
does not hold is for the spatially homogeneous models of Bianchi type class B.

---

[1]The line element has been turned to a simplified form without the shift vectors since it is always
possible to make a choice of the three out of the four gauge functions of the independent variable
x such that $N^i = 0$ at the expense of a more complicated spatial metric components.
[2]Even though the results above have been shown in [1] to hold for pure gravity, they can be easily
extended to include matter degrees of freedom as long as the total Lagrangian takes the form (1.3).

In the following, we are interested in the symmetries of the action (1.3) and the Lie-point symmetries of the Euler–Lagrange equations

$$E^0 = G_{\mu\nu}\dot{q}^\mu\dot{q}^\nu + 2N^2 V = 0 \tag{1.6a}$$

$$E^\kappa = \ddot{q}^\kappa + \Gamma^\kappa_{\mu\nu}\dot{q}^\mu\dot{q}^\nu - \frac{\dot{N}}{N}\dot{q}^\kappa + N^2 V^{,\kappa} = 0 \tag{1.6b}$$

where $V^{,\kappa} = G^{\rho\kappa} V_{,\rho}$.

1.1 Lagrangian Formalism

The usual approach in regular theories (i.e. without constraints) to find the generators of the point transformations that leave invariant the action is to consider the infinitesimal criterion of invariance. In the case of the minisuperspace models which are singular, this criterion has to be extended to include the lapse function at the same level as the q variables [1, 2]. The result is that the variational symmetries for the action (1.3) have an infinitesimal generator of the form

$$X = X_1 + X_2 \tag{1.7}$$

where

$$X_1 = \xi^\alpha(q)\frac{\partial}{\partial q^\alpha} - \tau(q) N\frac{\partial}{\partial N}, \tag{1.8a}$$

$$X_2 = \chi(t)\frac{\partial}{\partial t} + \chi(t)_{,t} N\frac{\partial}{\partial N}. \tag{1.8b}$$

provided that

$$\mathcal{L}_\xi G_{\mu\nu} = \tau(q)G_{\mu\nu} \qquad and \qquad \mathcal{L}_\xi V = -\tau(q)V \tag{1.9}$$

where $\mathcal{L}_\xi$ stands for the Lie derivative operator. The relations (1.9) is the definition of the conditional symmetries [3, 4]. In this definition, the ξ_I's can be conformal fields of the supermetric and not Killing fields and signifies that in order to have a variational symmetry for Lagrangians of this form, the vector ξ must be a simultaneous conformal Killing vector of both the potential and the supermetric with conformal factors of opposite signs. The vector X_1 represents the existing conditional symmetries encoded in the simultaneous conformal Killing fields of the potential V and the configuration space metric $G_{\mu\nu}$, while X_2 represents the time reparametrisation invariance encoded in the arbitrary function $\chi(t)$ remains an unrestricted function of time. This fact reflects the time reparametrisation invariance of the theory stemming out of the singular Lagrangian (1.3).

For the equations of motion (1.6) now, the Lie-point symmetries have a generator which can be decomposed as

$$X = X_1 + X_2 - \frac{2c}{1-n} Y, \tag{1.10}$$

where X_1 and X_2 are the generators (1.8) and

$$Y = q^\alpha \frac{\partial}{\partial q^\alpha} + \frac{1}{2} N \frac{\partial}{\partial N} \tag{1.11}$$

is the well-known scaling symmetry of the vacuum Einstein's equations [2]. These considerations lead to the conclusion that the Lie—point symmetries for the equations (1.6) in minisuperspace gravity are the ones with the infinitesimal generator X_1 which satisfy (1.9), plus the scaling symmetry generator Y and the time reparametrisation generator X_2.

1.2   Hamiltonian Formalism

The presence of a gauge symmetry, in our case being the reparametrisation invariance, becomes evident through the presence of constraints, especially in the Hamiltonian formalism. The canonical variables are chosen to be the dependent variables q^α and $N(x)$ together with their conjugate momenta

$$p_\alpha = \frac{\partial L}{\partial \dot{q}^\alpha} = \frac{1}{N} G_{\alpha\beta} \dot{q}^\beta, \tag{1.12a}$$

$$p_N = \frac{\partial L}{\partial \dot{N}} = 0. \tag{1.12b}$$

which obey the Poisson bracket relation $\{q^\alpha, p_b\} = \delta^a_b$. The relation (1.12b) is a primary constraint of the theory. The canonical Hamiltonian is given by

$$H = p_\gamma \dot{q}^\gamma - L = N \mathcal{H} = N \left(\frac{1}{2} G^{\alpha\beta}(q) p_\alpha p_\beta + V(q) \right), \qquad G_{\alpha\mu} G^{\mu\beta} = \delta^\beta_\alpha. \tag{1.13}$$

The consistency requirement that (1.12b) must be preserved by the x-evolution leads to the secondary constraint

$$\dot{p}_N = \{p_N, H\} \approx 0 \Rightarrow \mathcal{H} \approx 0 \tag{1.14}$$

and the algorithm is terminated since $\{\mathcal{H}, H\} = 0$. The two constraints are first class on account of the Poisson bracket $\{p_N, \mathcal{H}\} = 0$. They represent the x-

reparametrisation invariance of the action and reveal $N(x)$ as a Lagrange multiplier, defined by the quadratic constraint and not by the dynamical evolution equations involving the accelerations of the $n - 1$ $q^\alpha(x)$'s.

On the phase space, we can construct physical quantities as functions of the canonical variables. We consider the first integrals of motion which correspond to the invariance transformation generated by (1.7). These are

$$Q = \xi^\alpha p_\alpha - \chi(t) N \mathcal{H} \approx \xi^\alpha p_\alpha \tag{1.15}$$

and satisfy the $\frac{dQ}{dt} \approx 0$ by virtue of the constraint, thus they are constants of motion

$$Q_i = \kappa_i. \tag{1.16}$$

It is convenient to choose a parametrisation such that the superpotential becomes constant. This choice can always be performed by choosing a lapse parametrisation $\bar{N} = NV$. Under this change of N the scaled Lagrangian becomes

$$\bar{L} = \frac{1}{2\bar{N}} \bar{G}_{\kappa\lambda} \dot{q}^\kappa \dot{q}^\lambda - \bar{N} \tag{1.17}$$

with $\bar{G}_{\kappa\lambda} = V G_{\kappa\lambda}$ and trivially $\bar{V} = 1$. The advantage of this parametrisation is that we obtain $\mathcal{L}_\xi \bar{G}_{\mu\nu} = 0{=}0$ and $\mathcal{L}_\xi \bar{V} = 0$. This means that the simultaneous conformal Killing fields ξ are becoming Killing fields of both $\bar{G}_{\kappa\lambda}$ and $\bar{V}$. It is very interesting that the well-known scaling generator Y becomes just the homothetic Killing field of this metric

$$\mathcal{L}_Y \bar{G}_{\mu\nu} = \frac{n}{2} \bar{G}_{\mu\nu} \tag{1.18}$$

leading to a sort of integral of motion. In the following, we perform the quantisation in this parametrisation for the superpotential without loss of generality.

Finally, we mention that apart from the linear constants of motion, one can also find higher order generators, i.e. irreducible Killing tensor fields leading to new conserved quantities that do not reduce to the first order ones [5]. An important quantity is the Casimir invariant which commutes with all the Q_i's

$$\{Q_i, Q_{cas}\} = 0, \tag{1.19}$$

and is constructed by an element of the universal enveloping algebra spanned by ξ_i's. These can also be used for the reduction of the kinematical space during the canonical quantisation.

2 Quantisation

The first step for the canonical quantisation as introduced by Dirac [6] is the selection of the configuration variables and their conjugate momenta so that they are turned to operators. In our case, we select the variables q^a which are usually identified with the components of the spacetime metric (scale factors and the lapse function) and other matter degrees of freedom. Therefore we have

$$[\hat{q}^a, \hat{p}_b] = -i\delta^a_b \tag{2.1}$$

where it has been assumed that the Poisson brackets have been turned to operators according to $\{\cdot, \cdot\} \mapsto -i[\cdot, \cdot]$. For a general function of the fundamental variables, the quantisation is non-trivial since a choice of operator ordering has to be done, thus there is no unique representation from the classical function to the quantum correspondent.

The next step is the construction of a proper representation for the wave functional. In the Schrödinger representation we implement the relations

$$\hat{q}^a \Psi[q^a] = q^a \Psi[q^a], \tag{2.2}$$

$$\hat{p}_b \Psi[q^a] = -i\frac{\partial}{\partial q^b} \Psi[q^a]. \tag{2.3}$$

These relations do not define self-adjoint operators. The representation space consists of all the possible functionals and does not necessarily contain physical states, since there are gauge degrees of freedom. The issue of finding the physical states is very important. A reduction of the initial space of states is performed by imposing the constraints as conditions on the wave function. In our case, this means that we obtain the following relations:

$$\hat{p}_N \Psi = 0, \tag{2.4a}$$

$$\hat{\mathcal{H}} \Psi = 0, \tag{2.4b}$$

$$\hat{Q}_j \Psi = \kappa_j \Psi. \tag{2.4c}$$

In these relations, it has been assumed that the eigenvalues are the classical values of the integrals of motion/conserved quantities. In this approach, we also implement the linear quantities (2.4c) on the wave function, in order to reduce the kinematical configuration space. The quantum algebra of these linear quantities is

$$[\hat{\mathcal{H}}, \hat{Q}_j] = 0, \tag{2.5a}$$

$$[\hat{Q}_i, \hat{Q}_j] = c^k_{ij} \hat{Q}_k \tag{2.5b}$$

where we have assumed that it is isomorphic to the classical one. The last relation leads to the condition

$$[\hat{Q}_i, \hat{Q}_j]\Psi = ic^m \hat{Q}_{ij}\Psi \Rightarrow c_{ij}^m \kappa_m = 0. \qquad (2.6)$$

This equation indicates that not all operators $\hat{Q}_i$ can be imposed simultaneously on the wave function, thus leading to a choice of subalgebras which correspond to different admissible subsets of these operators.

The explicit form of these expression is highly dependent on the assumptions we make, since they are complicated expressions of the fundamental variables and an operator ordering has to be selected. In addition to that, in principle these operators are not self-adjoint and a choice of a measure for hermiticity and the existence of probability interpretation has to be implemented as well. Due to the conformal invariance of the supermetric, it is reasonable to ask the quantum theory is conformally invariant [3, 7]. This means that the kinetic part of $\hat{\mathcal{H}}$ must have the form of the conformal Laplacian i.e. the Laplacian corresponding to the reduced supermetric (i.e. of the reduced state space which is the result of imposing the quantum linear constraints equations) plus a multiple of the Ricci scalar. In the Schrödinger representation we thus have

$$\frac{1}{2}G^{ab} p_a p_b \rightarrow -\frac{1}{2}\Box_c^2 \qquad (2.7)$$

where

$$\Box_c^2 \equiv \Box^2 + \frac{(d-2)}{4(d-1)}R \qquad (2.8)$$

with d the dimension of the configuration space. Under this assumption, one can also find a proper measure for the construction of an inner product on the space of states. In this case, the measure has been chosen as $\mu = \sqrt{|\det G_{\alpha\beta}|} = \sqrt{|G|}$ [4, 8]. Under these conditions, the expressions for the Hermitian operators are

$$\hat{Q}_I \Psi = -\frac{i}{2\mu}\left(\mu \xi_I^\alpha \partial_\alpha + \partial_\alpha \mu \xi_I^\alpha\right)\Psi = \kappa_I \Psi, \qquad (2.9)$$

$$\hat{\mathcal{H}}\Psi = \left[-\frac{1}{2\mu}\partial_\alpha(\mu G^{\alpha\beta}\partial_\beta) + V(q)\right]\Psi = 0. \qquad (2.10)$$

2.1 Semiclassical Approach

The issue of deriving physical results from the solution of the quantum equations as well as their interpretation is an open issue, especially in quantum cosmology, where the usual interpretation of quantum theory cannot be applied. The main reasons are that one has to apply quantum mechanics of a single system as well as the issues related to unitarity and the problem of time. A way out of these problems is provided by the Bohmian interpretation [9–11], since it does not presuppose the classical domain and the trajectories in the configuration space are deterministic, thus the evolution of the corresponding geometry as well. Inserting the ansatz of the polar form of the wave function $\Psi(q) = \Omega(q)e^{iS(q)}$ where $\Omega(q)$ is the amplitude, $S(q)$ is the phase of the wave function, and replacing it to the Wheeler–DeWitt equation, the result is a modified Hamilton–Jacobi equation

$$\frac{1}{2}G^{\alpha\beta}\partial_\alpha S\partial_\beta S - \frac{1}{2}\frac{\Box\Omega}{\Omega} + V = 0, \tag{2.11}$$

and a second equation that in ordinary quantum theory is interpreted as the continuity equation

$$G^{\alpha\beta}\partial_\alpha S\partial_\beta\Omega + \frac{\Omega}{2\mu}\partial_\alpha(\mu G^{\alpha\beta}\partial_\beta S) = 0. \tag{2.12}$$

However, in this case probabilities are not defined and this equation cannot have the meaning of a continuity equation. The quantity $\partial_\alpha S$ can be identified with the momentum of the system, thus assuming that the classical definition

$$p_\alpha \equiv \frac{\partial L}{\partial \dot{q}^\alpha} = \partial_\alpha S \tag{2.13}$$

is still valid. We can also recognise the second term as a new potential term coming from quantum effects known as quantum potential,

$$\mathcal{Q}(q) \equiv -\frac{1}{2\Omega}\Box\Omega = -\frac{1}{2\mu}\partial_\alpha(\mu G^{\alpha\beta}\partial_\beta)\Omega. \tag{2.14}$$

When the quantum potential vanishes, the solution of this set of equations coincides with the classical one, while for a non-vanishing quantum potential this is not true.

In the following, the wave function is the solution obtained from each subalgebra obtained by the relation (2.6). This is either already in polar form or it is written in this way in some approximation limits; thus the phase function and the amplitude are immediately read from Ψ. As it is expected, for those models for which $\mathcal{Q} = 0$, the emerging semiclassical geometries are equivalent to the classical ones, while for non-zero potential we obtain a different solution. We can then study some important physical properties of the quantum solution, such as the existence of singularities, since the persistence of classical singularities at the quantum level remains a hot issue in modern quantum cosmology.

3 Example: Massless Field in the FLRW Universe

An interesting, explicit example of the FLRW geometry coupled to a scalar field, where we apply the canonical quantisation via conditional symmetries. We consider a massless scalar field coupled to a curved FLRW spacetime with line element

$$ds^2 = -N^2(t)dt^2 + a^2(t)\left(\frac{dr^2}{1 - kr^2} + r^2 d\theta^2 + r^2 \sin^2\theta d\varphi^2\right), \tag{3.1}$$

where $N(t)$ is the lapse function and $a(t)$ is the scale factor. Inserting this line element to the total action containing a gravitational and a matter field part and discarding a total time derivative, we find that the form of the Lagrangian, in the constant potential parametrisation is

$$L = n - \frac{36ka^2\dot{a}^2}{n} + \frac{3ka^4\dot{\phi}^2}{n}, \tag{3.2}$$

$n = 6kNa$ and with the corresponding Hamiltonian constraint and supermetric respectively

$$\mathcal{H} = -\frac{p_a^2}{72ka^2} + \frac{p_\phi^2}{12ka^4} - 1 \approx 0, \quad G_{\alpha\beta} = 6ka\begin{pmatrix} -12a & 0 \\ 0 & a^3 \end{pmatrix}. \tag{3.3}$$

This supermetric represents a flat two-dimensional superspace, admitting the following three symmetries and the homothecy

$$\xi_1 = \frac{e^{\phi/\sqrt{3}}}{a}\partial_a - \frac{2\sqrt{3}e^{\phi/\sqrt{3}}}{a^2}\partial_\phi, \quad \xi_2 = \frac{e^{-\phi/\sqrt{3}}}{a}\partial_a + \frac{2\sqrt{3}e^{-\phi/\sqrt{3}}}{a^2}\partial_\phi, \quad \xi_3 = \partial_\phi, \quad \xi_h = \frac{a}{4}\partial_a \tag{3.4}$$

where the numbered indices denote the Killing fields while h denotes the homothetic field. The corresponding first integrals of motion in the configuration space are

$$Q_1 = -\frac{12e^{\frac{\phi}{\sqrt{3}}}ka\left(6\dot{a} + \sqrt{3}a\dot{\phi}\right)}{n}, \quad Q_2 = \frac{12e^{-\frac{\phi}{\sqrt{3}}}ka\left(-6\dot{a} + \sqrt{3}a\dot{\phi}\right)}{n}, \tag{3.5a}$$

$$Q_3 = \frac{6ka^4\dot{\phi}}{n}, \quad Q_h = -\frac{18ka^3\dot{a}}{n} + \int dt\, n(t). \tag{3.5b}$$

The classical solution in any gauge always gives the existence of a singularity in the very early universe. We are interested in finding quantum corrections indicating that this singularity can be avoided. To this end, we promote the constraints as well as the first integrals Q_i to operators and impose them on the wave function. The eigenequations we find are

$$\hat{Q}_1\Psi = -\frac{ie^{\phi/\sqrt{3}}(-6\partial_\phi\Psi + \sqrt{3}a\partial_a\Psi)}{\sqrt{3}a^2} = \kappa_1\Psi, \tag{3.6}$$

$$\hat{Q}_2\Psi = -\frac{ie^{-\phi/\sqrt{3}}(6\partial_\phi\Psi + \sqrt{3}a\partial_a\Psi)}{\sqrt{3}a^2} = \kappa_2\Psi, \tag{3.7}$$

$$\hat{Q}_3\Psi = -i\partial_\phi\Psi = \kappa_3\Psi, \tag{3.8}$$

$$\hat{\mathcal{H}}\Psi = \frac{-144ka^4\Psi - 12\partial_{\phi\phi}\Psi + a(\partial_a\Psi + a\partial_{aa}\Psi)}{144ka^4} = 0, \tag{3.9}$$

where the measure is $\mu(a,\phi) = 6\sqrt{3}a^3k$. Because of the relation (2.6), it is not possible to simultaneously impose all the quantum operators $\hat{Q}_i$ to the wave function. We are thus led to consider only the abelian subalgebras, which in this case are the two-dimensional $(\hat{Q}_1, \hat{Q}_2)$ and the one-dimensionals $\hat{Q}_1, \hat{Q}_2, \hat{Q}_3$. The one-dimensional subalgebras spanned by the operators $\hat{Q}_1, \hat{Q}_2$ give solutions which are special cases of the two-dimensional case [4], while the subalgebra $\hat{Q}_1, \hat{Q}_2$ does not give quantum corrections, thus leading to the classical solution. In the case of the one-dimensional $\{\hat{Q}_3\}$, the solution of the equations (3.9) and (3.8) is

$$\Psi_{cl}(a,\phi) = e^{i\phi\kappa_3}(A_1 I_{-i\sqrt{3}\kappa_3}(6a^2) + B_1 I_{i\sqrt{3}\kappa_3}(6a^2)), \tag{3.10}$$

$$\Psi_{op}(a,\phi) = e^{i\phi\kappa_3}(A_2 J_{-i\sqrt{3}\kappa_3}(6a^2) + B_2 J_{i\sqrt{3}\kappa_3}(6a^2)), \tag{3.11}$$

for the closed and open case respectively. In order to write the wave function in polar form for the semiclassical analysis, approximation limits are taken, for small and large arguments of the Bessel functions. Using the simplifying assumption $A_1 = B_1$, $A_2 = B_2$ renders the, common for the two cases, wave function

$$\Psi_{sm} \approx c_1 e^{i\kappa_3\phi}\cos\ln a. \tag{3.12}$$

Similarly for the large values, assuming again $A_1 = B_1$, $A_2 = B_2$ the wave function becomes

$$\Psi_{la}^{cl} \approx \frac{e^{a^2}}{a}e^{i\kappa_3\phi}, \quad \Psi_{la}^{op} \approx \frac{\sin(6a^2)}{a}e^{i\kappa_3\phi}. \tag{3.13}$$

The quantum potential for small values does not vanish $\mathcal{Q}_{sm} = \frac{1}{144ka^4}$, while $\mathcal{Q}_{la}^{cl} = -\frac{1+4a^4}{144a^4k}$ and $\mathcal{Q}_{la}^{op} = \frac{144ka^4-1}{144ka^4}$. The phase function is $S = \kappa_3\phi$. The solution of the semiclassical equations with respect to (a, n) is

$$a = c, \quad n = \frac{6ka^4}{\kappa_3}\dot{\phi}, \tag{3.14}$$

and has a remaining freedom for the scalar field which we select to be such that the lapse function $N(t)$ of the semiclassical element is the same as for the classical, that is

$$\phi(t) = -\frac{8 \times 3^{3/4} t^{\sqrt{3}/2} \sqrt{-\frac{48kt^{2/\sqrt{3}}}{\kappa_1} - \frac{\kappa_1}{3} \kappa_1 \kappa_3^{3/2}} \left(-3 + \sqrt{1 + \frac{144kt^{2/\sqrt{3}}}{\kappa_1^2}} \, {}_2F_1\left(\frac{1}{2}, \frac{3}{4}; \frac{7}{4}; -\frac{144kt^{\frac{2}{\sqrt{3}}}}{\kappa_1^2}\right)\right)}{c^3(144kt^{2/\sqrt{3}}\kappa_1 + \kappa_1^3)}, \tag{3.15}$$

where ${}_2F_1 (a, b; c; d)$ is the Gauss hypergeometric function. Inserting the solution in the 4-dimensional element and after proper coordinate transformations the spacetime metric is written

$$ds^2 = -\frac{\lambda}{4\sqrt{T}(1 + T\epsilon)^3} dT^2 + \frac{1}{1 - \epsilon r^2} dr^2 + r^2 d\theta^2 + r^2 \sin^2 \theta d\varphi^2, \tag{3.16}$$

where the sign $(+)$ accounts for the closed case and $(-)$ for the open case while the identification $c^2 = \frac{\lambda^2}{16}$ has been considered in order for the constant λ coincide with that one in the classical metric. This spacetime has the interesting property of having constant Ricci scalar $R = 6k$, all higher derivatives of its Riemann tensor zero and constant all curvature scalars constructed from its Riemann tensor. Hence, there is no curvature and/or higher derivative curvature singularity.

4 Conclusions

In this review, we first described the variational and Lie-point symmetries of a reduced reparametrisation invariant action and its related Euler–Lagrange equations. We then used the symmetries coming from the relations (1.9) to the canonical quantisation of the system. The linear quantities lead to a selection rule (2.6) which prevents their simultaneous imposition on the wave function. Thus, we are led to choose subalgebras. Some of them lead to the classical solution but there are cases where we obtain quantum corrections. These, in the Bohmian interpretation we used, are indicated when the quantum potential in the quantum Hamilton–Jacobi equation does not vanish. The system we used as an example is the FLRW spacetime coupled to a massless scalar field, since there is of physical interest. This configuration has a classical singularity which was shown that it can be resolved under the assumptions we made for the construction of the quantum theory.

The motivation to promote the additional symmetries of the configuration space to operators and impose them on the wave function lies in the observation that, in the case of pure gravity and spatially homogeneous spacetimes, the group preserving the geometry of the spacetime is the time-dependent spatial diffeomorphisms group of automorphisms $Aut(G)$. This group can be decomposed into inner automorphisms

$Inn(G)$ and outer $Out(G) = Aut(G)/Inn(G)$. Its elements are

$$X_i = \lambda^\rho_{(i)\mu} g_{\rho\nu} \frac{\partial}{\partial g_{\mu\nu}} \tag{4.1}$$

where $\lambda^\beta_{(i)a} = (C^\beta_{(\rho)a}, \varepsilon^\beta_{(i)a})$ are the generators of the connected to the identity component of $Aut(G)$, with $C^\beta_{(\rho)a}$ related to the $Inn(G)$ and the linear constraint equations, $\mathcal{H}_i \approx 0$. The number of the members of this group is the dimension of the center of the Lie algebra; while the $\varepsilon^\beta_{(i)a}$ are related to the $Out(G) \equiv Aut(G)/Inn(G)$. It is natural to impose all the generators of $Aut(G)$ on the wave function during quantisation to reduce the kinematical space [12, 13]. The generalisation of this approach to the case of the gravity plus matter systems can be implemented through the notion of conditional symmetries, even though the strict correspondence between the generators of $Aut(G)$ and the conditional symmetries does not hold anymore. In this way, this method can also be extended to even more general systems with less symmetries and their quantisation.

Acknowledgements I would like to thank the organisers of the 1st Domoschool for their kind hospitality and the high level of lectures they provided during the school. This work was supported by the grant GAČR 14-37086G.

References

1. T. Christodoulakis, N. Dimakis, P.A. Terzis, Lie point and variational symmetries in minisuperspace Einstein gravity. J. Phys. A **47**, 095202 (2014). [1304.4359]
2. H. Stephani, M. MacCallum, *Differential Equations: Their Solution Using Symmetries* (Cambridge University Press, Cambridge, 1989)
3. K. Kuchar, Conditional symmetries in parametrized field theories. J. Math. Phys. **23**, 1647–1661 (1982)
4. T. Christodoulakis, N. Dimakis, P.A. Terzis, G. Doulis, T. Grammenos, E. Melas, et al., Conditional symmetries and the canonical quantization of constrained minisuperspace actions: the Schwarzschild case. J. Geom. Phys. **71**, 127–138 (2013). [1208.0462]
5. P.A. Terzis, N. Dimakis, T. Christodoulakis, A. Paliathanasis, M. Tsamparlis, Variational contact symmetries of constraint Lagrangians. arXiv:1503.00932
6. P.A.M. Dirac, *Lectures on Quantum Mechanics*. Belfer Graduate School of Science. Monographs (Belfer Graduate School of Science, New York, 1964)
7. T. Christodoulakis, J. Zanelli, Operator ordering in quantum mechanics and quantum gravity. Nuovo Cim. B **93**, 1–21 (1986)
8. T. Christodoulakis, J. Zanelli, Consistent algebra for the constraints of quantum gravity. Nuovo Cim. B **93**, 22–35 (1986)
9. D. Bohm, A suggested interpretation of the quantum theory in terms of hidden variables. 1. Phys. Rev. **85**, 166–179 (1952)
10. D. Bohm, A suggested interpretation of the quantum theory in terms of hidden variables. 2. Phys. Rev. **85**, 180–193 (1952)
11. D. Bohm, B. Hiley, Measurement understood through the quantum potential approach. Found. Phys. **14**, 255–274 (1984)

12. T. Christodoulakis, E. Korfiatis, G. Papadopoulos, Automorphism inducing diffeomorphisms and invariant characterization of Bianchi type geometries. Commun. Math. Phys. **226**, 377–391 (2002). [gr-qc/0107050]
13. T. Christodoulakis, T. Gakis, G. Papadopoulos, Conditional symmetries and the quantization of Bianchi type I vacuum cosmologies with and without cosmological constant. Class. Quant. Gravit. **19**, 1013–1026 (2002). [gr-qc/0106065]

Printed in the United States
By Bookmasters